UNDERSTANDING
MATHEMATICAL
PROOF

UNDERSTANDING MATHEMATICAL PROOF

John Taylor

Rowan Garnier

CRC Press
Taylor & Francis Group
Boca Raton   London   New York

CRC Press is an imprint of the
Taylor & Francis Group, an **informa** business

A CHAPMAN & HALL BOOK

CRC Press
Taylor & Francis Group
6000 Broken Sound Parkway NW, Suite 300
Boca Raton, FL 33487-2742

© 2014 by Taylor & Francis Group, LLC
CRC Press is an imprint of Taylor & Francis Group, an Informa business

No claim to original U.S. Government works

Printed on acid-free paper
Version Date: 20140131

International Standard Book Number-13: 978-1-4665-1490-4 (Paperback)

Visit the Taylor & Francis Web site at
http://www.taylorandfrancis.com

and the CRC Press Web site at
http://www.crcpress.com

Contents

Preface

The notion of 'proof' is central to mathematics; indeed, it is one of its unique features that distinguishes mathematics from other disciplines. In recent years, highly sophisticated software products have been developed, particularly so-called computer algebra systems and a variety of visualisation tools. As a result, some people suggested that mathematics is becoming more 'experimental' in nature and the notion of proof less salient. This has provoked vigorous denials from the mathematical community, which continues to maintain that proof is one of the key concepts that characterises the discipline. It is also one of the most difficult aspects of the subject both to teach and to master.

Our aims in this book are to describe the nature of mathematical proof, to explore the various techniques that mathematicians adopt in proving their results, and to offer advice and strategies for constructing proofs. Undergraduate mathematics students often experience difficulties both in understanding those proofs they come across in their courses and books and, more particularly, in constructing their own proofs. When first asked to write mathematical proofs, students of mathematics often struggle to know where to begin and where to end. In other words, students often find it challenging to know how to start an appropriate line of reasoning and knowing when they have 'done enough' to complete it. In short, constructing proofs may seem to a student to be something of a 'black art' known only to their teachers and the authors of their textbooks.

This book is intended primarily for undergraduate students of mathematics and other disciplines where reasoned arguments need to be developed and understood, such as engineering and computer science. We hope that this book will help improve students' ability to understand those proofs that they meet and will enhance their facility for constructing correct proofs of their own. Any mathematical proof has a definite logical structure and will be written with a certain audience in mind and with a particular style. In the 'zoo' of mathematical proofs, we may categorise the 'animals' (i.e., proofs themselves) by their logical types: direct proofs are placed in one enclosure, proofs by contradiction in another, proofs by mathematical induction in a third, and so on. To understand a proof, one needs to know which enclosure it belongs to; in other words, its underlying logical structure. One of our tasks is to describe the different 'enclosures' by cataloguing different types of proof. When it comes to writing proofs, a knowledge of the different kinds of proofs is important but is not, by itself, enough. So we also provide some hints and guidance on some of the approaches the proof writer may adopt to find an appropriate proof.

The book is structured as follows. Chapter 1 introduces the kind of reasoning that mathematicians use when writing their proofs and gives some example proofs to set the scene. In chapter 2, we introduce some basic logic in order to understand the structure both of individual mathematical statements and whole mathematical proofs. Much of the language of mathematics is based on the notions of sets and functions, which we introduce in chapter 3. In chapter 4, we dissect some proofs with a view to exposing some of the underlying features common to most mathematical proofs. In the last four chapters of the book, we explore in more detail different types of proof; to continue the analogy above, we explore in more detail each of the enclosures in the zoo of mathematical proofs.

RG and **JT**

September 2013

List of Figures

List of Tables

List of Symbols

Symbol	Description	Page
Logic		
$P \Leftrightarrow Q$	Biconditional connective: P if and only if Q	23
$P \Rightarrow Q$	Conditional connective: if P then Q	21
$P \wedge Q$	Conjunction: P and Q	18
$\Gamma \rightsquigarrow T$	Deduction in an axiom system: T can be deduced from the statements in Γ	168
$P_1, \ldots, P_n \vdash Q$	Deducible from: Q is deducible from $P_1, \ldots, P_n$	69
$P \veebar Q$	Exclusive disjunction: P or Q but not both	20
$P \vee Q$	Inclusive disjunction: P or Q	19
$\exists x \bullet P(x)$	Existential quantifier: there exists x such that $P(x)$	49
$P_1 \equiv P_2$	Logical equivalence: P_1 is logically equivalent to P_2	35
$P_1 \vDash P_2$	Logical implication: P_1 logically implies P_2	42
$\neg P$	Negation of a proposition P: not P	17
$\forall x \bullet P(x)$	Universal quantifier: for all x, $P(x)$	49
Sets and Functions		
$\lvert A \rvert$	Cardinality of set A	86
$A \times B$	Cartesian product of sets: $A \times B = \{(x,y) : x \in A \text{ and } y \in B\}$	103
$B(a,r)$	Closed ball in a metric space: $B(a,r) = \{x \in X : d(x,a) \leq r\}$	157
$[a,b]$	Closed interval $\{x \in \mathbb{R} : a \leq x \leq b\}$	90
$\bar{A}$	Complement of a set A	95
$g \circ f$	Composite of function f and function g	113
$A - B$	Difference of sets A and B	95
$m \mid n$	Divides: m divides n	187

Group Theory

Linear Algebra

Analysis

Chapter 1

Introduction

1.1 The need for proof

Mathematics has many different aspects, properties, and features, so it is probably impossible to answer succinctly the question 'what is mathematics?' There are, however, properties of the subject that spring readily to mind: most would agree that mathematics is abstract, rigorous, precise, and formal. No doubt, many would also add 'difficult' to that list. Mathematics is abstract because it deals with concepts rather than 'real' objects. For example, mathematics concerns itself with the number 3 rather than three apples or three buttons; it deals with 'idealised' triangles rather than imperfectly drawn 'real' ones, and so on. It is rigorous and precise because it does not accept vague or emotional arguments; rather, 'facts' about mathematical objects are only established by logical reasoning. The sum of angles in a Euclidean triangle is 180°, not because that would be a nice thing to have or because we want it to be so; we know the angle sum is 180° because we can prove this to be the case from the basic assumptions of Euclidean geometry. Mathematics is formal because mathematical objects are given explicit and precise definitions and mathematical reasoning has a definite and systematic structure.

Of course, people learn and understand mathematics at different 'levels', not all of which conform to the description given in the previous paragraph. When a young child learns about numbers, he or she may appreciate that a collection of two sweets when combined with a collection of three sweets produces a collection of five sweets without having any real concept of the abstract mathematical equation $2 + 3 = 5$. There are various theories that describe the different 'levels' of understanding mathematics. For example, the mathematics educator David Tall has described 'three worlds of mathematics' which he calls the 'embodied', 'proceptual' and 'formal' worlds [13]. The embodied world builds directly on our perceptions of the world, both the physical world and our mental world. Our concept of an idealised triangle, as an abstraction of real, physical triangles, belongs to this world. The proceptual world[1] is the world where symbols represent concepts and we apply processes to those sym-

---

[1] Tall coined this term from the words 'process' and 'concept'.

bols. Thus, for example, the representation of fractions as a/b and the rule $a/b + c/d = (ad + bc)/bd$ belong to this world. The formal world is one where formally expressed properties are used to specify mathematical structures. For example, some readers may be familiar with the mathematical notions of 'group', 'vector space', 'field', and so on; these belong to the formal world. Properties of these mathematical objects are deduced by formal proof rather than from direct experience. In this book, we will be operating (mostly) in the formal world, for it is here that the concept of mathematical proof belongs.

For mathematicians, Tall's formal world, where proof plays a central role, is the world of 'their' mathematics. In 1929, the eminent English mathematician G. H. Hardy articulated the special role that proof plays in mathematics when he wrote, 'It is generally held that mathematicians differ from other people in proving things, and that their proofs are in some sense grounds for their beliefs' [8]. Hardy is discussing the formal world of mathematics and he makes a critical observation that mathematicians differ from others in two crucial respects: in what they *do* (they prove things) and, more deeply, in their criteria for *belief* (the existence of proofs). In this way, the notion of proof is central to the formal world of mathematics.

If the notion of a mathematical proof is one of the defining features of mathematics, for those learning the subject, it is also one of its greatest challenges. In fact, there are (at least) three challenges for students of mathematics associated with the notion of proof. The first is to understand why mathematicians are 'obsessed' with their proofs. For instance, why is it that (pure) mathematics textbooks are peppered with proofs? A second challenge is to understand what constitutes a proof and, just as important, to understand when an argument does *not* constitute a proof. The final — and most difficult — challenge is to develop an ability to construct one's own proofs. In this book we aim to address each of these challenges.

A simple, if incomplete, description of a mathematical proof is 'a precise logical argument of a certain type that establishes a conclusion'. One of our aims is to explore this further to provide a fuller and more detailed understanding of what constitutes a proof. Of course, proofs are written by human beings, so there are many different styles and approaches adopted. Proofs are written for a variety of different audiences in various cultures and in many languages. A proof that may be appropriate for an audience at one level of mathematical development may not be appropriate for an audience with a deeper understanding of the subject. Nevertheless, we believe that there is something that is shared by (almost) all mathematical proofs. We shall attempt to delve beneath surface features such as style and language to examine the structure of the argument that lies beneath. This aspect of the book could be summarised as 'proofs dissected, examined, and their structure revealed.'

Our second goal is probably more difficult to achieve. Put simply, it is to show how to construct proofs. On one level, such a goal is unattainable. There is no magic formula which, if learned, will enable us to construct a proof of any result we care to choose. However, there are a variety of methods of proof that are appropriate in different contexts and any aspiring 'proof writer' needs to know these. We will also provide hints and guidance on how to approach the challenging task of finding a proof. Many teachers of mathematics will agree with the following advice to students. 'How do you stay in physical shape? Do exercise. How do you stay in mathematical shape? Do exercises!' For this reason, we present a variety of worked proofs and we provide numerous exercises for the reader to do (to stay in mathematical shape).

1.2 The language of mathematics

Like any specialised area of human activity, mathematics has its own language and commonly used terms that make it less accessible to the outsider. For example, two mathematicians in discussion may be heard uttering phrases similar to the following.

> '*I have a counter-example to your recent conjecture published in the Journal of Esoteric and Obscure Pure Mathematics.*'

> '*Ah, but your proof relies on Zorn's Lemma. Can you find a proof that is independent of the Axiom of Choice?*'

> '*If you factor out the nilpotent elements, your group is torsion-free.*'

Whereas these statements might make perfect sense to a professional mathematician, to an outsider they are almost certainly completely opaque. In this section, we introduce informally some of the language that is frequently used in mathematics texts. Many of the terms we introduce here will be considered in more depth later in the book.

Statements are the basic sentences of mathematics; they may be expressed in words or symbols or, frequently, using a combination of both. A statement may be true or false, or its truth may depend on the values of some variables that it contains. However, a statement does not express an opinion, emotion, or question. In this sense, we are using the term 'statement' to have a specific meaning [2] which differs from its common usage. In everyday language, the expression of an opinion, such as '*this rose is beautiful*', would often be referred to as a statement. The following are statements in the sense that we are using the term.

---

[2] We will define this more precisely in chapter 2; by a statement we mean either a proposition — see page 16 — or a propositional function — see page 46.

'17 *is a prime number*'; true.

'$\sqrt{2}$ *is rational*'; false — see section 6.5.

'$2^n - 1$ *is prime*'; truth depends on the value of n — see section 7.4.

'$n(n+1)$ *is even for every integer n*'; true – see example 1.4.

Mathematical arguments have **hypotheses** and a **conclusion**. The hypotheses are statements that are assumed to be true. A mathematical argument then proceeds to make deductions from the hypotheses until the final statement, the conclusion, is reached. As we shall discover, the hypotheses of an argument are assumed true for a variety of reasons. A hypothesis may be a statement that has previously been proved or it may be a basic assumption, or **axiom**, that underpins a whole branch of the subject. Alternatively, the argument may concern a conditional statement of the form 'if statement 1, then statement 2'. The argument may then have 'statement 1' as the hypothesis and 'statement 2' as the conclusion. Readers familiar with statistics, may have come across the idea that a hypothesis is a statement that is 'tested' based on some data; a statistical test will lead us to accept or reject the hypothesis with a certain level of confidence. This is not the case in mathematics; hypotheses are not tested, they are just assumed.

As in most disciplines, a branch of mathematics will have its own specialist terms. In mathematics, these will usually be precisely stated as **definitions**. A definition just gives a precise description of a word or phrase in terms of other, more basic words or phrases. Thus the phrase '*a prime number is an integer greater than 1 that has no factors other than 1 and itself*' provides a definition of 'prime number'. It is really saying that we may use 'prime number' as shorthand for 'an integer greater than 1 that has no factors other than 1 and itself'. Of course, we need to know what the more basic terms in a definition mean in order for the definition itself to make sense. For example, we would not expect most readers to obtain much meaning from the definition: *a simply connected 3-manifold is a topological space that is locally homeomorphic to $\mathbb{R}^3$ and whose fundamental group is trivial*. The 'problem' here is that the more basic terms such as 'topological space', 'locally homeomorphic' and so on are likely to be unfamiliar. Definitions are necessary in practice in order to keep the language of statements manageable. Consider, for example, the statement: *for all positive integers n, if $2^n - 1$ is prime, then n is prime.* (We will prove this later – see theorem 6.3.) Without the definition of 'prime number', we would need to write this in a much more cumbersome way, as follows: *for all positive integers n, if $2^n - 1$ is an integer greater than 1 that has no factors other than 1 and itself, then n is an integer greater than 1 that has no factors other than 1 and itself.* Clearly, the first way of stating the theorem is to be preferred. There is a danger with definitions, of course. Having too many definitions will make a piece of writing inaccessible as the reader may have to keep looking up the definitions of terms used and not be able to follow the 'flow' of the argument itself.

In mathematics, the truth of statements is established by providing a proof. A statement that has a proof is called a **theorem**. (We will not discuss here what constitutes a proof; that is for consideration elsewhere in this chapter and, indeed, elsewhere in this book.) Thus '17 *is a prime number*' is a theorem, albeit a rather trivial one, because it has a proof — we simply need to verify that each of the whole numbers $2, 3, \ldots, 16$ is not a factor of 17. A slightly more interesting theorem is the statement '*for all integers n, if n is odd, then n^2 is odd*'. In example 1.3, we give a proof of this statement and we are therefore justified in calling it a theorem.

A statement may fail to be a theorem for one of two reasons. It might be false and therefore no proof is possible. For example, the statement '$2^{256} - 1$ *is prime*' is false (see section 7.4) so, no matter how hard we try, we will not be able to find a proof. The other possibility is that we simply may not know whether or not the statement has a proof. A statement for which no proof or disproof is known is called a **conjecture**. An example of such a statement is: '*every even integer greater than 2 can be expressed as the sum of two primes*'. To date, no-one has been able either to prove the statement or to find an even integer greater than 2 that cannot be expressed as the sum of two primes. Actually, this is a very famous conjecture called Goldbach's Conjecture — see section 2.2.

Conjectures are not generally arbitrarily chosen statements for which a proof is not known. Most often, a conjecture is only stated as such if there is evidence to support it, usually in the form of examples where it is known to be true. To state something as a conjecture really means 'I think this is true but, so far, I have been unable to find a proof'. The status of a conjecture may change over time. Probably the most famous example of this is Fermat's Last Theorem. Pierre de Fermat (1601 – 1665) claimed to have discovered a 'remarkable proof' of the statement: '*for all integers $n > 2$, the equation $x^n + y^n = z^n$ has no solutions where x, y, and z are integers*'. However, Fermat never communicated his proof, so the statement should, more properly, have been known as Fermat's Conjecture. This remained a conjecture until the 1990s when it was proved by Andrew Wiles.[3] With the completion of the proof after over 300 years, the status of the statement changed from 'conjecture' to 'theorem'.

Sometimes a theorem may be called a 'lemma' or a 'corollary'. Although these are still theorems (a lemma or a corollary must still have a proof), the terminology indicates a slightly different 'status'. The proofs of some theorems are long and are best split into separate parts or phases. It may be sensible to prove some preliminary results first, separate from the main proof; then these preliminary results may be used within the main proof itself. In situations like this, the preliminary results are usually referred to as **lemmas**. Labelling

---

[3] The proof was originally announced in 1993, but a gap in the reasoning was subsequently found. Wiles and a colleague, Richard Taylor, were able to 'plug the gap' and complete the proof in 1994.

a theorem as a lemma probably indicates that the result is not particularly interesting in its own right and its principal purpose is as a 'stepping stone' towards what might be called the 'main' theorem. There are exceptions, of course; for example, Zorn's Lemma, mentioned at the beginning of this section, is an important result in its own right. In contrast to a lemma, a **corollary** generally follows a main theorem and is a relatively simple consequence of it. A corollary may simply be a special case of the preceding theorem or it may follow from it by a simple deduction. For example, a corollary of Pythagoras' theorem about right-angled triangles is that the diagonal of a square, with side length 1, is $\sqrt{2}$.

1.3 Reasoning

Human beings are constantly reasoning about their environment in order to make decisions and to guide actions. Mostly this reasoning will be informal and very often it will even be subconscious. Here are a few examples.

- *I know it takes me an hour to prepare, cook, and eat this evening's meal. I need to leave home at 8.30 pm to go out with my friends. So I need to start preparing my evening meal by 7.30 pm at the latest.*

- *The car approaching in that direction is signaling to turn off the road. The car approaching from the other direction is going slowly. Therefore it is safe for me to cross the road.*

- *When people catch a cold they cough and sneeze. If I am in close proximity to someone with a cold, I am likely to catch it. Mike is coughing and sneezing. So I should stay away from him.*

- *I know that whenever I drop an object, it will fall to the ground. If a glass falls to the ground, it will smash. So I must be very careful carrying this tray of glasses.*

Most human reasoning is based on experience, either our own direct experience or some shared, collective experience. For example, consider the following statement, related to the last example above.

If I drop this glass, then it will hit the floor and smash.

At some time or other, most of us will either have dropped a glass and seen it smash or observed someone else do so. Our reasoning here is therefore based on direct experience. Experience tells us that when we drop glasses they (a) fall to the floor and (b) when they hit the floor, they smash. Hence we deduce that if I drop *this glass*, then it will hit the floor and smash.

The kind of reasoning based on experience and experimentation is called **inductive reasoning**. Individual or collective experience tells us that a certain action (for example, dropping a glass) results in a particular consequence (the glass hits the floor and smashes) because:

- we have observed 'many' times that the certain action results in the particular consequence; and

- we have never — or almost never — observed the certain action *not* to result in the particular consequence.

This 'reasoning from experience' will not always be from direct, personal experience but may be based on collective experience of a group (or indeed, the collective experience of human kind). For example, neither of the authors of this book has ever sky-dived although we know people who have. Nevertheless, we believe that the action of sky-diving results in the sky-diver experiencing an adrenalin rush. This belief is based on the collective experience of those who have engaged in the activity of sky-diving. Similarly, imagine a scenario where a young child is playing with a pair of scissors. The child's mother might say:

> *Don't cut the lamp flex! If you cut the flex, then you will*
> *get electrocuted and it will hurt you.*

In this case, we hope that the statement is not based on the direct experience of the mother or child. However, the mother still 'knows' that cutting a live electric flex causes electrocution and that hurts! This is part of our shared collective experience; some people have experienced this directly and this experience is then shared more widely.

Inductive reasoning — reasoning based on (direct or indirect) experience and observation — is probably the basis for most human knowledge and belief. But it does not provide absolute certainty. (The next time I drop a glass it might not break.) This is why, for example, in criminal law, the test of guilt is 'proof beyond reasonable doubt' rather than proof beyond *all* doubt. In assembling a web of evidence — forensic, eye-witness, expert opinion, circumstantial, etc. — the criminal prosecutor can rarely, if ever, achieve absolute certainty of guilt.

Inductive reasoning is also the basis for acceptance of scientific theories. A scientific theory can be thought of as an intellectual framework (which is frequently mathematical in nature) that explains existing observations and predicts new ones. The more experiments that produce data consistent with the theory, the stronger our belief in the theory becomes, but no amount of experimental data will *prove* a theory to be correct. A single experiment (conforming to the established scientific criteria of reliability and repeatability) that conflicts with the theoretical framework can invalidate it. Indeed, scientific breakthroughs are frequently the result of experiments that do not 'fit' with existing theories. In this way, it is sometimes said that scientific theories are falsifiable but never provable.

The type of reasoning used in mathematics is of a different kind altogether. In mathematics, the framework for belief is based on **deductive reasoning**, where facts are deduced from others by a process of logical deduction. As we mentioned in section 1.1, mathematicians' criteria for belief is based on mathematical proof: a correct logical argument that proceeds by making logical deductions from a collection of premises to a conclusion. For a mathematician, a proof eliminates doubt: if we accept that the premises are true, then inevitably we must accept the conclusion to be true also. Consider, for example, Pythagoras' theorem:

> *In any right-angled triangle, the area of the square on the hypotenuse equals the sum of the areas of the squares on the other two sides.*

For mathematicians, this is true because there is a proof of it. Even if we have constructed many hundreds, or even thousands, of right-angle triangles, measured the lengths of the sides, and verified Pythagoras' Theorem for each triangle (and have never found a right-angled triangle for which the theorem fails), this evidence would not be sufficient for a mathematician. There are numerous results that are true for many thousands of examples, but which are not true in general. For the mathematician, therefore, only rigorous logical deduction is sufficient to guarantee the truth of a result in all cases. We explore this kind of deductive reasoning in more detail in the next section.

Although only deductive reasoning is allowed in establishing mathematical facts, inductive reasoning still has an important place in mathematics. In general, theorems begin life as conjectures and conjectures are formulated through inductive reasoning. Consider Pythagoras' theorem about right-angled triangles. The formulation of the theorem almost certainly came about by observing examples of right-angled triangles, calculating the squares of the lengths of the sides, and observing that, in each case, the square of the length of the hypotenuse equals the sum of the squares of the lengths of the other sides. Only once the statement of a theorem — more properly, the statement of a conjecture — has been formulated, can the search for a proof begin.

1.4   Deductive reasoning and truth

In mathematics, we require a standard of reasoning that is both precise — there should be no ambiguity — and logically rigorous — we may only draw conclusions that 'follow inevitably' from what we already know. Let us explore in a bit more detail what it means to say that one statement follows inevitably from others.

Example 1.1

Consider the following piece of reasoning.

> The square of any even integer is also even.
> 1234567893 is odd.
> ---
> Therefore $(1234567893)^2$ is odd.

Is this reasoning valid? Firstly, we need to explain the notation used. In a fragment of reasoning presented in this way, the (implicit) claim is that the statement below the line follows logically from the statements above the line.

To ask whether the reasoning is valid is *not* to ask whether or not the individual statements are true. Rather, it is asking whether the conclusion, which is the statement below the line following 'Therefore ...', *necessarily* follows from the previous statements. In this case, each of the three individual statements is true, but the reasoning itself is *not* valid. This is important but quite subtle. The point is that last statement '$(1234567893)^2$ is odd' is true but its truth *does not follow from* the truth of the previous statements. To see this, suppose we replace the term 'odd' with 'prime' and, to be consistent, we also replace the term 'even' (meaning 'not odd') with 'composite' (meaning 'not prime'). Then the reasoning is the following.

> The square of any composite integer is also composite.
> 1234567893 is prime.
> ---
> Therefore $(1234567893)^2$ is prime.

In this case the first two statements are true as before but the final statement is false. An obvious requirement of any piece of valid reasoning must be that, if the initial (above the line) statements are true, then the (below the line) conclusion must also be true. Hence this second reasoning fragment is clearly invalid. Each of the two reasoning fragments has the following form.

> If any integer has property P, then its square also has property P.
> A particular integer n does not have property P.
> ---
> Therefore n^2 does not have property P.

The second instance of this form of reasoning shows that it is not valid. The purported conclusion does not 'follow inevitably' from the given initial statements. The important point to note here is that whether a piece of reasoning is valid (or not) depends on the *form* of the reasoning fragment, not on the truth of the individual statements. In some cases, it will not be immediately or intuitively obvious whether a reasoning fragment is valid, so we will explore this further in chapter 2. For example, the following form of reasoning, which is closely related to that above, is valid.

> If any integer has property P, then its square also has property P.
> For a particular integer n, its square n^2 does not have property P.
> _____
> Therefore n does not have property P.

Example 1.2

Now consider the following piece of reasoning.

> For any prime integer n, the integer $2^n - 1$ is also prime.
> 23 is prime.
> _____
> Therefore $2^{23} - 1 = 8\,388\,607$ is prime.

In this case, the conclusion is false because $2^{23} - 1$ factorises as follows

$$2^{23} - 1 = 8\,388\,607 = 47 \times 178481.$$

Does this mean that the reasoning itself is invalid? In fact, the real reason why the conclusion is false is because one of the 'initial' statements is false. The statement

> '*if n is a prime integer then $2^n - 1$ is also prime*'

is false. Indeed, the example of $n = 23$ *shows* that this statement is false. There are, in fact, many examples of prime numbers n for which $2^n - 1$ is not prime. However, this does not mean that the reasoning itself is incorrect. Were it to be the case that each of the statements

> *for any prime positive integer n, the integer $2^n - 1$ is also prime*

and   23 *is prime*

was true, then it would necessarily be true that $2^{23} - 1$ would be prime.

The structure of the reasoning fragment may be described as follows.

> For any n, if n has property P, then n has property Q.
> A particular k has property P.
> _____
> Therefore k has property Q.

This is, indeed, a valid piece of reasoning. Again, this example emphasises that whether a piece of reasoning is valid depends on the structure of the fragment and not the truth of the individual statements.

As we have noted, establishing truth in mathematics depends on making valid logical deductions. However, the two previous examples show that the validity of a piece of reasoning does not, by itself, guarantee the truth of the conclusion. Equally, the truth of a conclusion does not mean that the reasoning that gave rise to it was valid. The important situation is when the initial statements are true *and* the reasoning is valid; then the conclusion is guaranteed to be true also. This is the situation in a mathematical proof which starts with true statements and makes a sequence valid of logical deductions from these until a conclusion is reached. Provided the starting statements are true and the reasoning is valid, then the conclusion must also be true. In the following section, we introduce two proofs to give a flavour of what is to come in the rest of the book.

1.5 Example proofs

We conclude this chapter by giving proofs of two simple results about the integers or whole numbers. In each case, readers will be able convince themselves of the truth of the result 'beyond reasonable doubt' by considering a selection of example integers and checking the result in each case. Our proofs will be short deductive proofs that, we hope, establish the results 'beyond *all* doubt'.

We shall examine the structure of mathematical proofs in more detail later but, in broad terms, a proof of a theorem must show that, given the truth of certain 'precursor' statements that will be assumed as part of the reader's 'background knowledge', the truth of the theorem inevitably follows. Normally, such a proof takes the form of a sequence of statements whose truth is guaranteed either because it is assumed background knowledge or because it follows from earlier statements in the proof itself. In other words, the truth of each statement in the proof is established using valid deductive reasoning from the truth of other statements whose truth is known or assumed.

Example 1.3
Consider the following theorem. In this example, we will give a proof of the theorem (which is why it is legitimate to call it a theorem). Although we will try not to labour this too much, we will examine the statement of the theorem and its proof in rather more detail than is normal in a mathematics text.

Theorem 1.1
For all integers n, if n is odd, then n^2 is odd.

Before we think about proving this theorem, we first need to understand precisely

(a) the overall logical structure of the statement and

(b) the meaning of the terms used.

We will explore the structure of individual statements such as this more fully in chapter 2. For now, it is sufficient to note that this is a *conditional statement* of the form

$$\text{if (statement 1), then (statement 2).}$$

We will explore conditional statements in more detail in chapter 2. For now, we need to note that a conditional statement is true provided statement 2 is true in all circumstances in which statement 1 is true. This means that we need to establish that 'n^2 *is odd*' is true in all circumstances that 'n *is odd*' is true.

The term 'integer' just means 'whole number'. What does it mean to say that an integer is *odd*? This probably seems like a curious question since we anticipate that all readers will know that the odd integers are

$$\ldots, -7, -5, -3, -1, 1, 3, 5, 7, \ldots \ .$$

However, knowing this list of the odd integers is not sufficient for our purposes. We need to establish that, whenever n is in this list of odd integers, its square n^2 is also in the list. What we really need is a *property* that defines what it means (for an integer) to be odd. Then we can establish that whenever n has this 'oddness property', then n^2 also has the 'oddness property'. The property we are seeking is the following.

Oddness property: An integer n is odd precisely when it can be expressed as $n = 2m + 1$ for some (other) integer m.

For example, 17 is odd because $17 = 2 \times 8 + 1$; -243 is odd because $-243 = 2 \times (-122) + 1$. Similarly, 26 is not odd because it cannot be expressed as $2m + 1$ where m is an integer. (We *can* express $26 = 2m + 1$, but then $m = 12.5$ is not an integer.)

Now that we have understood both the structure of the statement and the meaning of the terms used, we can give a proof.

Proof. Let n be an odd integer.

Then, by the oddness property, $n = 2m + 1$ for some integer m. Now

$$\begin{aligned}
n^2 &= (2m + 1)^2 \\
&= 4m^2 + 4m + 1 \\
&= 2(2m^2 + 2m) + 1 \\
&= 2M + 1 \quad \text{where } M = 2m^2 + 2m.
\end{aligned}$$

We have shown that $n^2 = 2M + 1$ where $M = 2m^2 + 2m$ is an integer. Hence, by the oddness property, n^2 is odd. □

Example 1.4

In this example, we will give alternate proofs of the following theorem.

Theorem 1.2

For all integers n, $n^2 + n$ is even.

As in the previous example, we need the following property that defines what it means for an integer to be even.

Evenness property: An integer n is even precisely when it can be expressed as $n = 2m$ for some (other) integer m.

Our first proof considers separately the two possibilities for n: either it is even or it is odd. Then it uses the evenness or oddness properties for n, together with a little algebra, to deduce the evenness property for $n^2 + n$.

Proof 1.

There are two cases to consider: n is even or n is odd.

Case 1: n is even.

Then $n = 2m$ for some integer m. Therefore

$$
\begin{aligned}
n^2 + n &= (2m)^2 + 2m \\
&= 4m^2 + 2m \\
&= 2(2m^2 + m), \quad \text{where } 2m^2 + m \text{ is an integer.}
\end{aligned}
$$

Therefore $n^2 + n$ is even in this case.

Case 2: n is odd.

Then $n = 2m + 1$ for some integer m. Therefore

$$
\begin{aligned}
n^2 + n &= (2m + 1)^2 + 2m + 1 \\
&= 4m^2 + 4m + 1 + 2m + 1 \\
&= 4m^2 + 6m + 2 \\
&= 2(2m^2 + 3m + 1), \quad \text{where } 2m^2 + 3m + 1 \text{ is an integer.}
\end{aligned}
$$

Therefore $n^2 + n$ is even in this case also.

In both cases $n^2 + n$ is even, which completes the proof. □

This proof is probably the most obvious way of establishing the given result, especially given our proof in the previous example. It is not unusual to break a proof down into separate cases and consider each separately. The only prerequisite knowledge for the proof is the evenness and oddness properties and some basic algebra. Compare this with the following, much shorter proof.

Proof 2.

Let n be an integer. Then $n^2 + n = n(n + 1)$.

Now n and $n + 1$ are consecutive integers, so one of them is even and the other is odd (but we don't know which is which). Hence their product is the product of an even and an odd integer, which is even.

Hence $n^2 + n = n(n + 1)$ is even.   □

This second proof is clearly much shorter, avoids the need for separate cases and requires much less algebra. However, it does require the clever observation that factorising $n^2 + n$ provides a helpful first step. This proof also assumes an additional piece of background knowledge that the product of an even and an odd integer is even. If we accept this as fact, then surely the second proof is simpler and, arguably, more elegant. Some readers might reasonably object that the proof is not complete as it has not established the 'missing' piece of background information. To overcome these objections, we could replace the penultimate sentence with the following.

> The product of an even and an odd integer is of the form $2p \times (2q + 1) = 2(p(2q + 1))$, for some integers p and q, which is even.

Exactly how much background information may be assumed and how much must be included explicitly in a proof will always be a matter of judgement by the person constructing the proof.

No doubt some readers will prefer our first proof and some will think the second one is better. That is as it should be — there is no 'right answer' to the question, which is the better proof?

Chapter 2

Logic and Reasoning

2.1 Introduction

In this chapter, we consider in more detail the notions of 'statements' and 'deductive reasoning' that were introduced in chapter 1. We aim to understand the structure of both individual statements and whole logical arguments. Since mathematical proofs are structured, logical arguments, a working knowledge of basic logic and logical reasoning will be important in understanding the structure and correctness of mathematical proofs.

As we have seen, some statements are about specific objects, such as '17 *is prime*'. Others concern whole classes of objects; for example, *'for all positive integers n, if $2^n - 1$ is prime, then n is prime'*. Some statements are 'atomic' and cannot be broken down further; others are 'compound ' and are built from simpler statements into more complicated structures. An example of the latter type of statement is: *'if n is even and m is odd, then the product mn is even'*. Understanding how statements are put together and how this affects their truth is one of the goals of this chapter. Similarly, recognising when and why a statement is not true is important. Consider the following imaginary dialogue, for example.

Person A: I've noticed that, whenever $x + y$ is positive, xy is also positive.

Person B: No, sorry, you're wrong because, for example, when $x = 1$ and $y = -2$ both $x + y$ and xy are negative.

Who, if anyone, is correct? In this case, neither person is correct. Person A's statement is certainly false. Although Person B's example is correct, this is not the negation of Person A's statement and so is not the reason why Person A is wrong.

In the previous chapter, we considered reasoning fragments that purported to show that one statement 'follows inevitably' from a collection of other statements. Later in this chapter we will examine both what it means to say one statement 'follows logically' from other statements and also when a piece of reasoning establishes this connection.

2.2   Propositions, connectives, and truth tables

Propositions are statements that are either true, T, or false, F. The truth or falsity of a proposition is called its **truth value**. Since there are only two truth values, a proposition that is not true is regarded as being false and a proposition that is not false is regarded as being true. We will generally use upper case letters, P, Q, R, etc. to denote propositions. Here are some examples of propositions together with their truth values where these can be determined.

1. *The Potomac river flows into the Chesapeake bay.*   T

2. *Brighton is the capital city of the United Kingdom.*   F

3. $2 + 2 = 4$.   T

4. *Today is my birthday.*   depends

5. *The digit in the 7th position after the decimal point*   T
   in the decimal expansion of π is 6.

6. *The digit in the 200 millionth position after the dec-*   ?
   imal point in the decimal expansion of π is 6.

7. *Every even integer greater than 2 can be expressed as*   unknown
   the sum of two primes.

Note that in some cases we will know the truth value of a proposition, but in others we may not. For example, in the first two statements above, we need to know some geography (of North America and the British Isles respectively) in order to know their truth values. The truth value of proposition 4, 'today is my birthday', depends on who utters the statement and when. If the statement is true when I utter it, then it will be false if I utter the statement 24 hours later. The two sentences involving the decimal expansion of π are propositions since they are definitely true or false. We can identify that proposition 5 is true using a calculator, spreadsheet, or other software to evaluate $\pi = 3.14159265\ldots$. Proposition 6 is a little more tricky. Although the 200 millionth decimal digit of π is known,[1] we (the authors) do not know its value. So, although we know that statement 6 is definitely true or false, we don't actually know which it is.

Statement 7 is a famous unsolved problem in mathematics. It is known as the Goldbach Conjecture and was first stated in 1742 by the German mathematician Christian Goldbach in correspondence with Leonhard Euler. Although it has been verified that every even integer up some vast number (at least

---

[1] In fact, π has been evaluated to several trillion digits.

10^{17}) is indeed expressible as the sum of two primes, it is still currently unknown whether *every* even integer greater than 2 can be so expressed. Hence, although statement 7 is definitely either true or false, as yet no one knows which.[2]

Sentences that cannot be viewed as true or false are not propositions. These include instructions, questions, exclamations, exhortations, demands, opinions, and so forth. Hence the following are not propositions.

1. *Keep off the grass.* an instruction

2. *Formal methods are vital for software development.* an opinion

3. *Will you come to my party?* a question

Negation

Given a proposition P, adding the prefix 'It is not the case that ... ' or inserting 'not' in an appropriate place in the sentence results in another proposition with the reverse truth value. The resulting proposition is called the **negation** of P, which we will denote $\neg P$. This is usually read as 'not P'. Other notations for negation are $\bar{P}$ and $\sim P$.

Example 2.1
Let P be the proposition '*Today is my birthday.*' Then the negation of P is the proposition '*It is not the case that today is my birthday.*' This is a rather unnatural sentence and a more natural way of expressing $\neg P$ is '*Today is **not** my birthday.*'

In example 2.1, we noted that there is more than one way to express negation. Whilst adding the prefix 'It is not the case that ... ' to P will always produce the negation $\neg P$, there is usually a more natural-sounding sentence for the negation. From the point of view of logic, what is important is that $\neg P$ is true when P is false and $\neg P$ is false when P is true. We summarise this in the following table, called the **truth table** for negation.

---

[2] In 2013, a related conjecture called Goldbach's Ternary Conjecture was proved by Harald Helfgott, a Peruvian mathematician working in Paris. This states that every odd integer greater than 5 is the sum of three primes. It is a weaker conjecture than Goldbach's original conjecture: if every even integer greater than 2 can be expressed as the sum of two primes, then adding 3 to each sum of two primes proves Goldbach's ternary conjecture. Even the weaker ternary conjecture was extraordinarily difficult to prove. Not only did it involve some highly sophisticated number theory, but it also required approximately 150,000 hours (17 years) of computer processing. In fact, this was part of a larger calculation that took approximately 440,000 hours (50 years) of computer processing. The experts still believe that a proof of the original Goldbach conjecture is a long way off.

$$\begin{array}{c|c} P & \neg P \\ \hline T & F \\ F & T \end{array}$$

Connectives and compound propositions

The examples of propositions given above are **simple propositions** because they make a single statement. By contrast, the proposition

Today is my birthday and I am having a party

is not a simple proposition since it makes two distinct statements. We view this proposition as being composed of the two simple propositions 'today is my birthday' and 'I am having a party' conjoined using the word 'and'. It is an example of a compound proposition, and its truth value depends on the truth values of the two component simple propositions. For example, if today is not my birthday, but nevertheless, I am still having a party, then the compound proposition is false.

In general, we may combine simple propositions together using **connectives** (such as 'and' and 'or') to form **compound propositions**. The truth value of a compound proposition depends on the truth values of its component parts as well as the connective(s) used. For example, the proposition above is only true when both the simple propositions 'today is my birthday' and 'I am having a party' are true; in all other cases, the compound proposition is false. There are five connectives that we will consider: conjunction, inclusive disjunction, exclusive disjunction, conditional, and biconditional.

Conjunction

The **conjunction** of two propositions is generally expressed by using '*and*' between the two propositions. If P and Q are propositions, then their conjunction is the proposition 'P and Q'. The conjunction is denoted $P \wedge Q$ and this is usually read as 'P and Q'. The truth values of the conjunction $P \wedge Q$ are defined by the following truth table.

$$\begin{array}{cc|c} P & Q & P \wedge Q \\ \hline T & T & T \\ T & F & F \\ F & T & F \\ F & F & F \end{array}$$

Note that there are four rows to the truth table because there are four possible combinations of truth values of P and Q. It is important to realise that the truth table defines what we mean by conjunction. Thus the conjunction $P \wedge Q$ has truth value T only when both the component propositions P and Q, sometimes called the **conjuncts**, have truth value T; in all other cases $P \wedge Q$ has truth value F. There are other ways of expressing conjunction in English other than using the word 'and'. Consider the following propositions.

1. *Sue went to the party last night* **but** *Jamie didn't.*
2. *Hamid passed the module* **whereas** *Mike did not.*
3. *Jim played football* **although** *he was suffering from a cold.*

In each case, the proposition is true only when both of its component simple propositions are true; hence, each proposition is the conjunction of its two components. The English sentences each have slightly different nuances compared with the corresponding sentences using 'and' to form the conjunction. For example, the third proposition hints at a little surprise that Jim played football while suffering from a cold. Any natural language, such as English, can express a vast range of subtlety, emotion, and nuance that we are unable to capture in logical expressions. However natural languages are also prone to ambiguity. It is for this reason that legal documents, for example, which seek to eliminate ambiguity, are written in a rather formulaic manner that does not follow the way most people speak. Although logical expressions are unable to capture subtlety, provided they are properly formulated, they can express statements precisely and unambiguously.

Disjunction

The **disjunction**, or **inclusive disjunction**, of two propositions is generally expressed by using '*or*' between the two propositions. If P and Q are propositions, then their disjunction is the proposition 'P or Q'. The disjunction is denoted $P \vee Q$ and this is read as 'P or Q'. The truth values of the conjunction $P \vee Q$ are defined by the following truth table.

P	Q	$P \vee Q$
T	T	T
T	F	T
F	T	T
F	F	F

Again, we regard the truth table as defining the meaning of disjunction. Thus $P \vee Q$ is false only when both component propositions (called the **disjuncts**) are false; if either or both of the disjuncts are true, then the disjunction is true. In particular, note that $P \lor Q$ is true when both P and Q are true. For example, the disjunction

> *Today is my birthday or I am having a party*

should surely be regarded as true when I am enjoying my birthday party. This is the so-called *inclusive* use of disjunction and it is the default interpretation in logic.

There is an alternative form of disjunction, called **exclusive disjunction**, that is denoted $P \veebar Q$. In this case, the (exclusive) disjunction of P and Q is false when both P and Q are true. The truth values of $P \veebar Q$ are defined by the following truth table.

P	Q	$P \veebar Q$
T	T	F
T	F	T
F	T	T
F	F	F

For example, the statement

> *Next week, I will work at home or visit a friend in Paris*

suggests an exclusive interpretation of 'or'. The two disjuncts, 'I will work at home' and 'I will visit a friend in Paris', appear to be mutually exclusive. In this case, we may interpret the proposition as true only when just one of the two disjuncts is true. However, even in this case, one could imagine a scenario where both conjuncts are true — suppose I live in Paris, for example. In logic, the default interpretation of 'or' is the inclusive disjunction. If we wish to ensure that a disjunction is given the exclusive interpretation, we would normally use the construction

> 'P or Q, but not both'.

This often leads to sentences which are somewhat unnatural in English. For example, *you will receive a fine or you have diplomatic immunity but not both.* However, since disjunctive statements can often be interpreted inclusively, we will add 'but not both' whenever the exclusive interpretation is the intended one.

Conditional

A conditional proposition is commonly denoted in English by a phrase of the form

> 'if ... then ...'.

Here are some typical examples.

1. *If the backspace key is pressed, then the cursor moves one place to the left.*

2. *If I leave my umbrella at home, then it will rain.*

3. *If you break the speed limit, you will receive a fine.*

4. *If n is a prime number, then n is not a perfect square.*

Note that the word 'then' is sometimes omitted from the phrase 'if ... then ...'. This is illustrated in the third proposition above.

If P and Q are propositions, then the **conditional** '*if P then Q*' is denoted $P \Rightarrow Q$. Here the proposition P is called the **antecedent** and the proposition Q is called the **consequent** of the conditional proposition. Note that $P \Rightarrow Q$ may also be expressed as '*P implies Q*'. The truth table for $P \Rightarrow Q$ is given below.

P	Q	$P \Rightarrow Q$
T	T	T
T	F	F
F	T	T
F	F	T

Note that a conditional proposition $P \Rightarrow Q$ is true in all cases except where the antecedent P is true and the consequent Q is false. It conforms to normal usage that the phrase 'if P, then Q' should be regarded as false in the case where P is true and Q is false. For example, suppose that, on a particular day, I leave my umbrella at home but it does not rain that day. In this case the statement

If I leave my umbrella at home, then it will rain

is certainly false. What is usually less easily appreciated is why the conditional $P \Rightarrow Q$ is true in each of the last two rows; in other words, $P \Rightarrow Q$ is true when P is false *regardless* of the truth value of Q. This is best understood by considering a suitable example. Consider the first conditional proposition listed above

If the backspace key is pressed, then the cursor moves one place to the left

and consider the cases when the backspace key (on a computer keyboard) is *not* pressed. The proposition makes no claims about what happens in this case. If the cursor remains in the same place, then the conditional proposition is true. Equally, if the cursor *does* move one place to the left (possibly caused by a mouse or touchpad, for example) then the conditional statement is also true (because it cannot be reasonably be regarded as false).

Similar reasoning can be applied to the other examples listed above. Consider, for example, the situation where I carry my umbrella with me when I leave home. Regardless of whether or not it then rains, the statement

> *If I leave my umbrella at home, then it will rain*

cannot be regarded as false, so must therefore be considered true.

There are other ways of expressing conditional propositions. As noted above, the phrase '*P implies Q*' is also denoted by $P \Rightarrow Q$, although this is used more frequently in mathematics than in everyday speech. Thus

> *n is a prime number implies n is not a perfect square*

is a natural-sounding sentence in mathematics, whereas

> *I leave my umbrella at home implies it will rain*

would sound a little curious in everyday conversation. It is for this reason that $P \Rightarrow Q$ is often read as 'if P, then Q' instead of 'P implies Q'. In mathematics, if $P \Rightarrow Q$ we also say that P is a **sufficient condition** for Q and that Q is a **necessary condition** for P.

It is also worth noting that $P \Rightarrow Q$ is sometimes denoted by the phrase 'Q if P' where the order of P and Q in the sentence is reversed. Thus

> *You will receive a fine if you break the speed limit*

is an alternative way of expressing '*If you break the speed limit, then you will receive a fine.*'

Unlike conjunction and disjunction, conditional statements are not symmetric in their component propositions. Conjunction and disjunction are symmetric in the sense that $P \wedge Q$ and $Q \wedge P$ have the same meaning as do $P \vee Q$ and $Q \vee P$. However, $P \Rightarrow Q$ and $Q \Rightarrow P$ are not equivalent logically. For example, the conditional propositions

> *If you break the speed limit, then you will receive a fine*

and

> *If you receive a fine, then you will break the speed limit*

have quite different meanings.

We can illustrate this generally by comparing the truth tables for $P \Rightarrow Q$ and $Q \Rightarrow P$ as shown below.

P	Q	$P \Rightarrow Q$	$Q \Rightarrow P$
T	T	T	T
T	F	F	T
F	T	T	F
F	F	T	T

This illustrates that $P \Rightarrow Q$ and $Q \Rightarrow P$ only have the same truth value when P and Q are either both true or both false. The conditional proposition $Q \Rightarrow P$ is called the **converse** of $P \Rightarrow Q$.

Biconditional

The **biconditional** connective is usually expressed in English by the phrase '...if and only if ...' usually between the two component propositions. An example is:

The backspace key is pressed if and only if the cursor moves one place to the left.

If P and Q are propositions, then the biconditional 'P if and only if Q' is denoted by $P \Leftrightarrow Q$. The truth table for the biconditional is given below. Note that the biconditional $P \Leftrightarrow Q$ is true when P and Q have the same truth values, both true or both false, and $P \Leftrightarrow Q$ is false when P and Q have opposite truth values.

P	Q	$P \Leftrightarrow Q$
T	T	T
T	F	F
F	T	F
F	F	T

As we have seen with other connectives, there are other ways of expressing a biconditional proposition. An alternative to 'P if and only if Q' is to use the phrase 'If and only if P, then Q' as in the following example.

If and only if the backspace key is pressed, then the cursor moves one place to the left.

However, since the truth values of $P \Leftrightarrow Q$ are symmetric in P and Q, we usually prefer the more symmetric way of expressing $P \Leftrightarrow Q$ as 'P if and only if Q'. In the case where $P \Leftrightarrow Q$ we also say that P is a necessary and sufficient condition for Q (and also that Q is a necessary and sufficient condition for P). The biconditional is not frequently used in everyday discourse, which is one reason why the linguistic constructions for it appear a little odd.

Truth tables

Using negation and the five connectives defined above, we can build more complicated compound propositions such as

$$(P \wedge Q) \Rightarrow \neg R, \quad \neg(P \vee Q) \wedge (P \vee S) \text{ and } (P \vee \neg Q) \Rightarrow (P \wedge Q).$$

In order to obtain the truth values for these more complicated propositions, we may build their truth tables column by column as we now illustrate.

Examples 2.2

1. Construct a truth table for $(P \vee \neg Q) \Rightarrow (P \wedge Q)$.

   Solution

   Firstly, note that there are two simple propositions, P and Q, involved in the compound proposition so, to account for the possible truth values, we will require four rows to the truth table. For the sake of consistency, we will always list the truth values in the following order.

P	Q	...
T	T	...
T	F	...
F	T	...
F	F	...

   Before we can find the truth values of this conditional proposition, we need to obtain the truth values of the antecedent $P \vee \neg Q$ and the consequent $P \wedge Q$. The antecedent also requires the negation of Q, so we build the next three columns of the truth table as follows.

P	Q	$\neg Q$	$P \vee \neg Q$	$P \wedge Q$
T	T	F	T	T
T	F	T	T	F
F	T	F	F	F
F	F	T	T	F

   Finally, we can complete the table using the truth values in the last two columns to evaluate the truth values of $(P \vee \neg Q) \Rightarrow (P \wedge Q)$. We can do this in two ways.

   Firstly, referring to the truth table of the conditional $P \Rightarrow Q$ on page 21, note that a conditional proposition is true in all cases where the antecedent is false. In our example, that situation only occurs in row 3 where $P \lor \neg Q$ is false; hence in row 3, $(P \lor \neg Q) \Rightarrow (P \land Q)$ has truth value T. Referring again to the truth table of $P \Rightarrow Q$, note that, when the antecedent is true, the truth values of a conditional agree with the truth values of its consequent. In our example, the antecedent is true in rows 1, 2 and 4 so, in these rows, the truth values of $(P \lor \neg Q) \Rightarrow (P \land Q)$ follow those of its consequent $P \land Q$.

P	Q	$\neg Q$	$P \lor \neg Q$	$P \land Q$	$(P \lor \neg Q) \Rightarrow (P \land Q)$
T	T	F	T	T	T
T	F	T	T	F	F
F	T	F	F	F	T
F	F	T	T	F	F

The alternative way of obtaining the final column of truth values is to note that the only occasion when a conditional is false is when its antecedent is true and its consequent is false. This occurs in rows 2 and 4, so $(P \lor \neg Q) \Rightarrow (P \land Q)$ is false in these rows and true otherwise.

Note that the vertical lines between columns are not significant to the meaning of the truth table. They may be inserted appropriately to break up the table and make it easier to read.

2. Construct a truth table for $(P \lor Q) \Rightarrow (Q \land R)$.

Solution

Since there are three simple propositions P, Q, and R involved in the compound proposition, we will need eight rows in the truth table to capture all the possible combinations of truth values. Again, for the sake of consistency and to ensure that we always list all eight different possibilities, we will always list the truth values in the order shown below.

P	Q	R	...
T	T	T	...
T	T	F	...
T	F	T	...
T	F	F	...
F	T	T	...
F	T	F	...
F	F	T	...
F	F	F	...

To complete the truth table for $(P \lor Q) \Rightarrow (Q \land R)$, we first need to construct columns containing the truth values of $P \vee Q$ and $Q \wedge R$. Then we can use the reasoning outlined in the previous example to obtain the truth values for the conditional $(P \vee Q) \Rightarrow (Q \wedge R)$. In this way, we obtain the following truth table.

P	Q	R	$P \vee Q$	$Q \wedge R$	$(P \vee Q) \Rightarrow (Q \wedge R)$
T	T	T	T	T	T
T	T	F	T	F	F
T	F	T	T	F	F
T	F	F	T	F	F
F	T	T	T	T	T
F	T	F	T	F	F
F	F	T	F	F	T
F	F	F	F	F	T

3. Construct a truth table for $(P \wedge \neg R) \Leftrightarrow (Q \wedge (P \vee R))$.

Solution

Building up the truth table column by column, we obtain the truth table shown below. Note that, so that the table does not become unwieldy, we have labelled $P \wedge \neg R$ as A (column 5) and $Q \wedge \neg(P \vee R)$ as B (column 7). This allows us to express the compound proposition $(P \wedge \neg R) \Leftrightarrow (Q \wedge (P \vee R))$ simply as $A \Leftrightarrow B$.

Observing the truth table for the biconditional proposition $P \Leftrightarrow Q$ on page 23, we may note that a biconditional is true when its component simple propositions have the same truth value and is false when its component simple propositions have the opposite truth values. This observation allows us to complete the final column of the truth table.

				A		B	
P	Q	R	$\neg R$	$P \wedge \neg R$	$P \vee R$	$Q \wedge (P \vee R)$	$A \Leftrightarrow B$
T	T	T	F	F	T	T	F
T	T	F	T	T	T	T	T
T	F	T	F	F	T	F	T
T	F	F	T	T	T	F	F
F	T	T	F	F	T	T	F
F	T	F	T	F	F	F	T
F	F	T	F	F	T	F	T
F	F	F	T	F	F	F	T

Logic and natural language

As we have seen, natural language is more expressive than propositional logic as it is able to express nuances that we cannot capture using logical expressions. For example, each of the two English sentences

> *Max played for the first team and he scored the wining try*

and

> *Max played for the first team although he was nursing an injury*

is logically a conjunction. However, the different way the sentences are expressed may carry nuances that would not be captured simply by saying that each is a conjunction $P \wedge Q$.

As we have noted, natural language is also prone to ambiguity. It is for this reason, for example, that the language of legal documents, such as contracts, often seems complicated and contrived as it attempts to eliminate ambiguity. Some sentences in English may have more than one completely different meaning. Consider, for example, the meaning of the sentence, '*She hit the man with a walking stick.*' Similarly, small changes in punctuation can change the meaning of English sentences. A well-known example is the phrase, '*Eats, shoots and leaves*'[3], which has quite different meanings with and without the comma. Of course, many comedians use the ambiguity of natural language to great humorous effect.

It should be clear, then, that it is not always straightforward 'translating' between logic and natural language. Using brackets, logical sentences can convey quite complicated propositions which natural language may sometimes struggle to express. For example, consider how we might distinguish in English between two similar sentences with logical structures $(P \wedge Q) \vee R$ and $P \wedge (Q \vee R)$ respectively.

Examples 2.3

1. Let P: *The sun is shining.*
             Q: *It is raining.*
             R: *There is a rainbow.*

   Translate each of the following propositions into symbolic form.

   (i) *If the sun is shining and it is raining, then there is a rainbow.*
   (ii) *It is not raining and the sun is shining.*
   (iii) *If there is no rainbow, then the sun is not shining.*

---

[3] This was taken by Lynne Truss as the title for her best-selling book on the importance of correct punctuation.

(iv) *Either it is not raining or, if the sun is shining, then there is a rainbow.*

(v) *There is not a rainbow and it is raining if and only if the sun is not shining.*

Solution

(i) There are two connectives in the sentence: a conditional 'if ... then ...' and a conjunction '... and ...'. However the 'top-level' structure of the sentence is a conditional and the conjunction only appears in the antecedent of the conditional proposition:

> If (*the sun is shining and it is raining*), *then* (*there is a rainbow*).

Therefore we may symbolise the proposition as

$$(P \wedge Q) \Rightarrow R.$$

(ii) This proposition is a conjunction where the first conjunct is the negation of Q. Hence we may symbolise it as

$$(\neg Q) \wedge P.$$

There is a convention in writing logic sentences where negation 'binds more tightly' than the other connectives. This means that the negation symbol $\neg$ only negates what is immediately to its right, so we may omit the brackets and symbolise the proposition as

$$\neg Q \wedge P.$$

Had we wanted to negate the whole conjunction $Q \wedge P$, *it is raining and the sun is shining*, then we would need to use brackets as follows: $\neg(Q \wedge P)$. This convention in logic is similar to one in arithmetic or algebra where the negative also binds more tightly than the binary operations of addition and multiplication. Thus, for example, $-3 + 5$ is interpreted as $(-3) + 5 = 2$ and not as $-(3 + 5) = -8$.

(iii) Again, the highest-level structure of the sentence is a conditional, where the antecedent is the negation of R and the consequent is the negation of P. Hence, using the convention mentioned in part (ii) above, we may symbolise this as

$$\neg R \Rightarrow \neg P.$$

(iv) The sentence contains both a disjunction '(either) ... or ...' and a conditional 'if ... then ...'. In order to symbolise the sentence, we need to decide two things: at the highest level, is the sentence a disjunction or a conditional and is the disjunction inclusive or exclusive? Firstly, it should be clear that the basic structure of the sentence is a disjunction with the conditional forming the second disjunct. The use of 'either ... or ...' — and also knowledge of when rainbows actually occur — may indicate that the disjunction is exclusive, in which case the sentence is symbolised as

$$\neg Q \veebar (P \Rightarrow R).$$

Alternatively, we may regard the use of 'either ... or ...' simply as a device to emphasise that the top level structure is a disjunction rather than a conditional. In this case, we would apply the convention that disjunctions are interpreted as inclusive unless they are explicitly denoted as exclusive and symbolise the sentence as

$$\neg Q \vee (P \Rightarrow R).$$

Either of these is a reasonable interpretation of the English language sentence and we leave the reader to decide which is their preferred interpretation. This again emphasises that natural language may sometimes be ambiguous and open to more than one interpretation.

(v) Clearly this sentence contains a conjunction '... and ...' and a biconditional '... if and only if ...'. However it is not obvious, from the English sentence, how to bracket the simple propositions, so

$$\neg R \wedge (Q \Leftrightarrow \neg P) \text{ and } (\neg R \wedge Q) \Leftrightarrow \neg P$$

are equally valid interpretations. This is a little problematic since we have no reason to choose one interpretation over the other and the two interpretations are logically different.

2. Let P: *Jo is on holiday.*
           Q: *Jo goes swimming.*
           R: *Jo studies logic.*

Translate each of the following propositions into naturally sounding English sentences.

(i) $P \veebar Q$

(ii) $Q \Rightarrow \neg R$

(iii) $R \vee (P \Rightarrow Q)$

(iv) $(R \wedge \neg Q) \Rightarrow \neg P$

(v) $(P \wedge Q) \vee R$

Solution

(i) This is an exclusive disjunction, so we need to ensure that this is made explicit in the natural language sentence. Hence we may express $P \veebar Q$ in English as:

> *Jo is on holiday or she goes swimming but not both.*

Note that, to keep the sentence as natural-sounding as possible, we have used the pronoun 'she' rather than repeat the proper noun 'Jo' twice in the same sentence.

(ii) This is a conditional where the only complication is that the consequent is a negated proposition. It can be expressed as:

> *If Jo goes swimming, then she doesn't study logic.*

(iii) The basic structure of the logic sentence is a disjunction where the second disjunct is a conditional. We will use the construction 'either ...or ...' to emphasise that the top-level structure of the sentence is a disjunction and write it as

> *Either Jo studies logic or, if she is on holiday, she goes swimming.*

Note that we have also omitted the word 'then' from the conditional statement that forms the second disjunct in order to emphasise further that the main connective in the sentence is a disjunction. Some people will prefer to omit 'either' as this hints at an exclusive disjunction and write the sentence instead as:

> *Jo studies logic or, if she is on holiday, she goes swimming.*

(iv) Here we have a conditional where the antecedent is a conjunction. We write this as:

> *If Jo studies logic and doesn't go swimming, then she is not on holiday.*

To emphasise that the principal structure of the sentence is conditional, we have omitted the pronoun 'she' in the second part of the conjunction 'Jo studies logic and she doesn't go swimming.'

(v) It can be tricky in written language to indicate the positioning of the brackets in sentences such as $(P \wedge Q) \vee R$. One way we can hint at the placement of the brackets in this case is to omit the pronoun 'she' in the conjunction and write

> *Jo is on holiday and goes swimming or she studies logic.*

Of course, this would not have been possible had the three simple propositions involved in the sentence $(P \wedge Q) \vee R$ referred to three different objects or individuals. Another option is to use the construction 'either ... or' to indicate the bracketing, as follows.

> *Either Jo is on holiday and she goes swimming or she studies logic.*

This is effective at indicating the bracketing, but there is a downside: some people are inclined to interpret 'either ... or' as an exclusive disjunction.

Tautologies and contradictions

Consider the truth table for the compound proposition $(P \wedge Q) \Rightarrow (P \vee Q)$ given below.

P	Q	$P \wedge Q$	$P \vee Q$	$(P \wedge Q) \Rightarrow (P \vee Q)$
T	T	T	T	T
T	F	F	T	T
F	T	F	T	T
F	F	F	F	T

Notice that, no matter what the truth values of the component simple propositions P and Q, the compound proposition $(P \wedge Q) \Rightarrow (P \vee Q)$ is always true. This means that the compound proposition is true by virtue of the *structure* of the proposition itself and not its content. We can see this by substituting particular propositions for P and Q. For example, suppose

> P: *Today is my birthday*
> Q: *I am happy.*

Then $(P \wedge Q) \Rightarrow (P \vee Q)$ is the proposition

> *If today is my birthday and I'm happy, then today is my birthday or I'm happy,*

which is clearly true irrespective of whether or not it really is my birthday or whether or not I really am happy. Such a compound proposition that is true by virtue of its structure is called a 'tautology'.

Clearly, the negation of a tautology will always have truth value F. Any compound proposition with this property is called a 'contradiction'. An example of a contradiction is the proposition $(\neg P \wedge Q) \wedge (P \vee \neg Q)$. This is shown in the following truth table.

P	Q	$\neg P$	$\neg Q$	$\neg P \wedge Q$	$P \vee \neg Q$	$(\neg P \wedge Q) \wedge (P \vee \neg Q)$
T	T	F	F	F	T	F
T	F	F	T	F	T	F
F	T	T	F	T	F	F
F	F	T	T	F	T	F

Definition 2.1

A compound proposition is a **tautology** if it is true for all combinations of truth values of its component (simple) propositions.

A compound proposition is a **contradiction** if it is false for all combinations of truth values of its component (simple) propositions.

Exercises 2.1

1. Construct truth tables for each of the following propositions.

   (i) $\neg(P \vee \neg Q)$

   (ii) $(P \wedge Q) \Leftrightarrow \neg Q$

   (iii) $P \veebar (Q \Rightarrow \neg P)$

   (iv) $(P \Rightarrow Q) \wedge (Q \Rightarrow P)$

   (v) $(P \Rightarrow R) \wedge (R \Rightarrow Q)$

   (vi) $(P \Rightarrow \neg Q) \wedge R$

   (vii) $P \vee \neg(R \Leftrightarrow Q)$

   (viii) $\neg(P \wedge Q) \Rightarrow (P \wedge \neg R)$

   (ix) $P \vee (Q \Leftrightarrow R)$

   (x) $(P \wedge R) \vee (\neg P \wedge \neg Q)$

2. The four propositions S, W, R, and T are defined as follows.

S:	The sun shines.
W:	The wind blows.
R:	The rain falls.
T:	The temperature rises.

   Translate each of the following propositions into natural-sounding English sentences.

   (i) $W \Rightarrow (\neg S \vee R)$

    (ii) $(W \wedge R) \Leftrightarrow \neg S$

   (iii) $(W \vee R) \wedge \neg T$

   (iv) $\neg(S \wedge W(\Rightarrow (R \vee \neg T)$

    (v) $(\neg R \veebar T) \Rightarrow (S \wedge \neg W)$

3. With the four propositions S, W, R, and T defined as in question 1 above, symbolise each of the following propositions.

    (i) The wind blows but the rain doesn't fall.

   (ii) If the sun shines and the wind doesn't blow, then the temperature rises.

   (iii) The temperature rises if and only if the sun shines and the rain doesn't fall.

   (iv) Either the sun shines and the temperature rises or the wind blows and the rain falls.

    (v) If the sun doesn't shine or the wind blows with rain falling, then the temperature doesn't rise.

4. Determine whether each of the following propositions is a tautology, a contradiction, or neither.

    (i) $P \wedge \neg P$

   (ii) $\neg P \vee P$

   (iii) $P \Rightarrow (P \vee Q)$

   (iv) $(P \wedge Q) \wedge \neg(P \vee Q)$

    (v) $(P \Rightarrow Q) \vee (Q \Rightarrow R)$

   (vi) $(P \Rightarrow Q) \wedge (Q \Rightarrow R)$

   (vii) $(P \wedge \neg P) \Rightarrow (Q \vee R)$

 (viii) $(P \Rightarrow \neg Q) \wedge (\neg R \Rightarrow P)$

5. A compound proposition A is called a **substitution instance** of a compound proposition B if A may be obtained from B by replacing each occurrence of one or more of its simple constituent propositions by another proposition; in making the substitution, every occurrence of a given simple proposition must be replaced by the same proposition.

For example, each of the following is a substitution instance of the proposition $(P \wedge R) \Rightarrow (Q \vee R)$:

Substitution instance	Substitution
$(P \wedge S) \Rightarrow (Q \vee S)$	$R \longrightarrow S$
$(P \wedge \neg R) \Rightarrow (Q \vee \neg R)$	$R \longrightarrow \neg R$
$(P \wedge (S \vee T)) \Rightarrow (Q \vee (S \vee T))$	$R \longrightarrow S \vee T$
$(\neg P \wedge R) \Rightarrow ((P \wedge Q) \vee R)$	$P \longrightarrow \neg P,\ Q \longrightarrow (P \wedge Q)$

(i) Which of the following propositions is a substitution instance of the compound proposition $(P \Rightarrow Q) \wedge \neg (P \vee R)$? If the proposition is a substitution instance, state the substitution that connects it to $(P \Rightarrow Q) \wedge \neg (P \vee R)$.

(a) $(\neg P \Rightarrow Q) \wedge \neg (\neg P \vee R)$

(b) $(P \Rightarrow R) \wedge \neg (P \vee Q)$

(c) $(P \Rightarrow Q) \wedge \neg (S \vee R)$

(d) $(P \Rightarrow \neg Q) \wedge \neg (P \vee (R \Leftrightarrow S))$

(e) $(P \Rightarrow Q) \vee \neg (P \wedge R)$

(f) $((R \Rightarrow S) \Rightarrow S) \wedge \neg ((R \Rightarrow S) \vee R)$

(ii) Explain why, if B is a tautology, then any substitution instance of B is also a tautology and similarly if B is a contradiction, then any substitution instance of B is also a contradiction.

(iii) Using part (ii) and the results of question 4, determine whether each of the following propositions is a tautology, a contradiction, or neither.

(a) $(P \vee Q) \wedge \neg (P \vee Q)$

(b) $\neg (P \Rightarrow Q) \vee (P \Rightarrow Q)$

(c) $P \Rightarrow (P \vee (Q \wedge R)$

(d) $(P \Rightarrow (Q \wedge S)) \vee ((Q \wedge S) \Rightarrow R)$

2.3   Logical equivalence and logical implication

On page 25, we derived the truth table for $(P \vee \neg Q) \Rightarrow (P \wedge Q)$. A simplified version of the truth table, without the columns that show how the truth values are built up, is repeated below.

P	Q	...	$(P \vee \neg Q) \Rightarrow (P \wedge Q)$
T	T	...	T
T	F	...	F
F	T	...	T
F	F	...	F

Looking at the table, we notice a curious fact: the truth values for $(P \vee \neg Q) \Rightarrow (P \wedge Q)$ are the same as those for Q. So, although the two propositions $((P \vee \neg Q) \Rightarrow (P \wedge Q)$ and $Q)$ are structurally quite different, they have the same logical outcome in terms of their truth values. In this case we say that the two propositions are 'logically equivalent'.

Definition 2.2

Two propositions P_1 and P_2 are **logically equivalent**, written $P_1 \equiv P_2$, if they have the same truth values for all combinations of truth values of their component simple propositions.

With this definition, we may write

$$[(P \vee \neg Q) \Rightarrow (P \wedge Q)] \equiv Q.$$

Note that logical equivalence $\equiv$ is a relationship between propositions (having the same truth values) and is *not* a new logical connective, since $P \equiv Q$ is not a new proposition.

Intuitively, we can think of logically equivalent propositions as, in a sense, saying the same thing in a different way. This will be very important when we come to proving mathematical propositions. As we shall see, if we need to prove a proposition P_1, sometimes it will be easier to prove a logically equivalent proposition P_2 because the structure of P_2 is easier to work with. But this is legitimate since P_1 and P_2 are true under exactly the same circumstances. Hence it will be important for us to establish certain logical equivalences that we will be able to use, where appropriate, when constructing mathematical proofs.

Examples 2.4

1. Show that $P \Rightarrow Q$ is logically equivalent to $\neg P \vee Q$.

   Solution

   We first construct a truth table giving the truth values for both $P \Rightarrow Q$ and $\neg P \vee Q$.

P	Q	$\neg P$	$\neg P \vee Q$	$P \Rightarrow Q$
T	T	F	T	T
T	F	F	F	F
F	T	T	T	T
F	F	T	T	T

   The fourth and fifth columns of the table show that $\neg P \vee Q$ and $P \Rightarrow Q$ have the same truth values for all combinations of truth values of P and Q. Therefore

   $$(\neg P \vee Q) \equiv (P \Rightarrow Q).$$

   There is a danger that expressions like this stating a logical equivalence can become complicated through the use of brackets. From now on we will generally omit the brackets around P_1 and P_2 in the expression $P_1 \equiv P_2$.

2. Show that $[P \Rightarrow (Q \wedge R)] \equiv [(P \Rightarrow Q) \wedge (P \Rightarrow R)]$.

Solution

As before, we construct a truth table showing the truth values of both $P \Rightarrow (Q \wedge R)$ and $(P \Rightarrow Q) \wedge (P \Rightarrow R)$.

P	Q	R	$Q \wedge R$	$P \Rightarrow (Q \wedge R)$	$P \Rightarrow Q$	$P \Rightarrow R$	$(P \Rightarrow Q)$ $\wedge (P \Rightarrow R)$
T	T	T	T	T	T	T	T
T	T	F	F	F	T	F	F
T	F	T	F	F	F	T	F
T	F	F	F	F	F	F	F
F	T	T	T	T	T	T	T
F	T	F	F	T	T	T	T
F	F	T	F	T	T	T	T
F	F	F	F	T	T	T	T

Comparing columns 5 and 8 of the truth table, we see that $P \Rightarrow (Q \wedge R)$ and $(P \Rightarrow Q) \wedge (P \Rightarrow R)$ have the same truth values. Hence

$$P \Rightarrow (Q \wedge R) \equiv (P \Rightarrow Q) \wedge (P \Rightarrow R).$$

There is a simple relationship between logically equivalent propositions and the notion of a tautology. Suppose that two compound propositions P_1 and P_2 are logically equivalent, $P_1 \equiv P_2$. Recall that a biconditional proposition is true precisely when the truth values of its two components are the same (page 23). Since the truth values of P_1 and P_2 are identical, it means that the biconditional proposition $P_1 \Leftrightarrow P_2$ will always have truth value T and is therefore a tautology. In other words, if we know that $P_1 \equiv P_2$, then it follows that the biconditional proposition $P_1 \Leftrightarrow P_2$ is a tautology. It works the other way round as well. Suppose that $P_1 \Leftrightarrow P_2$ is a tautology. Then all the truth values of $P_1 \Leftrightarrow P_2$ are T. Since a biconditional proposition is true precisely when its two components have the same truth values, this means that P_1 and P_2 are logically equivalent. In summary, we have established the following.

Theorem 2.1
Let P_1 and P_2 be two (compound) propositions. Then $P_1 \equiv P_2$ if and only if $P_1 \Leftrightarrow P_2$ is a tautology.

Example 2.5
Show that the proposition $\neg(P \wedge Q) \Leftrightarrow (\neg P \vee \neg Q)$ is a tautology. Deduce that $\neg(P \wedge Q)$ is logically equivalent to $\neg P \vee \neg Q$.

Solution

As always we begin by constructing the appropriate truth table.

P	Q	$\neg P$	$\neg Q$	$P \wedge Q$	A $\neg(P \wedge Q)$	B $\neg P \vee \neg Q$	$A \Leftrightarrow B$
T	T	F	F	T	F	F	T
T	F	F	T	F	T	T	T
F	T	T	F	F	T	T	T
F	F	T	T	F	T	T	T

The last column, containing the truth values of $\neg(P \wedge Q) \Leftrightarrow (\neg P \vee \neg Q)$, only has truth values T. Hence $\neg(P \wedge Q) \Leftrightarrow (\neg P \vee \neg Q)$ is a tautology. It therefore follows from theorem 2.1 that

$$\neg(P \wedge Q) \equiv \neg P \vee \neg Q.$$

This logical equivalence, which could be paraphrased as 'the negation of a conjunction is the disjunction of the negations', is one of two equivalences known collectively as 'De Morgan's Laws'. It should be noted that, if we only need to establish the logical equivalence, $\neg(P \wedge Q) \equiv \neg P \vee \neg Q$, it is simpler to do so directly by comparing their truth values — in columns 6 and 7 of the truth table above. Establishing that the biconditional $\neg(P \wedge Q) \Leftrightarrow (\neg P \vee \neg Q)$ is a tautology required an additional column in the truth table.

Standard logical equivalences

In ordinary algebra (or arithmetic), we are familiar with rules that allow us to simplify expressions. Here are some simple examples:

$$x + 0 = x, \ x - (-y) = x + y, \ \text{and} \ x(y + z) = xy + xz.$$

There are similar logical equivalences that will allow us to manipulate logical expressions much as we might manipulate algebraic ones. Some of the logical rules are comparable to their algebraic counterparts and some are unique to logic, as the following examples demonstrate.

Examples 2.6

1. **Distributive law of conjunction over disjunction**

   Show that $P \wedge (Q \vee R) \equiv (P \wedge Q) \vee (P \wedge R)$.

   Note that this law parallels the distributive law of multiplication over addition: $x(y + z) = (xy) + \cdot(xz)$.

Solution

Consider the following truth table.

P	Q	R	$Q \vee R$	$P \wedge (Q \vee R)$	$P \wedge Q$	$P \wedge R$	$(P \wedge Q)$ $\vee (P \wedge R)$
T	T	T	T	T	T	T	T
T	T	F	T	T	T	F	T
T	F	T	T	T	F	T	T
T	F	F	F	F	F	F	F
F	T	T	T	F	F	F	F
F	T	F	T	F	F	F	F
F	F	T	T	F	F	F	F
F	F	F	F	F	F	F	F

Since the truth values in columns 4 and 7 are identical, we have

$$P \wedge (Q \vee R) \equiv (P \wedge Q) \vee (P \wedge R).$$

2. **Absorption law of conjunction over disjunction**

   Show that $P \wedge (P \vee Q) \equiv P$.

   Note that this law has no counterpart in ordinary algebra.

   Solution

   The truth values for $P \wedge (P \vee Q)$ are given in the following truth table.

P	Q	$P \vee Q$	$P \wedge (P \vee Q)$
T	T	T	T
T	F	T	T
F	T	T	F
F	F	F	F

   The truth values for $P \wedge (P \vee Q)$ in column 4 are the same as those for P in column 1 so $P \wedge (P \vee Q) \equiv P$.

In table 2.1 below we list a number of standard logical equivalence laws, together with their usual names. Each of these may be established using truth tables as in our previous examples. In the table, 'true' stands for any tautology and 'false' stands for any contradiction.

As we noted above, these laws can be used to manipulate logical expressions

Logical Equivalence Laws

Idempotent Laws	$P \wedge P \equiv P$
	$P \vee P \equiv P$
Commutative Laws	$P \wedge Q \equiv Q \wedge P$
	$P \vee Q \equiv Q \vee P$
Associative Laws	$(P \wedge Q) \wedge R \equiv P \wedge (Q \wedge R)$
	$(P \vee Q) \vee R \equiv P \vee (Q \vee R)$
Absorption Laws	$P \wedge (P \vee Q) \equiv P$
	$P \vee (P \wedge Q) \equiv P$
Distributive Laws	$P \wedge (Q \vee R) \equiv (P \wedge Q) \vee (P \wedge R)$
	$P \vee (Q \wedge R) \equiv (P \vee Q) \wedge (P \vee R)$
Involution Law	$\neg(\neg P) \equiv P$
De Morgan's Laws	$\neg(P \wedge Q) \equiv \neg P \vee \neg Q$
	$\neg(P \vee Q) \equiv \neg P \wedge \neg Q$
Identity Laws	$P \vee \text{false} \equiv P$
	$P \wedge \text{true} \equiv P$
	$P \vee \text{true} \equiv \text{true}$
	$P \wedge \text{false} \equiv \text{false}$
Complement Laws	$P \vee \neg P \equiv \text{true}$
	$P \wedge \neg P \equiv \text{false}$
	$\neg \text{false} \equiv \text{true}$
	$\neg \text{true} \equiv \text{false}$

TABLE 2.1: Logical equivalence laws.

without having to resort to truth tables. This is illustrated in the following example where we compare the solution using the laws with that using truth tables. Note that for logic expressions that involve, say, four or more simple propositions, truth tables are large and somewhat unwieldy.

Example 2.7

Show that $P \wedge [(Q \vee (P \wedge R)] \equiv P \wedge (Q \vee R)$.

Solution 1

For our first solution, we manipulate the expression on the left-hand side using the standard logical equivalences.

$$
\begin{aligned}
P \wedge [(Q \vee (P \wedge R)] &\equiv P \wedge [(Q \vee P) \wedge (Q \vee R)] && \text{Distributive law} \\
&\equiv [P \wedge (Q \vee P)] \wedge (Q \vee R) && \text{Associative law} \\
&\equiv [P \wedge (P \vee Q)] \wedge (Q \vee R) && \text{Commutative law} \\
&\equiv P \wedge (Q \vee R) && \text{Absorption law}
\end{aligned}
$$

Solution 2

As an alternative, we could draw up the following truth table.

P	Q	R	$P \wedge R$	$Q \vee (P \wedge R)$	$P \wedge [Q \vee (P \wedge R)]$	$Q \vee R$	$P \wedge (Q \vee R)$
T	T	T	T	T	T	T	T
T	T	F	F	T	T	T	T
T	F	T	T	T	T	T	T
T	F	F	F	F	F	F	F
F	T	T	F	T	F	T	F
F	T	F	F	T	F	T	F
F	F	T	F	F	F	T	F
F	F	F	F	F	F	F	F

The truth values in columns 6 and 8 are identical, so $P \wedge [(Q \vee (P \wedge R)] \equiv P \wedge (Q \vee R)$.

The standard logical equivalence laws listed in the table above describe what we might call the 'algebra' of propositions. In addition to these laws, we now introduce three more logical equivalence laws that will be used when we come to consider mathematical proofs.

Contrapositive Law	$P \Rightarrow Q \equiv \neg Q \Rightarrow \neg P$.

The following truth table establishes the logical equivalence since the truth values in columns 5 and 6 are identical. The proposition $\neg Q \Rightarrow \neg P$ is called the **contrapositive** of the conditional $P \Rightarrow Q$. Hence the contrapositive law says that a conditional is logically equivalent to its contrapositive.

P	Q	$\neg P$	$\neg Q$	$P \Rightarrow Q$	$\neg Q \Rightarrow \neg P$
T	T	F	F	T	T
T	F	F	T	F	F
F	T	T	F	T	T
F	F	T	T	T	T

Contradiction Law $(\neg P \Rightarrow \text{false}) \equiv P$.

Recall that false denotes any contradiction; that is, any compound proposition that only has truth value F. The truth table that establishes the logical equivalence only requires two rows, as follows.

P	$\neg P$	false	$\neg P \Rightarrow \text{false}$
T	F	F	T
F	T	F	F

The truth values in the last column are the same as those for P in the first column so $(\neg P \Rightarrow \text{false}) \equiv P$.

Biconditional Law $P \Leftrightarrow Q \equiv (P \Rightarrow Q) \wedge (Q \Rightarrow P)$.

Recall that the proposition $Q \Rightarrow P$ is called the converse of $P \Rightarrow Q$ (page 23). The biconditional law therefore states that a biconditional proposition is logically equivalent to the conjunction of the corresponding conditional and its converse. The following truth table establishes the law since the truth values in columns 3 and 6 are the same.

P	Q	$P \Leftrightarrow Q$	$P \Rightarrow Q$	$Q \Rightarrow P$	$(P \Rightarrow Q) \wedge (Q \Rightarrow P)$
T	T	T	T	T	T
T	F	F	F	T	F
F	T	F	T	F	F
F	F	T	T	T	T

Logical implication

When constructing mathematical proofs, we will want to know that each statement in a proof 'follows logically' from previous statements. In other words, we will wish to ensure that the truth of a statement in a proof is guaranteed by the truth of the statements that come before it in the proof. When dealing with propositions, this notion is called 'logical implication'.

Let P_1 and P_2 be two (possibly compound) propositions such that P_2 is true in all cases where P_1 is true. Then we say that P_1 **logically implies** P_2, written $P_1 \vDash P_2$. This notion may be extended to several propositions. Let $P_1, P_2, \ldots, P_n$ and Q be propositions. We say that the propositions $P_1, P_2, \ldots, P_n$ **logically imply** the proposition Q, written $P_1, P_2, \ldots, P_n \vDash Q$, provided Q is true in all those cases where *each* of the propositions $P_1, P_2, \ldots, P_n$ is true. For the sake of clarity of notation, we will also write $\{P_1, P_2, \ldots, P_n\} \vDash Q$ in this case. Since the conjunction $P_1 \wedge P_2 \wedge \ldots \wedge P_n$ is true precisely when each of the propositions $P_1, P_2, \ldots, P_n$ is true, $\{P_1, P_2, \ldots, P_n\} \vDash Q$ is equivalent to saying $P_1 \wedge P_2 \wedge \ldots \wedge P_n \vDash Q$.

Examples 2.8

1. Show that $(P \Rightarrow Q) \wedge (Q \Rightarrow R) \vDash (P \Rightarrow R)$.

   Solution

   A truth table giving the truth values of both $(P \Rightarrow Q) \wedge (Q \Rightarrow R)$ and $P \Rightarrow R$ is given below.

P	Q	R	$P \Rightarrow Q$	$Q \Rightarrow R$	$(P \Rightarrow Q) \wedge (Q \Rightarrow R)$	$P \Rightarrow R$
T	T	T	T	T	**T**	T
T	T	F	T	F	F	F
T	F	T	F	T	F	T
T	F	F	F	T	F	F
F	T	T	T	T	**T**	T
F	T	F	T	F	F	T
F	F	T	T	T	**T**	T
F	F	F	T	T	**T**	T

   There are four rows where $(P \Rightarrow Q) \wedge (Q \Rightarrow R)$ is true — rows 1, 5, 7, and 8. In this case, these rows have been highlighted by emboldening the truth value T although we will not routinely do this. In each of these rows, $P \Rightarrow R$ is also true. Hence $P \Rightarrow R$ is true in all cases where $(P \Rightarrow Q) \wedge (Q \Rightarrow R)$ is true so

   $$(P \Rightarrow Q) \wedge (Q \Rightarrow R) \vDash (P \Rightarrow R).$$

Note that $P \Rightarrow R$ is also true in some situations where $(P \Rightarrow Q) \wedge (Q \Rightarrow R)$ is false, namely those cases given by rows 3 and 6. However we only require $P \Rightarrow R$ to be true in all cases where $(P \Rightarrow Q) \wedge (Q \Rightarrow R)$ is true; what happens when $(P \Rightarrow Q) \wedge (Q \Rightarrow R)$ is false is of no significance here.

2. Show that $\{P \vee Q, P \Rightarrow R, Q \Rightarrow R\} \vDash R$.

Solution

We need to show that in those cases where $P \vee Q$, $P \Rightarrow R$ and $Q \Rightarrow R$ are all true, then so too is R. The following truth table gives the truth values of $P \vee Q$, $P \Rightarrow R$ and $Q \Rightarrow R$.

P	Q	R	$P \vee Q$	$P \Rightarrow R$	$Q \Rightarrow R$
T	T	T	T	T	T
T	T	F	T	F	F
T	F	T	T	T	T
T	F	F	T	F	T
F	T	T	T	T	T
F	T	F	T	T	F
F	F	T	F	T	T
F	F	F	F	T	T

The propositions $P \vee Q$, $P \Rightarrow R$ and $Q \Rightarrow R$ are all true only in rows 1, 3, and 5. In each of these rows, R is also true. Therefore R is true in all those cases where *each* of the propositions $P \vee Q$, $P \Rightarrow R$ and $Q \Rightarrow R$ is true, so

$$\{P \vee Q, P \Rightarrow R, Q \Rightarrow R\} \vDash R.$$

Exercises 2.2

1. Using truth tables, establish each of the following logical equivalences.

   (i) $\neg(P \vee Q) \equiv \neg P \wedge \neg Q$

   (ii) $P \vee (P \wedge Q) \equiv P$

   (iii) $P \Rightarrow \neg Q \equiv \neg(P \wedge Q)$

   (iv) $(P \wedge Q) \vee R \equiv (P \vee R) \wedge (Q \vee R)$

   (v) $(P \vee Q) \Rightarrow R \equiv (P \Rightarrow R) \wedge (Q \Rightarrow R)$

   (vi) $P \Rightarrow (Q \Rightarrow R) \equiv (P \wedge Q) \Rightarrow R$

   (vii) $((P \Rightarrow Q) \Rightarrow R) \equiv ((\neg P \Rightarrow R) \wedge (Q \Rightarrow R))$.

(viii) $\neg(P \wedge Q) \wedge R \equiv (\neg P \wedge R) \vee (\neg(Q \vee \neg R))$

(ix) $(P \wedge Q) \Rightarrow R \equiv (P \Rightarrow R) \wedge (Q \Rightarrow \neg P)$

(x) $(P \Leftrightarrow Q) \vee R \equiv [(P \Rightarrow (Q \vee R)) \wedge (Q \Rightarrow (P \vee R))]$

2. Using truth tables, establish each of the following logical implications.

   (i) $(P \wedge Q) \vDash (P \vee Q)$

   (ii) $(P \Leftrightarrow Q) \wedge Q \vDash P$

   (iii) $\{P \Rightarrow Q, P \vee \neg Q\} \vDash \neg(P \vee Q)$

   (iv) $\{P \Rightarrow (Q \vee R), P \wedge R\} \vDash (P \vee R)$

   (v) $\{P \vee (Q \wedge R), R \Rightarrow P\} \vDash (\neg Q \Rightarrow R)$

   (vi) $\{P \Rightarrow Q, (P \wedge Q) \Rightarrow R\} \vDash (P \Rightarrow R)$

   (vii) $\{P \Rightarrow R, Q \Rightarrow R, \neg P \Rightarrow Q\} \vDash (\neg P \vee R)$

3. Use the logical equivalence laws on page 39 to establish each of the following logical equivalences.

   (i) $\neg P \vee (P \wedge Q) \equiv \neg P \vee Q$

   (ii) $\neg(P \vee Q) \vee (\neg P \wedge Q) \equiv \neg P$

   (iii) $\neg(\neg P \wedge Q) \wedge (P \vee \neg Q) \equiv (P \vee \neg Q)$

   (iv) $P \vee (Q \wedge (P \vee \neg Q)) \equiv P$

   (v) $P \wedge [(P \wedge Q) \vee \neg P] \equiv (P \wedge Q)$

   (vi) $(P \wedge Q) \vee (\neg P \vee R) \equiv (\neg P \vee Q) \vee R$

   (vii) $\neg(P \wedge \neg(Q \wedge R)) \equiv (\neg P \vee Q) \wedge \neg(P \wedge \neg R)$

   (viii) $[(P \wedge Q) \vee (R \wedge S)] \equiv [(P \vee R) \wedge (P \vee S) \wedge (Q \vee R) \wedge (Q \vee S)]$

4. Show that $P \Rightarrow Q$ is logically equivalent to $\neg P \vee Q$. This is sometimes called the **Law of Material Implication**.

   Using the law of material implication and the other logical equivalence laws on page 39, establish each of the following.

   (i) $\neg Q \wedge (P \Rightarrow Q) \equiv \neg(P \wedge Q)$

   (ii) $P \Rightarrow (P \wedge Q) \equiv (P \Rightarrow Q)$

   (iii) $(P \wedge Q) \Rightarrow R \equiv P \Rightarrow (Q \Rightarrow R)$

   (iv) $(P \vee Q) \Rightarrow R \equiv (P \Rightarrow R) \wedge (Q \Rightarrow R)$

   (v) $P \Rightarrow (Q \vee R) \equiv (P \Rightarrow Q) \vee (P \Rightarrow R)$

2.4   Predicates and quantification

In propositional logic, we cannot represent the fact that several propositions may be referring to the same object or have the same subject matter. For example, we would need to denote the propositions

> *Sarah is an athlete*
> *Sarah is physically fit*
> *All athletes are physically fit*

using different letters (P, Q, R say), and then the fact that they are making statements about the same individuals or properties of individuals would be lost.

A proposition of the form '*Sarah is an athlete*' may be regarded as having two components. One is the object or individual — 'Sarah' in this case. The other is the property which the object or individual is said to possess — being an athlete, in this example. If we wish to be able to capture the fact that two propositions refer to the same object or refer to the same property, then we will need a notation for propositions that refer specifically to the objects and their properties.

A **predicate** describes a property of one or more objects or individuals. We will use uppercase letters to denote predicates. The following are examples of predicates.

> $R :$ ...*is red.*
> $A :$ ...*is an athlete.*
> $F :$ ...*is physically fit.*
> $T :$ ...*cannot be tolerated under any circumstances.*

A predicate can be 'turned into' a proposition by supplying an appropriate object or individual. With the predicates defined above, we have the following propositions.

> R(this rose) : *This rose is red.*
> A(Sarah) : *Sarah is an athlete.*
> F(Sarah) : *Sarah is physically fit.*
> T(John): *John cannot be tolerated under any circumstances.*

The general notation for representing propositions using objects and predicates should be clear from these examples. If P is a predicate and a is an object or individual, then we use $P(a)$ to represent the proposition '*The object or individual a has the property P*'. More compactly,

$$P(a) : a \text{ has property } P.$$

We may negate propositions denoted in this way and build them into more complex propositions using connectives as we have done previously. Using the notation above, here are some examples.

$$\neg R(\text{this rose}) : \quad \textit{This rose is not red.}$$
$$A(\text{Sarah}) \wedge A(\text{Dan}) : \quad \textit{Sarah is an athlete and Dan is an athlete.}$$
$$\text{(or \textit{Sarah and Dan are athletes})}$$
$$A(\text{Sarah}) \Rightarrow F(\text{Sarah}) : \quad \textit{If Sarah is an athlete, then she is physically fit.}$$

So far in this section we have simply introduced a slightly more expressive notation for propositions that allows us to represent separately the object being referred to and the property that it is claimed to possess. There are various ways in which we can exploit this new notation. Suppose we replace a specific object or individual by a variable. Then we obtain expressions like the following.

$$R(x) : \quad x \text{ is red.}$$
$$A(x) : \quad x \text{ is an athlete.}$$
$$F(y) : \quad y \text{ is physically fit.}$$

These expressions are not propositions since they cannot be declared true or false. They are called 'propositional functions' by analogy with 'numerical' functions. Readers familiar with functions (see chapter 3) will recognise an expression such as $f(x) = x^2 + 2$ as (partly) defining a function. Here, x is a numerical variable. Substituting a value for x gives a value for the expression; for example, when $x = 3$ we have $f(3) = 3^2 + 2 = 11$. It is similar with propositional functions. For the propositional function $A(x) : x$ is an athlete, substituting an object such as 'Sarah' for the variable x gives a proposition $A(\text{Sarah})$: Sarah is an athlete.

More formally, a **propositional function** is a logical expression which depends on the value of one or more variables. Let P be some predicate which we may read as '*... has property P*'. Then $P(x)$ is the propositional function '*x has property P*'.

As we have seen, one way to convert a propositional function $P(x)$ into a proposition is to substitute (the name of) a particular object or individual for the variable x. If a is a particular object or individual, then $P(a)$ is a proposition. Thus the propositional function $P(x)$ represents a whole family of possible propositions, one for each possible object or individual which may be substituted for the variable x. The relationship between predicates, propositional functions, and propositions is illustrated in the following example.

Predicate:	$M :$	*...is my friend.*
Propositional function:	$M(x) :$	*x is my friend.*
Propositions:	$M(\text{Elaine}) :$	*Elaine is my friend.*
	$M(\text{Jack}) :$	*Jack is my friend.*

It is also important to note that propositional functions behave in the similar way to propositions. We can negate propositional functions and use all the standard connectives between propositional functions. Using the predicates defined previously, we have the following examples.

$$\neg R(x) : \quad x \text{ is not red.}$$
$$R(x) \wedge R(y) : \quad x \text{ is red and } y \text{ is red.}$$
$$(\text{or } x \text{ and } y \text{ are red.})$$
$$A(x) \Rightarrow F(x) : \quad \text{If } x \text{ is an athlete, then } x \text{ is physically fit.}$$
$$R(x) \vee A(y) : \quad x \text{ is red or } y \text{ is an athlete.}$$

Example 2.9
The following predicates are defined.

$$M : \quad \text{...is my friend.}$$
$$R : \quad \text{...is rich.}$$
$$F : \quad \text{...is famous.}$$
$$B : \quad \text{...is boring.}$$

Symbolise each of the following propositions.

(i) *Jimmy is my friend but he's not rich.*

(ii) *Sue and Mark are rich.*

(iii) *If Peter is my friend, then he is not boring.*

(iv) *Mark is rich or famous but not both.*

(v) *If Jane is rich and not boring, then she is my friend.*

Solution

(i) Recall that 'but' signifies a conjunction (see page 19). Hence the proposition is symbolised as

$$M(\text{Jimmy}) \wedge \neg R(\text{Jimmy}).$$

Note that we only use connectives between propositions (or propositional functions) and *not* between predicates themselves. Thus '$(M \wedge \neg R)(\text{Jimmy})$' is *not* a well-formed proposition.

(ii) This is a conjunction of the propositions *Sue is rich* and *Mark is rich* and so is:
$$R(\text{Sue}) \wedge R(\text{Mark}).$$

In the same way that we do not use connectives between predicates, so it is also incorrect to use connectives between objects. Hence the expression '$R(\text{Sue} \wedge \text{Mark})$' is also *not* a well-formed proposition.

(iii) This is a conditional where both the antecedent and the consequent refer to Peter:
$$M(\text{Peter}) \Rightarrow \neg B(\text{Peter}).$$

(iv) This is a (rare) example of an exclusive disjunction:
$$R(\text{Mark}) \veebar F(\text{Mark}).$$

(v) The top-level structure of the proposition is that of a conditional:
$$(R(\text{Jane}) \wedge \neg B(\text{Jane})) \Rightarrow M(\text{Jane}).$$

Quantifiers

Using predicates, or propositional functions, we can identify when propositions refer to the same individual or the same property. However, we are still unable to represent (except using a single letter) propositions such as the following.

> *All athletes are physically fit.*
> *Some roses are red.*

In general, if $P(x)$ is a propositional function, then the following statements are propositions since each can be declared true or false.

> *For all x, $P(x)$*   or   *All x have property P.*
> *For some x, $P(x)$*   or   *Some x have property P.*

These propositions are **quantified propositional functions** and the phrases 'for all' and 'for some' are called **quantifiers**. A proposition that asserts that all objects (in a certain class or category) have a specified property is said to be **universally quantified**. Universally quantified propositions may be recognised by phrases such as 'for all ...', 'for every ...' or 'for each ...'. A proposition that asserts that some objects (in a certain class or category) have a specified property is said to be **existentially quantified**. Existentially quantified propositions may be recognised by phrases such as 'some ...', 'there exists ...' or 'at least one ...'.

As usual, in natural language there are different ways of expressing quantified propositions. The important point is that, however it is expressed, a universally quantified proposition is one that is true when all the specified objects have the specified property. Similarly, an existentially quantified proposition is one that is true when *at least one* of the specified objects has the specified property. Existentially quantified statements are often expressed in the plural: *some roses are red.* This could have been expressed as: *some rose is red.* In either case, the proposition is true when there exists an example of a rose that is red. Here are some further examples of quantified propositions expressed in a variety of ways.

> Universally quantified propositions
> *All fish swim in water.*
> *Every bird can fly.*
> *Each apple in this batch is bruised.*

> Existentially quantified propositions
> *Some birds cannot fly.*
> *There exist black swans.*
> *At least one prime number is even.*

We need a way of symbolising quantified propositions. The only missing pieces in the notation we have so far developed are symbols for the quantifiers 'all' and 'some' which we now introduce.

The **universal quantifier** 'for all' is denoted $\forall$ and a universally quantified propositional function is denoted $\forall x \bullet P(x)$ or just $\forall x \; P(x)$. Here the 'bullet' is being used simply as punctuation to separate the quantified variable 'for all x' from the propositional function $P(x)$ being quantified. We will use this notation, but many authors omit the bullet. The following summarises how we symbolise a universally quantified proposition and the different ways that the symbols may be read.

> $\forall x \bullet P(x)$:   for all x, $P(x)$
> for every x, $P(x)$
> for each x, $P(x)$

The **existential quantifier** 'there exists' is denoted $\exists$ and an existentially quantified propositional function is denoted $\exists x \bullet P(x)$ or just $\exists x \; P(x)$. Again the bullet is an optional, but useful, piece of punctuation. The symbol $\exists$ is most commonly read as 'there exists' and, in this case, we may read the bullet as 'such that'. This is illustrated in the following summary of an existentially quantified proposition and the different ways that the symbols may be read.

> $\exists x \bullet P(x)$:   there exists x, such that $P(x)$
> for some x, $P(x)$
> for at least one x, $P(x)$

Examples 2.10

1. The following predicates are defined.

M :   ... *is my friend.*
P :   ... *is Peter's friend.*
U :   ... *is prone to unruly behaviour.*
S :   ... *should be shunned.*

Symbolise each of the following propositions.

(i) *All of Peter's friends are prone to unruly behaviour.*

(ii) *Some of my friends are also friends of Peter's.*

(iii) *If all of Peter's friends are prone to unruly behaviour, then some of them should be shunned.*

(iv) *Some of Peter's and my friends are prone to unruly behaviour.*

(v) *All of my friends are prone to unruly behaviour and some of them should be shunned.*

(vi) *None of Peter's friends are my friends.*

Solution

(i) We first paraphrase this into a 'half-way house' between English and symbolic logic:

> *For all x, if x is Peter's friend, then x is prone to unruly behaviour.*

From this we can readily symbolise the proposition as:

$$\forall x \bullet P(x) \Rightarrow U(x).$$

Most people find the somewhat unnatural-sounding half-way house to be a useful device in understanding the structure of the quantified sentence. With some practise, however, it is usually possible to be able to symbolise many sentences without resorting to this.

(ii) In this case, the 'half-way house' between English and symbolic logic might be expressed as:

> *For some x, x is my friend and (also) x is Peter's friend.*

Hence the symbolic version is:

$$\exists x \bullet M(x) \wedge P(x).$$

(iii) In this case the proposition is a conditional 'if ... then ...' where the antecedent is universally quantified and the consequent is existentially quantified. We begin by symbolising the antecedent and consequent separately.

Antecedent
Half-way house: For all x, if x is Peter's friend, then x is prone to unruly behaviour.
Symbolic expression: $\forall x \bullet P(x) \Rightarrow U(x)$

Consequent
Half-way house: For some x, x is Peter's friend and x should be shunned.
Symbolic expression: $\exists x \bullet P(x) \wedge S(x)$

Now we can symbolise the whole sentence as

$$(\forall x \bullet P(x) \Rightarrow U(x)) \Rightarrow (\exists x \bullet P(x) \wedge S(x)).$$

(iv) This is similar to (ii) above in the sense that the sentence may be paraphrased as saying some objects with the first property also have the second property. In this case, the first property is a conjunction so that the half-way house 'pseudo-English' statement is:

> *For some x, (x is my friend and x is Peter's friend) and x is prone to unruly behaviour.*

Hence the symbolic version is

$$\exists x \bullet (M(x) \wedge P(x)) \wedge U(x).$$

By the associative law for conjunction, $(P \wedge Q) \wedge R \equiv P \wedge (Q \wedge R)$ (see page 39), so the brackets around $M(x) \wedge P(x)$ may be omitted and we can write this more simply as

$$\exists x \bullet M(x) \wedge P(x) \wedge U(x).$$

(v) This is similar to (iii) above in the sense that the proposition contains two 'sub-propositions', one universally quantified and the other existentially quantified. However, in this case the two component quantified propositions are conjoined. If we separately symbolise 'all of my friends are prone to unruly behaviour' and 'some of my friends should be shunned' and then form their conjunction, we obtain the following.

$$(\forall x \bullet M(x) \Rightarrow U(x)) \wedge (\exists x \bullet M(x) \wedge S(x))$$

(vi) The phrase 'None of ...' signifies the negation of 'Some of ...'; in other words, 'none of ...' is equivalent to 'it is not the case that some of ...'. So we first symbolise the sentence 'some of Peter's friends are my friends' and then negate the result. This gives:

$$\neg \exists x \bullet P(x) \wedge M(x).$$

It is worth noting that there is a convention being applied here: to negate a quantified propositional function, we simply write the negation symbol in front of the quantifier

$$\neg\,(\exists x \bullet P(x)) \equiv \neg\exists x \bullet P(x)$$
$$\neg\,(\forall x \bullet P(x)) \equiv \neg\forall x \bullet P(x).$$

2. Symbolise each of the following propositions.

   (i) *All politicians wear suits and talk too much.*

  (ii) *Not everybody approves of capital punishment.*

 (iii) *Everyone was curious but no one was prepared to ask.*

Solution

In these examples the predicates are not already defined so we need, in each case, to define the predicates that we wish to use.

  (i) There are three predicates involved in the proposition which we need to define.

     Let P : ... is a politician

          S : ... wears a suit

          T : ... talks too much.

     Then the half-way house is:

> For all x, if x is a politician, then (x wears a suit and x talks too much).

     Hence we symbolise the proposition as

$$\forall x \bullet P(x) \Rightarrow (S(x) \wedge T(x)).$$

  (ii) This is the negation of the proposition 'everybody approves of capital punishment'. The term 'everybody ...' is a little awkward because it really means 'all people ...'. Hence, one of the predicates we will need will be '... is a person'.

     Let P : ... is a person

          A : ... approves of capital punishment.

     Then the half-way house for the proposition 'everybody approves of capital punishment' is

> For all x, if x is a person then x approves of capital punishment,

     which we would symbolise as $\forall x \bullet P(x) \Rightarrow A(x)$. Hence the original negated proposition 'not everybody approves of capital punishment' is

$$\neg\forall x \bullet P(x) \Rightarrow A(x).$$

(iii) In this sentence, 'everyone' and 'no one' seem to be referring to a specific group of individuals This may have already been made clear in the context in which the sentence was uttered. One of the predicates we will need, therefore, is '...is a person (belonging to the specific group being referred to)'.

The phrase 'no one ...' represents 'it is not the case that someone ...'; in other words, 'no one ...' signals the negation of the existential quantifier.

Let P :   ...is a person (belonging to the specific group being referred to)

      C :   ...was curious

      A :   ...was prepared to ask.

Then we can symbolise 'everyone was curious' as $\forall x \bullet P(x) \Rightarrow C(x)$. We may also symbolise 'someone was prepared to ask' as $\exists x \bullet P(x) \wedge A(x)$. Hence the given proposition 'everyone was curious but no one was prepared to ask' may be represented as follows.

$$(\forall x \bullet P(x) \Rightarrow C(x)) \wedge (\neg \exists x \bullet P(x) \wedge A(x))$$

Note that the previous examples all build on the following two general patterns for symbolising universally and existentially quantified propositions.

$$\text{All } Ps \text{ are } Qs : \quad \forall x \bullet P(x) \Rightarrow Q(x)$$
$$\text{Some } Ps \text{ are } Qs : \quad \exists x \bullet P(x) \wedge Q(x)$$

Universe of Discourse

Consider again the proposition from example 2.10.2 (iii) above.

Everyone was curious but no one was prepared to ask.

When symbolising this, we noted that it appeared that the proposition was most likely about some specific group of people. Furthermore, we needed a predicate that referred to membership of the particular group. There is an alternative approach in cases like this where the proposition is referencing some known collection of objects or individuals. Instead of defining the predicate

P: ...is a person (belonging to the specific group being referred to).

we could specify that the variable(s) will only range over the people in the specific group. To do this we define the 'universe of discourse' to be the collection of people in the specific group. Then any variable x will refer only to people in the group. This gives a more natural and more efficient way of representing the proposition, as follows.

Let   Universe :   people in the specific group being referred to

C :   ...was curious

A :   ...was prepared to ask.

Then we may symbolise the proposition as

$$(\forall x \bullet C(x)) \wedge (\neg \exists x \bullet A(x)).$$

Compare this with the previous symbolisation

$$(\forall x \bullet P(x) \Rightarrow C(x)) \wedge (\neg \exists x \bullet P(x) \wedge A(x)).$$

The saving, in both space and mental effort, comes from defining the particular group of people once (as the universe) rather than having to refer to it (via a predicate) every time we wish to refer to elements of the group.

In general, for a given propositional function $P(x)$, the **universe of discourse** is the set from which we may select an object or individual to substitute for x. As we have seen, defining a universe of discourse can simplify the symbolisation of quantified propositional functions. If a universe of discourse is not defined, then we must assume that any object or individual may be substituted for x. In principle, we may define separate universes for each propositional variable. In practice, however, we will usually have a single universe for all the propositional variables under consideration. However, when there are two different kinds of objects under consideration, it may make sense to define two separate universes.

Example 2.11
Defining suitable universes as appropriate, symbolise each of the following propositions.

(i) *All of my friends are either rich or famous.*

(ii) *All athletes who can lift heavy weights cannot run fast.*

(iii) *Everybody was too hot and someone fainted.*

(iv) *If everyone was generous, then no one would go hungry.*

Solution

(i) We define the universe to be 'my friends' and define predicates:

R :   ...is rich

A :   ...is famous.

Then we may paraphrase the proposition as:

> *For all x (who is my friend), x is rich or x is famous.*

Hence we may symbolise as: $\forall x \bullet R(x) \vee F(x)$.

(ii) Let   Universe :   athletes
         H :   ...can lift heavy weights
         R :   ...can run fast.

Then we may paraphrase the proposition as:

> *For all (athletes) x, if x can lift heavy weights, then
> x cannot run fast.*

Symbolically, we have: $\forall x \bullet H(x) \Rightarrow \neg R(x)$.

(iii) Again it appears that the sentence refers to a specific group of people, so we define this group as the universe.

Let   Universe :   people in the specific group being referred to
         H :   ...was too hot
         F :   ...fainted.

Then the sentence is a conjunction of two propositions:

> *For all x, x was too hot and for some x, x fainted.*

Therefore the symbolic expression is: $(\forall x \bullet H(x)) \wedge (\exists x \bullet F(x))$.

(iv) Let   Universe :   people
         G :   ...was generous
         H :   ...goes hungry.

The proposition is a conditional where the antecedent and consequent are both (separately) quantified propositions. The antecedent, 'everyone was generous', is straightforward to symbolise: $\forall x \bullet G(x)$.

Recall that 'no one ...' signifies the negation of 'someone ...'. Hence the consequent may be symbolised as: $\neg \exists x \bullet H(x)$.

Putting these together, we may symbolise the given proposition as:

$$(\forall x \bullet G(x)) \Rightarrow (\neg \exists x \bullet H(x)).$$

Two-place predicates

The predicates we have considered so far are **one-place predicates**. To convert each such predicate to a proposition requires the name of a single member of the universe of discourse or quantification over a single variable.

There are predicates which are relational in nature and require two objects to convert them to propositions. Consider, for example, the predicate 'belongs to'. This describes the relationship of 'belonging' and requires two objects to become a proposition: '*this cat belongs to me*', for example. Predicates like this that require two objects to convert them to propositions are called **two-place predicates** or, sometimes, **relational predicates**.

When denoting two-place predicates, we will follow a similar notation to one-place predicates and place the objects after the predicate symbol. However, this time there will be a pair of objects. Thus if we define

$$B: \quad `\dots belongs\ to\ \dots`$$

then the proposition '*this cat belongs to me*' is denoted

$$B(\text{this cat}, \text{me}).$$

There are two key points to note in relation to the objects required for a two-place predicate. Firstly, their order is important. For example, the two propositions

'*this cat belongs to me*' and '*I belong to this cat*'

have quite different meanings. Similarly, changing the order of the objects can change the truth value of the resulting proposition. For example, '17 is greater than 3' is a true proposition, whereas '3 is greater than 17' is false.

The second point to note is that the two objects may be drawn from different universes. For example, consider the predicate

$$C: \quad `\dots is\ the\ capital\ city\ of\ \dots`$$

To convert this into a meaningful proposition, the first object needs to be a city and the second object should be a country (or other entity, such as a state or region, that has a capital city). Thus

> *Paris is the capital city of France*
> *Brighton is the capital city of the United Kingdom*
> *Little Rock is the capital city of Arkansas*

are all properly defined propositions (of which two are true and one is false). However

> *this banana is the capital city of the Eiffel tower*, and
> *a herring is the capital city of Sarah*

are simply not meaningful propositions.

Given a two-place predicate P, we may form a **two-variable propositional**

function $P(x, y)$ by supplying two variables x and y. We have seen that replacing both variables with objects or individuals gives a proposition. Replacing just one of the variables with an object produces a single-variable propositional function. For example,

$$B(x, y) : \quad x \text{ belongs to } y$$

is a two-variable proposition. Replacing one of the variables gives single-variable propositional functions such as

$$B(\text{this cat}, y) : \quad \textit{this cat belongs to } y$$
$$B(x, \text{me}) : \qquad \textit{x belongs to me.}$$

We may now quantify the remaining variable to obtain a proposition. For example, assuming a universe of people for the second variable:

$$\exists y \bullet B(\text{this cat}, y) : \quad \textit{this cat belongs to someone}$$
$$\forall x \bullet B(x, \text{me}) : \qquad \textit{everything belongs to me.}$$

In general, if $P(x, y)$ is a two-variable propositional function and a and b are objects (of the appropriate type), then each of the following are propositions:

$$P(a, b)$$
$$\forall x \bullet P(x, b)$$
$$\exists x \bullet P(x, b)$$
$$\forall y \bullet P(a, y)$$
$$\exists y \bullet P(a, y)$$

Finally, we may also convert a two-variable propositional function $P(x, y)$ into a proposition by quantifying over both of the variables. There are eight ways in which this may be done.

$$\forall x \, \forall y \bullet P(x, y) \qquad \forall y \, \forall x \bullet P(x, y)$$
$$\exists x \, \forall y \bullet P(x, y) \qquad \forall y \, \exists x \bullet P(x, y)$$
$$\forall x \, \exists y \bullet P(x, y) \qquad \exists y \, \forall x \bullet P(x, y)$$
$$\exists x \, \exists y \bullet P(x, y) \qquad \exists y \, \exists x \bullet P(x, y)$$

Note that, when different quantifiers are used, their order is significant. For example, consider the following propositions where the universe for both variables is the set of positive integers $\mathbb{Z}^+$.

$$\forall m \, \exists n \bullet m < n \qquad \exists n \, \forall m \bullet m < n$$

The first of these, $\forall m \, \exists n \bullet m < n$, asserts that, for every positive integer m, there is a positive integer n that is greater than it. This is clearly a true proposition because there is no largest positive integer. The second proposition, $\exists n \, \forall m \bullet m < n$, asserts that there exists a positive integer n that is larger than every positive integer m. This is just as clearly a false proposition again because there is no largest positive integer. In this case, changing the order of the quantifiers has changed the truth value of the proposition.

Examples 2.12

1. The following one-variable and two-variable propositional functions are given.

$$
\begin{aligned}
B(x, y) &: \quad x \text{ belongs to } y \\
D(x, y) &: \quad x \text{ detests } y \\
L(x, y) &: \quad x \text{ loves } y \\
C(x) &: \quad x \text{ is a cat} \\
F(x) &: \quad x \text{ is fierce} \\
P(x) &: \quad x \text{ is a person}
\end{aligned}
$$

Write each of the propositions symbolised below as natural-sounding English sentences.

(i) $\forall x \bullet C(x) \Rightarrow L(\text{Max}, x)$

(ii) $\exists x \bullet C(x) \wedge F(x) \wedge B(x, \text{Chloe})$

(iii) $\forall x \, \exists y \bullet P(x) \Rightarrow (C(y) \wedge L(x, y))$

(iv) $\forall x \, \forall y \bullet (C(x) \wedge F(x)) \Rightarrow (P(y) \Rightarrow D(y, x))$

(v) $\exists x \, \forall y \bullet P(x) \wedge ((C(y) \wedge B(y, x)) \Rightarrow L(x, y))$

(vi) $\exists x \, \exists y \bullet P(x) \wedge C(y) \wedge \neg F(y) \wedge D(x, y)$

Solution

(i) The literal translation of the proposition is 'for all x, if x is a cat, then Max loves x'. An idiomatic version of this is '*Max loves all cats*' or simply '*Max loves cats*'.

(ii) The 'pseudo English' translation of the proposition is 'for some x, x is a cat and x is fierce and x belongs to Chloe'. A natural-sounding sentence for this would be '*there is a fierce cat that belongs to Chloe*' or, better, '*Chloe has a fierce cat*'.

(iii) Since the first propositional function $P(x)$ does not involve the variable y, the quantifier $\exists y$ may be moved past $P(x)$. In other words,

$$
\begin{aligned}
&\forall x \, \exists y \bullet P(x) \Rightarrow (C(y) \wedge L(x, y)) \\
\equiv \quad &\forall x \bullet P(x) \Rightarrow (\exists y \bullet C(y) \wedge L(x, y))
\end{aligned}
$$

and the second logical sentence is easier to translate. Literally it means 'for all x, if x is a person, then there exists y such that y is a cat and x loves y'. As a natural-sounding sentence we might say '*everyone loves some cat*'.

(iv) The literal 'half-way house' translation is 'for all x and for all y, if x is a cat and x is fierce, then if y is a person, y detests x. In other words, '*all people detest all fierce cats*' or, more naturally, '*everyone detests fierce cats*'.

(v) Again, since this is quite a complicated expression, we may move the quantifier $\forall y$ past the propositional function $P(x)$ since it does not involve the variable y and re-write the logical sentence as:

$$\exists x \, \forall y \bullet P(x) \wedge ((C(y) \wedge B(y,x)) \Rightarrow L(x,y))$$
$$\equiv \quad \exists x \bullet P(x) \wedge (\forall y \bullet (C(y) \wedge B(y,x)) \Rightarrow L(x,y)).$$

This second sentence can be translated as 'for some x, x is a person and, for all y, if y is a cat and y belongs to x, then x loves y'. As an English sentence, we might write '*there is someone who loves all the cats that belong to them*' or, more naturally, '*someone loves all their cats*'.

(vi) Literally, this translates into 'pseudo English' as 'there exists x and there exists y such that x is a person and y is a cat and y is not fierce and x detests y'. As an idiomatic sentence: '*someone detests a cat that is not fierce*'.

2. Let the universe of discourse (for both variables) be 'people' and define the following two-variable propositional functions.

$$
\begin{aligned}
K(x,y) &: \quad x \text{ knows } y \\
T(x,y) &: \quad x \text{ is taller than } y \\
Y(x,y) &: \quad x \text{ is younger than } y
\end{aligned}
$$

Symbolise each of the following propositions.

(i) Jess knows everyone.

(ii) Someone knows everyone.

(iii) Everyone knows Ahmed.

(iv) Everyone knows someone.

(v) Everyone knows someone who is taller than them.

(vi) Everyone Misha knows is taller than him.

(vii) Sam is taller than everyone (who is) younger than him.

(viii) Everyone is taller than all those people who are younger than them.

Solution

(i) We may rewrite the sentence as, 'for all y, Jess knows y'. Hence we may symbolise it as: $\forall y \bullet K(\text{Jess}, y)$.

Note that the name of the variable is not important; we chose y as it appeared in the second place in the propositional function, but any variable name would be acceptable. Thus each of the following are also acceptable, even if the third is not one we would normally choose!

$$\forall x \bullet K(\text{Jess}, x), \ \forall t \bullet K(\text{Jess}, t), \ \forall wallaby \bullet K(\text{Jess}, wallaby).$$

(ii) This proposition is similar to the previous one except that we are now asserting the existence of someone who knows everyone rather than stating that this 'all knowing' person is Jess. This means that we simply need to take the previous proposition and replace 'Jess' with an existentially quantified variable.

In pseudo-English we have: there exists x such that for all y, x knows y. Hence in symbols the proposition is: $\exists x \, \forall y \bullet K(x, y)$.

Note that this example is much easier to symbolise once we have already attempted part (i). This provides a useful tool in symbolising propositions involving two quantifiers: first replace one quantifier with a named object or individual and symbolise the resulting proposition that now only has a single quantifier. Secondly, replace the named individual with an appropriately quantified variable.

(iii) This is similar to the first proposition except that the named individual is the person 'known' rather than the person 'knowing'. We may symbolise as $\forall x \bullet K(x, \text{Ahmed})$.

(iv) Following on from example (iii), we simply need to replace 'Ahmed' with an existentially quantified variable $\exists y$. Thus we may symbolise the proposition as $\forall x \, \exists y \bullet K(x, y)$. In the original sentence, 'everyone knows someone', the universal quantification precedes the existential quantification and this is mirrored in the symbolic version.

There is, however, an important point to make in relation to the proposition in part (iii). In that case, the proposition asserts that everyone knows the same specific individual, namely Ahmed. For the proposition 'everyone knows someone' to be true, it is not necessarily the case that everyone knows *the same* person. For example, although I might know Ahmed, you may not, but you might know Julie instead. So long as everyone in the universe has someone that they know, the proposition is true.

(v) This proposition is an extension of that in part (iv). We simply need to add that the person y, who is known by x, is taller than x. Hence we may symbolise as:

$$\forall x \, \exists y \bullet K(x, y) \wedge T(y, x).$$

(vi) In our half-way house language we may write this as: 'for all x, if Misha knows x, then x is taller than Misha'. Hence the fully symbolised version is: $\forall x \bullet K(\text{Misha}, x) \Rightarrow T(x, \text{Misha})$.

(vii) Firstly, we may write in literal form as 'for all x, if x is younger than Sam, then Sam is taller than x'. Therefore we may symbolise the proposition as: $\forall x \bullet Y(x, \text{Sam}) \Rightarrow T(\text{Sam}, x)$.

(viii) Comparing this with the previous proposition, we just need to replace 'Sam' with a universally quantified variable. Hence the proposition is: $\forall y \, \forall x \bullet Y(x, y) \Rightarrow T(y, x)$.

3. Let the universe of discourse be 'people' and consider the two-variable propositional function

$$L(x, y): \text{ '}x \text{ loves } y\text{'.}$$

Write down English sentences representing each of the following possible ways of doubly quantifying $L(x, y)$.

   (i) $\forall x \, \forall y \bullet L(x, y)$

  (ii) $\forall x \, \exists y \bullet L(x, y)$

 (iii) $\exists y \, \forall x \bullet L(x, y)$

 (iv) $\exists x \, \forall y \bullet L(x, y)$

  (v) $\forall y \, \exists x \bullet L(x, y)$

 (vi) $\exists x \, \exists y \bullet L(x, y)$

Solution

(i) Everyone loves everyone.

(ii) In our half-way house language, this would be 'for all x there exists y such that x loves y' or just 'every x loves some y'. In natural-sounding language we would say 'everyone loves someone'.

(iii) Again, in the half-way house language, this is 'there exists y such that, for all x, x loves y'. In order to preserve the order in which the variables are mentioned, we note that 'x loves y' can be replaced with 'y is loved by x', so that the half-way house expression becomes 'some y is loved by every x'. More naturally, this is 'someone is loved by everyone'.

(iv) This time the 'unnatural' expression is 'there exists x such that, for all y, x loves y' or 'some x loves all y'. More naturally, we might say 'someone loves everyone'.

(v) Replacing 'x loves y' with 'y is loved by x' as in part (iii), we have 'for all y, there exists x such that y is loved by x' or 'every y is loved by some x'. More naturally, 'everyone is loved by someone'.

(vi) Someone loves somebody.

Note that (ii) and (iii) are logically different propositions and they differ only in the ordering of the quantifiers. The same is true of are propositions (iv) and (v). This illustrates the fact we have noted previously — when we have different quantifiers, the order in which they are written is significant.

Negation of Quantified Propositional Functions

At first glance, the proposition '*no athletes are physically fit*' might suggest the negation of the proposition '*all athletes are physically fit*'. However, recall that the negation of a proposition is false when the proposition is true and is true when the proposition is false. Therefore '*no athletes are physically fit*' is not the negation of '*all athletes are physically fit*' because, in the case where some athletes are fit and some are not, both propositions are false. The proposition '*all athletes are physically fit*' is false when there is at least one athlete who is not physically fit. Hence the negation of '*all athletes are physically fit*' is '*some athletes are not physically fit*'. If we define the universe to be 'athletes' and we let $P(x)$ denote '*x is physically fit*', then, in symbols, the negation of $\forall x \bullet P(x)$ is $\exists x \bullet \neg P(x)$,

$$\neg \forall x \bullet P(x) \equiv \exists x \bullet \neg P(x).$$

Similarly, the existentially quantified proposition '*some athletes are physically fit*' is false in the unlikely event that all athletes are not physically fit. In other words, the negation of '*some athletes are physically fit*' is the proposition '*all athletes are not physically fit*'. In symbols,

$$\neg \exists x \bullet P(x) \equiv \forall x \bullet \neg P(x).$$

Note that, in each case, to negate the quantified proposition, the quantifier changes (from $\forall$ to $\exists$ or vice versa) and the propositional function itself is negated. Thus, each of the two logical equivalences above conform to the general description (or 'meta-rule') that says,

> *to negate a quantified propositional function, change the*
> *quantifier and negate the propositional function.*

There is nothing special about the context — that is, athletes and their fitness — of the previous examples. So the two logical equivalences above, or the meta-rule, give the following general rule for negating quantified propositions.

Rule for negating quantified propositions

$$\neg \forall x \bullet P(x) \equiv \exists x \bullet \neg P(x).$$

$$\neg \exists x \bullet P(x) \equiv \forall x \bullet \neg P(x).$$

This rule is particularly helpful for negating quantified statements where more than one quantifier is involved. Consider, for example, the proposition 'someone knows everyone'. In exercise 2.12.2 (ii), we symbolised this as

$$\exists x \, \forall y \bullet K(x, y)$$

where $K(x, y)$ is 'x knows y' defined over the universe of people. We may ask: what is the negation of this proposition? To answer this, we move the negation through the quantifiers in two stages, applying the rule for negating quantified propositions at each stage, as follows.

$$\neg \exists x \, \forall y \bullet K(x, y) \quad \equiv \quad \forall x \, \neg \forall y \bullet K(x, y)$$
$$\equiv \quad \forall x \, \exists y \bullet \neg K(x, y).$$

Turning this back into English, we can first write is as; 'for all x there exists y such that x does not know y'. In idiomatic form, we may write '*everyone has someone they don't know*'.

On page 53, we introduced the following two general forms of quantified propositions.

All Ps are Qs : $\forall x \bullet P(x) \Rightarrow Q(x)$

Some Ps are Qs : $\exists x \bullet P(x) \wedge Q(x)$

We conclude this chapter by considering the negation of each of these propositions. Consider first, the negation of 'all Ps are Qs': $\neg \forall x \bullet P(x) \Rightarrow Q(x)$. Applying the rule for negating quantified propositions, we have:

$$\neg \forall x \bullet P(x) \Rightarrow Q(x) \equiv \exists x \bullet \neg (P(x) \Rightarrow Q(x)).$$

To understand what this means, we need to consider the propositional function $\neg (P(x) \Rightarrow Q(x))$. In exercise 2.2.4 we established the following logical equivalence (called the Law of Material Implication).

$$P \Rightarrow Q \equiv \neg P \vee Q.$$

Therefore, using De Morgan's law and the involution law, we have

$$\neg (P \Rightarrow Q) \equiv \neg (\neg P \vee Q) \equiv P \wedge \neg Q.$$

Applying this to the quantified proposition above shows that

$$\neg \forall x \bullet P(x) \Rightarrow Q(x) \equiv \exists x \bullet P(x) \wedge \neg Q(x).$$

Now we may paraphrase $\exists x \bullet P(x) \wedge \neg Q(x)$ as 'some Ps are not Qs'. Hence the negation of 'all Ps are Qs' is 'some Ps are not Qs'.

Now consider the negation of 'some Ps are Qs': $\neg \exists x \bullet P(x) \wedge Q(x)$. Again applying the rule for negating quantified propositions, we have:

$$\neg \exists x \bullet P(x) \wedge Q(x) \equiv \forall x \bullet \neg(P(x) \wedge Q(x)).$$

Applying De Morgan's law and the law of material implication given in exercise 2.2.4, we have

$$\neg(P \wedge Q) \equiv \neg P \vee \neg Q \equiv P \Rightarrow \neg Q.$$

Applying this to the quantified proposition shows that

$$\neg \exists x \bullet P(x) \wedge Q(x) \equiv \forall x \bullet P(x) \Rightarrow \neg Q(x).$$

Now we may paraphrase $\forall x \bullet P(x) \Rightarrow \neg Q(x)$ as 'all Ps are not Qs'. Hence the negation of 'some Ps are Qs' is 'all Ps are not Qs' or 'every P is not a Q'.

Exercises 2.3

1. Translate each of the following propositions into symbolic form using one-place or two-place predicates. Define the predicates used and, where appropriate, define a universe of discourse.

    (i) *All students like to party.*

    (ii) *Some footballers are overpaid but are not talented.*

    (iii) *Everyone who went to the auction bought something.*

    (iv) *Someone shouted 'Fire!' and everyone panicked.*

    (v) *All celebrities give interviews but not all of them participate in reality TV.*

    (vi) *Not all athletes can run fast.*

    (vii) *No one likes people who are rude.*

    (viii) *If all Graham's friends came to the party, then some of them would have to stand.*

    (ix) *All of my friends believe in global warming but some of them drive large cars.*

    (x) *Everyone applauds someone who is courageous.*

(xi) *No one who went to the races was cold but some of them lost money.*

(xii) *If all politicians are honest, then none of them would receive illegal payments.*

2. The following propositional functions are defined. The universe of discourse is 'people'.

$$C(x): \quad x \text{ is clever}$$
$$H(x): \quad x \text{ is honest}$$
$$L(x): \quad x \text{ is likeable}$$
$$S(x): \quad x \text{ is smiles a lot}$$
$$D(x): \quad x \text{ is prone to depression}$$

Write each of the following propositions in natural-sounding English sentences.

(i) $S(\text{Julie}) \wedge D(\text{Julie})$

(ii) $(C(\text{Pete}) \wedge H(\text{Pete})) \Rightarrow L(\text{Pete})$

(iii) $\forall x \bullet L(x) \Rightarrow S(x)$

(iv) $\exists x \bullet C(x) \wedge \neg H(x)$

(v) $\forall x \bullet (\neg L(x) \vee \neg S(x)) \Rightarrow D(x)$

(vi) $\neg \forall x \bullet L(x) \wedge H(x)$

(vii) $\neg \forall x \bullet D(x) \Rightarrow S(x)$

(viii) $\neg \exists x \bullet L(x) \wedge \neg H(x)$

3. The following two-variable propositional functions are defined. The universe of discourse for the variable x is 'students' and the universe of discourse for the variable y is 'courses'.

$$T(x, y): \quad x \text{ takes } y.$$
$$E(x, y): \quad x \text{ enjoys } y.$$
$$P(x, y): \quad x \text{ passes } y.$$

Symbolise each of the following propositions.

(i) *Carl passes every course he takes.*

(ii) *Every student who takes Statistics enjoys it.*

(iii) *Some students who take Statistics do not pass.*

(iv) *There are students who take courses which they do not enjoy.*

(v) *Some students pass every course they take.*

(vi) *If Gemma passes Statistics, then any student who takes Statistics passes it.*

(vii) *There are some courses that are passed by all students who take them.*

(viii) *If all students take courses that they don't enjoy, then no student passes every course they take.*

4. This question uses the same universes of discourse and propositional functions as defined in question 3. For each of the following propositions written symbolically:

   (a) write the proposition as a natural-sounding English sentence;

   (b) negate the symbolic form and apply the rule for negating quantified propositions to move the negation through the quantifier(s);

   (c) write the resulting negated proposition as a natural-sounding English sentence.

   (i) $\forall x \bullet E(x, \text{Logic})$

   (ii) $\exists x \bullet \neg P(x, \text{Logic})$

   (iii) $\forall y \bullet T(\text{Poppy}, y) \Rightarrow E(\text{Poppy}, y)$

   (iv) $\exists y \bullet T(\text{Poppy}, y) \land \neg E(\text{Poppy}, y)$

   (v) $\forall x \exists y \bullet P(x, y)$

   (vi) $\neg \exists x \forall y \bullet P(x, y)$

   (vii) $\exists x \forall y \bullet T(x, y) \Rightarrow P(x, y)$

   (viii) $(\exists x \forall y \bullet \neg T(x, y)) \Rightarrow (\neg \forall x \exists y \bullet P(x, y))$

2.5 Logical reasoning

In chapter 1, we noted that the kind of reasoning that is deployed in mathematics is, or should be, both precise and logically rigorous. In section 2.3, we introduced the notion of logical implication which captures what we mean by 'logically rigorous'. Recall that a collection of propositions $P_1, P_2, \ldots, P_n$ logically implies another proposition Q, denoted $\{P_1, P_2, \ldots, P_n\} \vDash Q$, provided Q is true in all those cases where *each* of the propositions $P_1, P_2, \ldots, P_n$ is true. We were able to verify the relation of logical implication using truth tables. All of our examples were restricted to situations where there were at most three simple propositions involved. There was an obvious reason for this: a truth table for a compound proposition that involves n simple propositions has 2^n rows. So when there are more than three or four simple propositions involved, evaluating a truth table becomes unwieldy and impractical.

In this section we give an alternative way of establishing logical implication. We give a set of simple rules for deducing a 'conclusion' proposition from 'premise' propositions such that the premise propositions logically imply the conclusion proposition. To illustrate this, using truth tables we can show $P, P \Rightarrow Q \vDash Q \vee R$. Using our 'deduction rules' we will be able to establish this in two steps:

- one deduction rule allows us to deduce the conclusion Q from premises P and $P \Rightarrow Q$;

- a second deduction rule allows us to deduce the conclusion $Q \vee R$ from the premise Q.

We will summarise this deduction as a sequence of propositions where each proposition is either a premise or can be deduced from earlier propositions in the list using a deduction rule.

1. P premise
2. $P \Rightarrow Q$ premise
3. Q from 1,2 using first deduction rule above
4. $Q \vee R$ from 3 using second deduction rule above

A deduction rule will be written in two lines, with the premises of the rule on the first line and the conclusion on the second. Thus the two rules used in the deduction above are written as follows.

$$\frac{P, \; P \Rightarrow Q}{Q} \qquad \text{and} \qquad \frac{P}{P \vee Q}$$

Note that the second of these is written in terms of P and Q rather than Q and R, which were actually used in the deduction above. In applying a deduction rule we may substitute any propositions for those in the rule provided, of course, that the same proposition is substituted throughout for a particular proposition that appears in the rule. For example, given propositions $P \wedge Q$ and $(P \wedge Q) \Rightarrow (R \vee S)$, we may apply the first deduction rule above to deduce $R \vee S$.

We now list a set of rules that will allow us to make the deductions that we need. In each case the premise propositions logically imply the conclusion. As an illustration, in example 2.8.1, we showed that $\{P \Rightarrow Q, Q \Rightarrow R\} \vDash (P \Rightarrow R)$, and this is the basis for the deduction rule called hypothetical syllogism. This means that, as we apply the rules in making deductions, the conclusion of a sequence of deductions will always be logically implied by the premises.

Deduction rules for propositional logic

Simplification	$$\dfrac{P \wedge Q}{P}$$
Addition	$$\dfrac{P}{P \vee Q}$$
Conjunction	$$\dfrac{P, \; Q}{P \wedge Q}$$
Disjunctive syllogism	$$\dfrac{P \vee Q, \; \neg P}{Q}$$
Modus ponens	$$\dfrac{P, \; P \Rightarrow Q}{Q}$$
Modus tollens	$$\dfrac{P \Rightarrow Q, \; \neg Q}{\neg P}$$
Hypothetical syllogism	$$\dfrac{P \Rightarrow Q, \; Q \Rightarrow R}{P \Rightarrow R}$$
Resolution (Res)	$$\dfrac{P \vee Q, \; \neg P \vee R}{Q \vee R}$$
Constructive dilemma	$$\dfrac{P \Rightarrow Q, \; R \Rightarrow S, \; P \vee R}{Q \vee S}$$

In addition to these rules, we also allow the replacement of a proposition with any logically equivalent proposition. For example, if we have $\neg(P \wedge Q)$ in a deduction, then we may add the proposition $\neg P \vee \neg Q$ since $\neg(P \wedge Q) \equiv \neg P \vee \neg Q$ (De Morgan's rules, page 39).

These deduction rules are not independent of one another, and they are not a 'minimal set' of rules. Equally, there are other rules that we could have included — see exercise 2.4.2. What is important about the rules is that they are sufficient to be able to deduce any logical implication. The following definition will help make this a little more precise.

Definition 2.3
Let $P_1, P_2, \ldots, P_n, Q$ be propositions. We say that Q is **deducible from**

$P_1, P_2, \ldots, P_n$, written $P_1, P_2, \ldots, P_n \vdash Q$ or $\{P_1, P_2, \ldots, P_n\} \vdash Q$, if there is a list of propositions, the last of which is Q, such that each proposition is either one of the $P_1, P_2, \ldots, P_n$ or is obtained from earlier propositions in the list by applying one of the deduction rules or is logically equivalent to an earlier proposition in the list.

The list of propositions, together with justifications for each proposition belonging to the list, is called a **deduction** of Q from $P_1, P_2, \ldots, P_n$. We refer to the propositions $P_1, P_2, \ldots, P_n$ as the **premises** of the deduction and the proposition Q as the **conclusion** of the deduction.

The fact that, in each of our deduction rules the premises logically imply the conclusion, means that we will only be able to derive propositions that are logically implied by the premises. In other words, if Q is deducible from $P_1, P_2, \ldots, P_n$, then $P_1, P_2, \ldots, P_n$ logically imply Q:

$$\text{if } \{P_1, P_2, \ldots, P_n\} \vdash Q \text{ then } \{P_1, P_2, \ldots, P_n\} \vDash Q.$$

Logicians refer to this property as **soundness** and we can say that our deduction rules are **sound**. In addition, our rules are sufficient to be able to deduce any logical implication. In other words, if $P_1, P_2, \ldots, P_n$ logically imply Q, then Q is **deducible from** $P_1, P_2, \ldots, P_n$ using our deductions rules:

$$\text{if } \{P_1, P_2, \ldots, P_n\} \vDash Q \text{ then } \{P_1, P_2, \ldots, P_n\} \vdash Q.$$

Logicians refer to this property as **completeness**, and we can say that our deduction rules are **complete**. Completeness is, in fact, rather rare in logic and it is only for simple systems such as propositional logic that we can find a complete set of deduction rules.

Examples 2.13

1. Show that $\{(P \Rightarrow Q) \wedge (R \Rightarrow S), R\} \vDash Q \vee S$.

   Solution

   By the soundness property, it is sufficient to show that $Q \vee S$ can be deduced from the propositions $P \Rightarrow Q, R \Rightarrow S$, and R; symbolically, $\{(P \Rightarrow Q) \wedge (R \Rightarrow S), R\} \vdash Q \vee S$.

   Here is one way of completing the deduction.

1.	$(P \Rightarrow Q) \wedge (R \Rightarrow S)$	premise
2.	R	premise
3.	$(R \Rightarrow S) \wedge (P \Rightarrow Q)$	1. Equivalence: commutative law
4.	$R \Rightarrow S$	3. Simplification
5.	S	2, 4. Modus ponens
6.	$S \vee Q$	5. Addition
7.	$Q \vee S$	6. Equivalence: commutative law

Note that the inclusion of each proposition in the deduction has been justified either by noting that it is a premise or by explaining how it follows from previous propositions using the deduction rules. In the case where we introduce a proposition that is logically equivalent to an existing proposition in the deduction, we indicate this by 'equivalence' followed by a reason for the logical equivalence.

In each case in this deduction, the logical equivalence was the commutative law (for $\wedge$ or $\vee$) — see page 39. The reason we need this is that our deduction laws are quite precise. For example, from $P \wedge Q$ the Simplification deduction rule allows us to deduce P but not Q so, if we wish to deduce Q, we first need to deduce the logically equivalent proposition $Q \wedge P$. An alternative approach would have been to introduce two versions of the simplification rule, one where we may deduce P and one where we may deduce Q.

Is is frequently the case that there are several different ways of deducing a conclusion from a collection of premises. The following is an alternative deduction to the one given above. In this case we have given the abbreviated form of the rules used at each step.

1.	$(P \Rightarrow Q) \wedge (R \Rightarrow S)$	premise
2.	R	premise
3.	$(R \Rightarrow S) \wedge (P \Rightarrow Q)$	1. Equivalence: commutative law
4.	$P \Rightarrow Q$	1. Simplification
5.	$R \Rightarrow S$	3. Simplification
6.	$R \vee P$	2. Addition
7.	$P \vee R$	3. Equivalence: commutative law
8.	$Q \vee S$	4, 5, 7. Constructive Dilemma

2. Show that $\{(P \vee Q) \wedge (Q \vee R), \neg Q\} \vDash P \wedge R$.

Solution

As in the previous example, we give a deduction of $P \wedge R$ from the two premises $(P \vee Q) \wedge (Q \vee R)$ and $\neg Q$.

1.	$(P \vee Q) \wedge (Q \vee R)$	premise
2.	$\neg Q$	premise
3.	$P \vee Q$	1. Simplification
4.	$Q \vee P$	3. Equivalence: commutative law
5.	P	2, 4. Disjunctive syllogism
6.	$(Q \vee R) \wedge (P \vee Q)$	1. Equivalence: commutative law
7.	$Q \vee R$	6. Simplification
8.	R	2, 7. Disjunctive syllogism
9.	$P \wedge R$	5, 8. Conjunction

Arguments in Propositional Logic

In section 1.4, we considered briefly the nature of logical arguments. We are now in a position to be more precise about what constitutes a logical argument. An **argument** in propositional logic comprises a collection of propositions $P_1, P_2, \ldots, P_n$, called **premises**, and a proposition Q called the **conclusion**. An argument with premises $P_1, P_2, \ldots, P_n$ and conclusion Q is **valid** if the premises logically imply the conclusion

$$\{P_1, P_2, \ldots, P_n\} \vDash Q.$$

A **formal deduction**, or **formal proof**, of the validity of an argument is a deduction that shows

$$\{P_1, P_2, \ldots, P_n\} \vdash Q.$$

Examples 2.14

1. Consider the validity of the following argument.

   If Mark is correct, then unemployment will rise, and if Ann is correct then, there will be a hard winter. Ann is correct. Therefore unemployment will rise or there will be a hard winter.

   Solution

   Before we can test the validity of the argument, we need to understand the structure of the argument. We first symbolise the propositions involved.

Let	M:	*Mark is correct*
	U:	*Unemployment will rise*
	A:	*Ann is correct*
	H:	*There will be a hard winter.*

   There are two premises of the argument: $(M \Rightarrow U) \wedge (A \Rightarrow H)$ and A.[4] The conclusion is signalled by the word 'therefore'; in symbols, the conclusion is $U \vee H$.

   To show that the argument is correct, we need to establish that

   $$\{(M \Rightarrow U) \wedge (A \Rightarrow H), A\} \vDash U \vee H.$$

   In example 2.13.1, we established that

   $$\{(P \Rightarrow Q) \wedge (R \Rightarrow S), R\} \vDash Q \vee S$$

---

[4] It may be worth noting that, had the first sentence been split in two, *If Mark is correct, then unemployment will rise. If Ann is correct, then there will be a hard winter*, then we would regard the argument as having three premises $M \Rightarrow U$, $A \Rightarrow H$, and A. This does not affect the validity or otherwise of the argument because from $P \wedge Q$ we may deduce both P and Q and, conversely, from both P and Q we may deduce $P \wedge Q$.

by giving a deduction of $Q \vee S$ from $(P \Rightarrow Q) \wedge (R \Rightarrow S)$ and R. This is the same as the deduction that we wish to establish except that M, U, A, and H have been relabelled P, Q, R, and S, respectively. Clearly, the labelling of the propositions does not matter so the deduction given in example 2.13.1 can also be used to show that $\{(M \Rightarrow U) \wedge (A \Rightarrow H), A\} \vDash U \vee H$, so the argument is valid. For future reference, translating the deduction given in example 2.13.1 gives the following.

1.	$(M \Rightarrow U) \wedge (A \Rightarrow H)$	premise
2.	A	premise
3.	$(A \Rightarrow H) \wedge (M \Rightarrow U)$	1. Equivalence: commutative law
4.	$A \Rightarrow H$	3. Simplification
5.	H	2, 4. Modus ponens
6.	$H \vee U$	5. Addition
7.	$U \vee H$	6. Equivalence: commutative law

2. Provide a formal deduction of the validity of the following argument.

If Sam had taken my advice or he'd had his wits about him, then he would have sold his house and moved to the country. If Sam had sold his house, Jenny would have bought it. Jenny did not buy Sam's house. Therefore Sam did not take my advice.

Solution

We first symbolise the propositions as follows.

Let	A:	*Sam took my advice*
	W:	*Sam had his wits about him*
	H:	*Sam sold his house*
	C:	*Sam moved to the country*
	J:	*Jenny bought Sam's house.*

Then the argument has premises: $(A \vee W) \Rightarrow (H \wedge C)$, $H \Rightarrow J$ and $\neg J$. The conclusion is $\neg A$.

When attempting to discover a deduction, it often helps both to work forwards from the premises and also to work backwards from the conclusion. In other words, we may ask 'in order to be able to deduce the conclusion, what do we first need to deduce?'

In this example, we may deduce $\neg A$ if we can first deduce $\neg A \wedge \neg W \equiv \neg (A \vee W)$. Now we may deduce $\neg (A \vee W)$ from the first premise using modus tollens, provided that we can deduce $\neg (H \wedge C) \equiv \neg H \vee \neg C$. The full deduction is the following where the first few steps establish $\neg H \vee \neg C$.

1. $(A \lor W) \Rightarrow (H \land C)$ premise
2. $H \Rightarrow J$ premise
3. $\neg J$ premise
4. $\neg H$ 2, 3. Modus tollens
5. $\neg H \lor \neg C$ 4. Addition
6. $\neg(H \land C)$ 5. Equivalence: De Morgan's law
7. $\neg(A \lor W)$ 1, 6. Modus tollens
8. $\neg A \land \neg W$ 7. Equivalence: De Morgan's law
9. $\neg A$ 8. Simplification

3. Consider the validity of the following argument.

If Mark is correct, then unemployment will rise and if Ann is correct, then there will be a hard winter. Anne is correct. Therefore, unemployment will rise and there will be a hard winter.

Solution

This is almost the same argument as the one given in Example 1 above. The only difference is that the conclusion has been strengthened to be a conjunction. In other words, this argument concludes that unemployment will rise *and* there will be a hard winter, whereas the argument in example 1 claimed that only one of these events would occur.

With the symbolism used in example 1, the premises are $(M \Rightarrow U) \land (A \Rightarrow H)$ and A (as before) and the conclusion is $U \land H$. If we look more closely at the derivation in example 1, we can see that the reason why we were able to conclude $U \lor H$ was that we earlier deduced H. However this will not allow us to conclude $U \land H$.

At this point we may start thinking that the argument is not valid; in other words, that the premises do not logically imply the conclusion. In general, to show that P does not logically imply Q, $P \nVdash Q$, we need to show that Q is not true in all cases where P is true. This means we need to find a situation where P is true and Q is false.

Returning to the argument, to show that it is not valid, we need to find a situation where all of the premises are true but the conclusion is false. Now A must be true since it is a premise and H must be true since the deduction in Example 1 shows that H may be deduced from the premises. In order to make the conclusion false, we need to suppose that U is false. Consider the premise $(M \Rightarrow U) \land (A \Rightarrow H)$. The second conjunct $A \Rightarrow H$ is true since both A and H are true. If U is false, the first conjunct will be true only if M is also false.

So, suppose that:

M is false: Mark is not correct
U is false: unemployment will not rise
A is true: Ann is correct
H is true: there will be a hard winter.

Then both premises are true but the conclusion is false. Therefore the argument is not valid.

4. Consider the validity of the following argument.

If Rory didn't get the job, then he didn't become a lawyer. Rory didn't get the job and he took up golf. Rory became a lawyer or he didn't take up golf. Therefore Rory won the Open Championship.

Solution

We first symbolise the propositions as follows.

Let J: *Rory got the job*
         L: *Rory become a lawyer*
         G: *Rory took up golf*
         C: *Rory won the Open Championship.*

Then the argument has premises: $\neg J \Rightarrow \neg L$, $\neg J \wedge G$, and $L \vee \neg G$. The conclusion is C.

At this stage we may think that there is something strange about the argument as the conclusion C does not feature at all in the premises. Thus, were we able to provide a deduction of the argument, we would also be able to provide a deduction for any argument with these premises and *any conclusion whatsoever*. In fact, it is possible to provide a deduction as follows.

1.	$\neg J \Rightarrow \neg L$	premise
2.	$\neg J \wedge G$	premise
3.	$L \vee \neg G$	premise
4.	$\neg J$	2. Simplification
5.	$\neg L$	1, 4. Modus ponens
6.	$\neg G$	3, 5. Disjunctive syllogism
7.	$G \wedge \neg J$	2. Equivalence: commutative law
8.	G	7. Simplification
9.	$G \vee C$	8. Addition
10.	C	6, 9. Disjunctive syllogism

This deduction gives further grounds for concern since we have deduced both G (line 8) and its negation $\neg G$ (line 6). The reason for this is that the premises are inconsistent in the sense that it is not possible for them all to be true simultaneously. This could be verified by drawing up a truth table for the premises and showing that there is no combination of truth values of J, L, and G for which the premises are all true — see exercise 2.4.5.

Whenever we have a situation like this, we may deduce any proposition in the manner shown using addition (line 9 above) followed by disjunctive syllogism (line 10 above). Beginning with inconsistent premises, it is possible to deduce any proposition from them.

Arguments in Predicate Logic

It is relatively easy to extend the idea of a formal deduction to arguments in predicate logic. An argument in predicate logic will still have premises $P_1, P_2, \ldots, P_n$ and conclusion Q, but now some or all of these may be quantified propositional functions. In addition to the deduction rule in propositional logic, we also need rules that allow us to eliminate or introduce quantifiers.

For example, suppose that $\forall x \bullet P(x)$ is a premise of an argument. Then we may deduce the proposition $P(a)$ for any object a in the appropriate universe. We call this the 'quantifier elimination' rule for the universal quantifier $\forall$. The 'quantifier introduction' rule for the universal quantifier is a little more subtle. This is because we will usually not be able to deduce $P(a)$ for *each and every* element a of the universe. The quantifier introduction rule says that, if we have established $P(a)$ for an *arbitrary* element of the universe, then we may deduce $\forall x \bullet P(x)$. By 'arbitrary' we just mean that a has no special attributes other than being a member of the particular universe.

The corresponding rules for elimination and introduction of the existential quantifier are similar; the essential difference is that the object a of the universe is a specific one rather than an arbitrary element.

Deduction rules for predicate logic

Universal quantifier elimination ($\forall$-elimination)
From $\forall x \bullet P(x)$, we may deduce $P(a)$ for any a in the universe.

Universal quantifier introduction ($\forall$-introduction)
From $P(a)$ where a is an arbitrary element of the universe, we may deduce $\forall x \bullet P(x)$.

Existential quantifier elimination ($\exists$-elimination)
From $\exists x \bullet P(x)$, we may deduce $P(a)$ for some particular a in the universe.

Existential quantifier introduction ($\exists$-introduction)
From $P(a)$ where a is a particular element of the universe, we may deduce $\exists x \bullet P(x)$.

With these rules, in addition to the deduction rules for propositional logic, we can give formal deductions of arguments in predicate logic. The basic strategy is to apply quantifier elimination rules at the beginning of the deduction to obtain non-quantified propositions; then apply appropriate deduction rules from propositional logic; finally, apply appropriate quantifier introduction rules, where necessary. In the last step, we will need to be careful to note whether we have deduced $Q(a)$, say, for an arbitrary or a particular element a of the universe. In the former case, we may deduce $\forall x \bullet Q(x)$ and in the latter case we may deduce $\exists x \bullet Q(x)$.

Examples 2.15

1. Give a formal deduction of the validity of the following argument.

   All athletes are physically fit. Sarah is an athlete. Therefore Sarah is physically fit.

   Solution

   We first define the universe to be 'people' and symbolise the predicates as follows.

   Let A: ...*is an athlete*
            F: ...*is physically fit.*

   Then the argument has premises: $\forall x \bullet A(x) \Rightarrow F(x)$ and $A(\text{Sarah})$. The conclusion is $F(\text{Sarah})$.

1.	$\forall x \bullet A(x) \Rightarrow F(x)$	premise
2.	$A(\text{Sarah})$	premise
3.	$A(\text{Sarah}) \Rightarrow F(\text{Sarah})$	1. $\forall$-elimination
4.	$F(\text{Sarah})$	2, 3. Modus ponens

   Note that, in this case the conclusion is a non-quantified proposition, so we do not need a quantifier introduction step at the end.

2. Give a formal deduction of the validity of the following argument.

   All elephants are mammals. Some elephants are playful. Therefore some mammals are playful.

   Solution

   We first define the universe to be 'living beings' and symbolise the predicates as follows.

   Let E: ...*is an elephant*
            M: ...*is a mammal*
            P: ...*is playful.*

Then the argument has premises: $\forall x \bullet E(x) \Rightarrow M(x)$, $\exists x \bullet E(x) \wedge P(x)$. The conclusion is $\exists x \bullet M(x) \wedge P(x)$.

At the beginning of the argument after listing the premises, we may apply quantifier elimination to each premise. However, we need to be careful. If we apply $\forall$-elimination to the first premise, we deduce $E(a) \Rightarrow M(a)$ for an arbitrary element of the universe. However we cannot assume that this element a is also one for which $E(a) \wedge P(a)$ is true. Applying $\exists$-elimination to the premise $\exists x \bullet E(x) \wedge P(x)$ allows us to deduce $E(b) \wedge P(b)$ for some particular element b of the universe, but we don't know that it is the element a in the proposition $E(a) \Rightarrow M(a)$.

The way around this is to apply $\exists$-elimination first to deduce $E(a) \wedge P(a)$ for some particular element a of the universe. Then applying $\forall$-elimination to the first premise, we can infer $E(a) \Rightarrow M(a)$ for this a. However we need to be aware that a is not arbitrary as it was 'obtained' from an $\exists$-elimination step; therefore, at the end of the argument, we may only apply $\exists$-introduction and not $\forall$-introduction.

The full deduction is the following.

1.	$\forall x \bullet E(x) \Rightarrow M(x)$	premise
2.	$\exists x \bullet E(x) \wedge P(x)$	premise
3.	$E(a) \wedge P(a)$	2. $\exists$-elimination
4.	$E(a) \Rightarrow M(a)$	1. $\forall$-elimination
5.	$E(a)$	3. Simplification
6.	$M(a)$	4, 5. Modus ponens
7.	$P(a) \wedge E(a)$	3. Equivalence: commutative law
8.	$P(a)$	7. Simplification
9.	$M(a) \wedge P(a)$	6, 8. Conjunction
10.	$\exists x \bullet M(x) \wedge P(x)$	9. $\exists$-introduction

3. Give a formal deduction of the validity of the following argument.

   Everyone supports a football team or plays hockey. Everyone enjoys exercise or they don't play hockey. Therefore everyone who doesn't support a football team enjoys exercise.

   Solution

   We first define the universe to be 'people' and symbolise the predicates as follows.

Let	F:	... *supports a football team*
	H:	... *plays hockey*
	E:	... *enjoys exercise*

   Then the argument has premises: $\forall x \bullet F(x) \vee H(x)$, $\forall x \bullet E(x) \vee \neg H(x)$. The conclusion is $\forall x \bullet \neg F(x) \Rightarrow E(x)$.

In this case, when applying ∀-elimination to the premises we obtain non-quantified propositions with an arbitrary element of the universe. Hence, at the end of the deduction, we are able to apply ∀-introduction. The full deduction is the following.

1.	$\forall x \bullet F(x) \vee H(x)$	premise
2.	$\forall x \bullet E(x) \vee \neg H(x)$	premise
3.	$F(a) \vee H(a)$	1. ∀-elimination
4.	$E(a) \vee \neg H(a)$	2. ∀-elimination
5.	$H(a) \vee F(a)$	3. Equivalence: commutative law
6.	$\neg H(a) \vee E(a)$	4. Equivalence: commutative law
7.	$F(a) \vee E(a)$	5, 6. Resolution
8.	$\neg(\neg F(a)) \vee E(a)$	7. Equivalence: involution law
9.	$\neg F(a) \Rightarrow E(a)$	8. Equivalence: material implication law
10.	$\forall x \bullet \neg F(x) \Rightarrow E(x)$	9. ∀-introduction

---

Exercises 2.4

1. Find a deduction from the premises to the conclusion in each of the following cases.

   (i) $\{P \Rightarrow Q, P \wedge R\} \vdash Q$

   (ii) $\{P \Rightarrow \neg Q, Q \vee R\} \vdash P \Rightarrow R$

   (iii) $\{P \Leftrightarrow Q, \neg(P \wedge Q)\} \vdash \neg P \wedge \neg Q$

   (iv) $(P \wedge Q) \Rightarrow (S \wedge T), Q \wedge P\} \vdash S$

   (v) $\{(Q \vee S) \Rightarrow R, Q \vee P, \neg R\} \vdash P$

   (vi) $\{Q, \neg S, (P \wedge R) \Rightarrow S\} \vdash Q \vee \neg R$

   (vii) $\{(P \vee Q) \Rightarrow (R \wedge S), P\} \vdash R$

   (viii) $\{P \vee Q, R \Rightarrow \neg Q, \neg P, (\neg R \wedge Q) \Rightarrow S\} \vdash S$

   (ix) $\{P \Rightarrow \neg Q, Q \vee (R \wedge S)\} \vdash P \Rightarrow (R \wedge S)$

   (x) $\{(P \Rightarrow Q) \wedge (R \Rightarrow Q), S \Rightarrow (P \vee R), S\} \vdash Q$

2. Show that each of the following deductions can be made from the deduction rules given on page 67. In each case the premises of the deduction are the propositions above the line and the conclusion of the deduction is the proposition below the line.

   Note that this means that each of these rules could also be considered a deduction rule and could therefore be added to our list of rules.

(i) Rule of biconditional introduction

$$\frac{P \Rightarrow Q, \ Q \Rightarrow P}{P \Leftrightarrow Q}$$

(ii) Rule of biconditional elimination

$$\frac{P \Leftrightarrow Q}{P \Rightarrow Q}$$

(iii) Rule of case analysis 1

$$\frac{P \vee Q, \ P \Rightarrow R, \ Q \Rightarrow R}{R}$$

(iv) Rule of case analysis 2

$$\frac{P \Rightarrow R, \ Q \Rightarrow R}{(P \vee Q) \Rightarrow R}$$

3. Give formal deductions of the validity of each of the following arguments.

(i) If Mary drinks wine or eats cheese, then she gets a headache. She is drinking wine and eating chocolate. Therefore Mary will get a headache.

(ii) If you get a degree or you get a good job, then you will be successful and happy. You get a good job. Therefore you will be happy.

(iii) If the battery is flat or the car is out of petrol, then it won't start and I'll be late for work. The car is out of petrol or the battery is flat. Therefore I'll be late for work.

(iv) Either the project wasn't a success or Sally did not invest her inheritance. If she were sensible, then Sally would invest her inheritance. The project was a success. If Sally wasn't sensible and didn't invest her inheritance, then she is broke. Therefore Sally is broke.

(v) The murder was committed either by A or by both B and C. If A committed the murder, then the victim was poisoned. Therefore either C committed the murder or the victim was poisoned.

(vi) Peter is either brave or brainy and he is either brainy or bald. Peter is not brainy. Therefore he is brave and bald.

(vii) If it is useful, then I'll keep it, and if it is valuable, then I'll keep it. If it belonged to Ben, then it is useful or valuable. It belonged to Ben. So I'll keep it.

(viii) If it doesn't rain, then I'll go shopping. If I go shopping, then if I don't take an umbrella, it will rain. If I go by car, then I won't take an umbrella. So it will rain or I won't go by car.

(ix) If ghosts are a reality, then there are spirits roaming the Earth, and if ghosts are not a reality, then we do not fear the dark. Either we fear the dark or we have no imagination. We do have an imagination and ghosts are a reality. Therefore there are spirits roaming the Earth.

(x) If Tim committed the crime, then he'll flee the country and we'll never see him again. If we see Tim again, then he is not Tom's friend. Hence, if Tim committed the crime or he's Tom's friend, then we'll never see him again.

4. Show that each of the following arguments are not valid.

   (i) If Mary drinks wine and eats cheese, then she gets a headache. She is drinking wine or eating chocolate. Therefore Mary will get a headache.

   (ii) Either the project wasn't a success or Sally invested her inheritance. If she were sensible, then Sally would invest her inheritance. The project was a success. If Sally wasn't sensible and didn't invest her inheritance, then she is broke. Therefore Sally is broke.

   (iii) If it doesn't rain, then I'll go shopping. If I go shopping, then if I don't take an umbrella, it will rain. If I go by car, then I won't take an umbrella. So it will rain and I won't go by car.

   (iv) If ghosts are a reality, then there are spirits roaming the Earth or if ghosts are not a reality, then we do not fear the dark. Either we fear the dark or we have no imagination. We do have an imagination or ghosts are a reality. Therefore there are spirits roaming the Earth.

   (v) If Tim committed the crime, then he'll flee the country and we'll never see him again. If we see Tim again, then he is not Tom's friend. Hence if he's Tom's friend, then we'll never see Tim again.

5. Draw up a truth table, with simple propositions labelled J, L, and G, giving the truth values of $\neg J \Rightarrow \neg L$, $\neg J \wedge G$, and $L \vee \neg G$.

   Hence show that the premises in example 2.14.4 are inconsistent.

6. Provide a formal deduction of the validity of each of the following arguments in predicate logic.

   (i) Some people are good-looking and rich. Everyone who is rich is dishonest. Therefore there are people who are good-looking and dishonest.

   (ii) Some people are good-looking and rich. Everyone who is rich is dishonest. Therefore not everyone who is good-looking is honest.

   (iii) All even numbers are rational and divisible by two. Some even numbers are divisible by four. Hence some numbers are divisible by two and four.

(iv) All numbers which are integers are even or odd. All numbers which are integers are even or non-zero. Some numbers are integers. Therefore there are numbers that are either even or they are odd and non-zero.

 (v) All animals with feathers are not aquatic. There are aquatic animals that live in the sea. So there are animals that live in the sea and don't have feathers.

(vi) Some functions are continuous and differentiable. All functions which are continuous are defined for all values of x. Therefore some functions which are defined for all values of x are differentiable.

(vii) Everything which is enjoyable and cheap is harmful to one's health. All holidays are enjoyable. There are holidays which are not harmful to one's health. Therefore some things are not cheap.

(viii) There are no polynomials which are not differentiable functions. All differentiable functions are continuous. Therefore all polynomials are continuous.

Chapter 3

Sets and Functions

3.1 Introduction

In the previous chapter, we discussed the elements of logic which we may regard as providing a framework for constructing and understanding mathematical statements. Much of the language of mathematics, however, is described in terms of sets and functions, which we introduce in this chapter. If logic provides the 'backbone' of mathematics, then sets and functions put flesh on the bone.

We imagine that many of our readers will have met sets and functions previously, but possibly will not have covered all the material that we will introduce here or maybe not as formally as our approach. The material of this chapter will be assumed as background knowledge in the rest of the book. Some readers who have met sets and functions may choose to ignore this chapter on first reading, referring back to it as and when necessary.

In the next three sections, we first introduce sets, how they may be defined and their properties. We then look at operations on sets and building new sets from old. Finally, we introduce an important construction called the Cartesian product. In the last two sections of the chapter, we consider functions and some of their properties.

3.2 Sets and membership

The notion of a 'set' is one of the basic concepts of mathematics — some would say *the* basic concept. We will not give a precise, formal definition of a set but simply describe a **set** as a well-defined collection of objects called **elements**. The elements contained in a given set need not have anything in common (other than the obvious common attribute that they all belong to the given set). Equally, there is no restriction on the number of elements allowed in a set; there may be an infinite number, a finite number, or even no elements at all.

There is, however, one restriction we insist upon: given a set and an object, we should be able to decide (in principle at least — it may be difficult in practice) whether or not the object belongs to the set. This is what we meant above by a 'well-defined' collection: given an object a and a set A, it must be unambiguous whether or not a belongs to A. Thus, 'the set of all tall people' is *not* a set, whereas 'the set of all people taller than 2 m' is a set.

Clearly, a concept as general as this has many familiar examples as well as many frivolous ones. For example, a set could be defined to contain the Mona Lisa, London Bridge, and the number e (although what 'use' this set has is hard to envisage). This is a **finite set** because it contains a finite number of elements. The set containing all the positive, even integers is clearly well-defined. This is an **infinite set** as it contains infinitely many elements.

Notation

We shall generally use uppercase letters to denote sets and lowercase letters to denote elements. It is not always possible to adhere to this convention however; for example, when the elements of a particular set are themselves sets. The symbol $\in$ denotes 'belongs to' or 'is an element of'. Thus

$$a \in A \text{ means (the element) } a \text{ belongs to (the set) } A$$

and

$$a \notin A \text{ means } \neg(a \in A) \text{ or } a \text{ does not belong to } A.$$

Defining Sets

Sets can be defined in various ways. The simplest is by listing the elements enclosed between curly brackets or 'braces' { }.

Examples 3.1

1. $A = \{\text{Mona Lisa}, \text{London Bridge}, e\}$.

   This is the rather odd set containing three elements described above.

2. $B = \{2, 4, 6, 8, \ldots\}$

   This is the infinite set described above. We clearly cannot list *all* the elements. Instead we list enough elements to establish a pattern and use '...' to indicate that the list continues indefinitely.

3. $C = \{1, \{1, 2\}\}$

   This set has *two* elements, the number 1 and the set $\{1, 2\}$. This illustrates that sets can themselves be elements of other sets.

4. $D = \{1, 2, \ldots, 100\}$

   This set contains the first 100 positive integers. Again we use '...' to indicate that there are elements in the list which we have omitted, although in this case only finitely many are missing.

5. $E = \{\ \}$

   This set contains no elements. It is called the **empty set** or **null set**. The empty set is usually denoted $\varnothing$.

When listing elements, the order in which elements are listed is not important. Thus, for example $\{1, 2, 3, 4\}$ and $\{4, 3, 2, 1\}$ define the same set. Also, any repeats in the listing are ignored so that $\{1, 2, 2, 3, 3, 3, 4, 4, 4, 4\}$ also describes the same set. To be precise, two sets are **equal** if and only if they contain the same elements; that is, $A = B$ if $\forall x \bullet x \in A \Leftrightarrow x \in B$ is a true proposition, and conversely.

Listing the elements of a set is impractical except for small sets or sets where there is a pattern to the elements such as B and D in examples 3.1 above. An alternative is to define the elements of a set by a property or predicate (see section 2.4). More precisely, if $P(x)$ is a single-variable propositional function, we can form the set whose elements are all those objects a (and only those) for which $P(a)$ is a true proposition. A set defined in this way is denoted

$$A = \{x : P(x)\}.$$

This is read: the set of all x such that $P(x)$ (is true).

Note that 'within A' — that is, if we temporarily regard A as the universe of discourse — the quantified propositional function $\forall x \bullet P(x)$ is a true statement.

Examples 3.2

1. The set B in example 3.1 above could be defined as

   $B = \{n : n \text{ is an even, positive integer}\}$, or

   $B = \{n : n = 2m, \text{ where } m > 0 \text{ and } m \text{ is an integer}\}$,

   or, with a slight change of notation,

   $B = \{2m : m > 0 \text{ and } m \text{ is an integer}\}$.

   Note that although the propositional functions used are different, the same elements are generated in each case.

2. The set D in example 3.1 above could be defined as

   $D = \{k : k \text{ is an integer and } 1 \leq k \leq 100\}$.

3. The set $\{1, 2\}$ could alternatively be defined as $\{x : x^2 - 3x + 2 = 0\}$. We sometimes refer to the set $\{1, 2\}$ as the **solution set** of the equation $x^2 - 3x + 2 = 0$.

4. The empty set $\varnothing$ can be defined in this way using any propositional function $P(x)$ which is true for no objects x. Thus, for example,

$$\varnothing = \{x : x \neq x\}.$$

Note that, if $P(x)$ and $Q(x)$ are propositional functions which are true for precisely the same objects x, then the sets they define are equal, $\{x : P(x)\} = \{x : Q(x)\}$. For example, the two solution sets $\{x : (x-1)^2 = 4\}$ and $\{x : (x+1)(x-3) = 0\}$ are equal, since the two propositional functions $P(x) : (x-1)^2 = 4$ and $Q(x) : (x+1)(x-3) = 0$ are true for precisely the same values of x, namely -1 and 3.

Universal set

Sometimes we wish to 'build' our sets *only* from some larger 'context set' called the **universal set**, denoted $\mathscr{U}$. The universal set is simply the set of all objects being considered at the current time. The universal set plays much the same role as the universe of discourse does in logic. In chapter 2, we noted that we could define the universe of discourse to be what was convenient for the particular context. So it is with sets. For example, if we are interested in studying the properties of soccer players in a particular league, we might define

$$\mathscr{U} = \{\text{soccer players in the particular league}\}.$$

Then the set

$$A = \{x : x \text{ has scored at least three goals}\}$$

denotes only those players in the particular league who have scored at least three goals.

Sometimes the role of the universe is made explicit when defining sets using predicates with the notation

$$\{x \in \mathscr{U} : P(x)\},$$

which may be read as 'the set of all x in $\mathscr{U}$ such that $P(x)$'.

Cardinality

If A is a finite set, its **cardinality**, $|A|$, is the number of (distinct) elements which it contains. If A has an infinite number of elements, we say it has **infinite cardinality**, and write $|A| = \infty$.[1]

Other notations commonly used for the cardinality of A are $n(A)$, $\#(A)$, and $\bar{\bar{A}}$.

---

[1] There is a more sophisticated approach to cardinality of infinite sets which allows different infinite sets to have different cardinality. Thus 'different sizes' of infinite sets can be distinguished. In this theory the set of integers has different cardinality from the set of numbers, for example. For further details, see Garnier and Taylor [6], for example.

Examples 3.3

1. Let $A = \{0, 1, 2, 3, 4\}$. Then $|A| = 5$.

2. Let $A = \{1, 2, 2, 3, 3, 3, 4, 4, 4, 4\}$. Then $|A| = 4$. This is because we ignore repeats when listing the elements of sets, so $A = \{1, 2, 3, 4\}$.

3. $|\varnothing| = 0$ since the empty set $\varnothing$ contains no elements.

4. Generalising from Example 1, if $A = \{0, 1, \ldots, n\}$, then $|A| = n + 1$.

5. Let A be the set of positive even integers, $A = \{2, 4, 6, 8, \ldots\}$. Then $|A| = \infty$.

6. Let $A = \{1, 2, \{1, 2\}\}$. Then $|A| = 3$ because A contains three elements: the number 1, the number 2, and the set $\{1, 2\}$.

   Determining the cardinality of sets, some of whose elements are also sets, can be tricky. In this context, it may help to think of sets as some kind of abstract paper bag. In this example, the set A is a paper bag that, when we look inside, contains two numbers and another paper bag; that is, three elements in total.

7. Let $A = \{\{1, 2, 3, 4\}\}$. Then $|A| = 1$ because A contains a single element, namely the set $\{1, 2, 3, 4\}$.

   Using the paper bag analogy, the set A is a paper bag that contains, as its single element, another paper bag.

8. Let $A = \{\{1, 2\}, \{3, 4\}, \{1, 2\}\}$. In this case, we have $|A| = 2$.

   Initially, we might think that $|A| = 3$ since A contains three sets (paper bags). However two of these are the same set, $\{1, 2\}$, so we must ignore the repeated element. Hence $A = \{\{1, 2\}, \{3, 4\}\}$, which clearly contains two elements, so $|A| = 2$.

Subset

The set B is a **subset** of the set A, denoted $B \subseteq A$, if every element of B is also an element of A. Symbolically, $B \subseteq A$ if $\forall x \bullet x \in B \Rightarrow x \in A$ is true, and conversely. If B is a subset of A, we say that A is a **superset** of B, and write $A \supseteq B$.

Note that every set A is a subset of itself, $A \subseteq A$. Any other subset of A is called a **proper subset** of A. The notation $B \subset A$ is used to denote 'B is a proper subset of A'. Thus $B \subset A$ if and only if $B \subseteq A$ and $B \neq A$.

It should also be noted that $\varnothing \subseteq A$ for every set A. This is because the definition above is satisfied in a trivial way: the empty set has no elements, so certainly each of them belongs to A. Alternatively, for any object x, the proposition $x \in \varnothing$ is false, which means that the conditional $(x \in \varnothing) \Rightarrow (x \in A)$ is true.

Examples 3.4

1. $\{2, 4, 6, \ldots\} \subseteq \{1, 2, 3, \ldots\} \subseteq \{0, 1, 2, \ldots\}$. Of course, we could have used the proper subset symbol $\subset$ to link these three sets instead.

2. Similarly: $\{$women$\} \subseteq \{$people$\} \subseteq \{$mammals$\} \subseteq \{$creatures$\}$;
   $\{$Catch 22$\} \subseteq \{$novels$\} \subseteq \{$works of fiction$\}$;
   $\{$Mona Lisa$\} \subseteq \{$paintings$\} \subseteq \{$works of art$\}$; etc.

   Again, in each of these we could have used $\subset$ instead.

3. Let $X = \{1, 2, \{3, 4\}\}$. Then $\{1, 2\} \subseteq X$ because each element in $\{1, 2\}$ is also an element of X. However $\{3, 4\}$ is not a subset of X, which we can denote by $\{3, 4\} \not\subseteq X$, because the elements of $\{3, 4\}$ — 3 and 4 — are not themselves elements of X. Using the paper bag analogy introduced in example 3.3.6, the set X is a paper bag containing two numbers, 1 and 2, and the paper bag $\{3, 4\}$. Thus $\{3, 4\}$ is an *element* of X, $\{3, 4\} \in X$, but not a subset of X, $\{3, 4\} \not\subseteq X$. However, since $\{3, 4\}$ is an *element* of X, the set containing $\{3, 4\}$ as its only element is a subset of X, $\{\{3, 4\}\} \subseteq X$.

   Care clearly needs to be taken to distinguish between set membership and subset, particularly when a set has elements which are themselves sets. In particular, note that $a \in A$ is true if and only if $\{a\} \subseteq A$ is true.

Recall that two sets are equal, $A = B$, if and only if they contain the same elements: for all x, $x \in A$ if and only if $x \in B$. In section 2.3 (page 41), we introduced the following logical equivalence, which we called the Biconditional Law:

$$P \Leftrightarrow Q \equiv (P \Rightarrow Q) \wedge (Q \Rightarrow P).$$

Thus, the statements

$$x \in A \Leftrightarrow x \in B \quad \text{and} \quad (x \in A \Rightarrow x \in B) \wedge (x \in B \Rightarrow x \in A)$$

are true for precisely the same objects x. Saying that the first of these is true for all x means $A = B$. Saying that the second of these is true for all x means both $A \subseteq B$ (from $x \in A \Rightarrow x \in B$) and $B \subseteq A$ (from $x \in B \Rightarrow x \in A$). Therefore $A = B$ means the same thing as $A \subseteq B$ and $B \subseteq A$. We summarise this as a theorem.

Theorem 3.1

Two sets A and B are equal if and only if both $A \subseteq B$ and $B \subseteq A$.

Examples 3.5

1. Show that $\{x : 2x^2 + 5x - 3 = 0\} \subseteq \{x : 2x^2 + 7x + 2 = 3/x\}$.

Solution

Let $A = \{x : 2x^2 + 5x - 3 = 0\}$ and $B = \{x : 2x^2 + 7x + 2 = 3/x\}$.

We need to show that every element of A is an element of B. The equation $2x^2 + 5x - 3 = 0$ has solutions $x = \frac{1}{2}$ and $x = -3$, so $A = \{\frac{1}{2}, -3\}$.

When $x = \frac{1}{2}$, $2x^2 + 7x + 2 = \frac{1}{2} + \frac{7}{2} + 2 = 6 = 3/x$, so $\frac{1}{2} \in B$.

When $x = -3$, $2x^2 + 7x + 2 = 18 - 21 + 2 = -1 = 3/x$, so $-3 \in B$.

Therefore every element of A is an element of B, so $A \subseteq B$.

2. Let $A = \{\{1\}, \{2\}, \{1, 2\}\}$ and let B be the set of all non-empty subsets of $\{1, 2\}$. Show that $A = B$.

Solution

$A \subseteq B$ since each of the three elements of A is a non-empty subset of $\{1, 2\}$ and therefore an *element* of B.

$B \subseteq A$ since every non-empty subset of $\{1, 2\}$ (i.e., every element of B) is contained in A.

Using theorem 3.1, we conclude that $A = B$.

3. Suppose that $A \subseteq B$ and let C be the set defined by

$$C = \{x : x \in A \vee x \in B\}.$$

Show that $C = B$.

Solution

To show that C and B are equal, we will show that $B \subseteq C$ and $C \subseteq B$.

Consider $x \in B$. Then $x \in A \vee x \in B$ is true, so $x \in C$. Thus every element of B also belongs to C, so $B \subseteq C$.

Now consider $x \in C$. Then, by the definition of C, either $x \in A$ or $x \in B$ (or both). However, if $x \in A$, then it follows that $x \in B$ also, because $A \subseteq B$. Therefore, in either case we can conclude $x \in B$. This shows that every element of C also belongs to B, so $C \subseteq B$.

We have now shown $B \subseteq C$ and $C \subseteq B$, so theorem 3.1 allows us to conclude that $B = C$.

Special sets of numbers

The following is a list of some special sets of numbers that are frequently used as universal sets.

$\mathbb{N} = \{0, 1, 2, 3, \ldots\}$ the set of **natural numbers**.

$\mathbb{Z} = \{\ldots, -2, -1, 0, 1, 2, \ldots\}$ the set of **integers**.

$\mathbb{Q} = \{p/q : p, q \in \mathbb{Z}$ and $q \neq 0\}$ the set of **rational numbers**.

$\mathbb{R} =$ the set of **real numbers**; real numbers can be thought of as corresponding to points on a number line or as numbers written as (possibly infinite) decimals.

$\mathbb{C} = \{x+iy : x, y \in \mathbb{R}$ and $i^2 = -1\}$ the set of **complex numbers**.

Clearly the following subset relations hold amongst these sets:

$$\mathbb{N} \subseteq \mathbb{Z} \subseteq \mathbb{Q} \subseteq \mathbb{R} \subseteq \mathbb{C}.$$

Also frequently used are $\mathbb{Z}^+$, $\mathbb{Q}^+$, and $\mathbb{R}^+$, the sets of *positive* integers, rational numbers, and real numbers, respectively. Note that $\mathbb{N}$ is *not* equal to $\mathbb{Z}^+$ since 0 belongs to the former but not the latter. In addition, we shall sometimes use $\mathbb{E}$ and $\mathbb{O}$ to denote the sets of even and odd integers, respectively:

$$\mathbb{E} = \{2n : n \in \mathbb{Z}\} = \{\ldots, -4, -2, 0, 2, 4, \ldots\}$$
$$\mathbb{O} = \{2n+1 : n \in \mathbb{Z}\} = \{\ldots, -3, -1, 1, 3, 5, \ldots\}.$$

Finally, we will need a notation for intervals. Let a and b be real numbers such that $a < b$. The **open interval** (a, b) is $(a, b) = \{x \in \mathbb{R} : a < x < b\}$. The **closed interval** $[a, b]$ is $[a, b] = \{x \in \mathbb{R} : a \leq x \leq b\}$. The notation also extends to 'half-open' intervals $[a, b) = \{x \in \mathbb{R} : a \leq x < b\}$ and $(a, b] = \{x \in \mathbb{R} : a < x \leq b\}$.

Exercises 3.1

1. List the elements of each of the following sets, using the '...' notation where necessary:

   (i) $\{x : x$ is an integer and $-2 < x < 5\}$

   (ii) $\{x : x$ is a positive (integer) multiple of three$\}$

   (iii) $\{x : x = y^2$ and y is an integer$\}$

   (iv) $\{x : (3x - 1)(x + 2) = 0\}$

   (v) $\{x : x \geq 0$ and $(3x - 1)(x + 2) = 0\}$

   (vi) $\{x : x$ is an integer and $(3x - 1)(x + 2) = 0\}$

   (vii) $\{x : x$ is a positive integer and $(3x - 1)(x + 2) = 0\}$

   (viii) $\{x : 2x$ is a positive integer$\}$

2. Define each of the following sets using a predicate; that is, write each set in the form $\{x : P(x)\}$ for some suitable predicate P.

(i) $\{3, 4, 5, 6, 7, 8, 9, 10\}$

(ii) $\{2, 4, 6, 8, \ldots, 100\}$

(iii) The set of all odd integers.

(iv) $\{1, 4, 9, 16, 25, 36, 49, \ldots\}$

(v) $\{1, 2, 4, 8, 16, 32, 64, \ldots\}$

(vi) $\{2, 7, 12, 17, 22, 27, 32, 37, \ldots\}$

(vii) $\{a, b, c, d, e, f, \ldots, x, y, z\}$

(viii) The set of integers which can be written as the sum of the squares of two integers.

(ix) The set of positive integers that are powers of 2.

(x) The set of positive integers that are powers of some prime number.

3. State whether each of the following statements is true or false.

(i) $\{1, 2\} \in \{1, 2, 3, 4\}$

(ii) $\{1, 2\} \subseteq \{1, 2, 3, 4\}$

(iii) $\{1, 2\} \in \{\{1, 2\}, \{3, 4\}\}$

(iv) $\{1, 2\} \subseteq \{\{1, 2\}, \{3, 4\}\}$

(v) $0 \in \varnothing$

(vi) $\varnothing \in \{\{1, 2\}, \{3, 4\}\}$

(vii) $\varnothing \subseteq \{\{1, 2\}, \{3, 4\}\}$

(viii) $\varnothing \in \{\varnothing, \{\varnothing\}\}$

4. Determine the cardinality of each of the following sets:

(i) $\{2, 4, 6, 8, 10\}$

(ii) $\{x : x$ is an integer and $2/3 < x < 17/3\}$

(iii) $\{x : \sqrt{x}$ is an integer$\}$

(iv) $\{x \in \mathbb{Z} : x^2 \leq 2\}$

(v) $\{x \in \mathbb{R} : x^2 \leq 2\}$

(vi) $\{2, 4, \{6, 8\}, 10\}$

(vii) $\{2, 4, \{\{6, 8\}, 10\}\}$

(viii) $\{\{2\}, \{4\}, \{6\}, \{8\}, \{10\}\}$

(ix) $\{1, \{1\}, \{\{1\}\}, \{\{\{1\}\}\}\}$

(x) $\{\varnothing, \{\varnothing\}, \{\{\varnothing\}\}\}$.

5. In each of the following cases, state whether $x \in A$, $x \subseteq A$, both, or neither.

    (i) $x = \{1\}$; $A = \{1, 2, 3\}$

    (ii) $x = \{1\}$; $A = \{\{1\}, \{2\}, \{3\}\}$

    (iii) $x = \{1\}$; $A = \{1, 2, \{1, 2\}\}$

    (iv) $x = \{1, 2\}$; $A = \{1, 2, \{1, 2\}\}$

    (v) $x = \{1\}$; $A = \{\{1, 2, 3\}\}$

    (vi) $x = 1$; $A = \{\{1\}, \{2\}, \{3\}\}$

6. Given that $X = \{a, b, c, d\}$, list the elements of each of the following sets:

    (i) $\{A : A \subseteq X \text{ and } |A| = 3\}$

    (ii) $\{A : A \subseteq X \text{ and } |A| = 2\}$

    (iii) $\{A : A \text{ is a proper subset of } X\}$

    (iv) $\{A : A \subseteq X \text{ and } b \in A\}$.

7. Let $\mathscr{U} = \{x : x \text{ is an integer and } 1 \leq x \leq 12\}$. In each of the following cases, determine whether $A \subseteq B$, $B \subseteq A$, both, or neither.

    (i) $A = \{x : x \text{ is odd}\}$ $B = \{x : x \text{ is a multiple of } 5\}$

    (ii) $A = \{x : x \text{ is even}\}$ $B = \{x : x^2 \text{ is even}\}$

    (iii) $A = \{x : x \text{ is even}\}$ $B = \{x : x \text{ is a power of } 2\}$

    (iv) $A = \{x : 3x + 1 > 10\}$ $B = \{x : x^2 > 20\}$

    (v) $A = \{x : \sqrt{x} \in \mathbb{Z}\}$ $B = \{x : x \text{ is a power of } 2 \text{ or } 3\}$

    (vi) $A = \{x : \sqrt{x} \leq 3\}$ $B = \{x : x \text{ is a perfect square}\}$

3.3 Operations on sets

Venn–Euler diagrams [2] are a useful visual representation of sets. In these diagrams, sets are represented as regions in the plane and elements which belong to a given set are placed inside the region representing it. Frequently, all the sets in the diagram are placed inside a box which represents the universal

---

[2] These diagrams are more commonly called just 'Venn diagrams' after John Venn, the nineteenth-century English mathematician. In fact, diagrams such as figure 3.2 are more properly called 'Euler diagrams' after Leonhard Euler who first introduced them in 1761. Although both Venn and Euler had precise rules for constructing their diagrams, today the term 'Venn diagram' is used informally to denote any diagram that represents sets by regions in the plane.

set $\mathscr{U}$. If an element belongs to more than one set in the diagram, the two regions representing the sets concerned must overlap and the element is placed in the overlapping region.

Example 3.6

Let $\quad \mathscr{U} = \{1, 2, 3, \ldots, 12\},$
$$A = \{n : n \text{ is even}\} = \{2, 4, 6, 8, 10, 12\},$$
$$B = \{n : n \text{ is prime}\} = \{2, 3, 5, 7, 11\}.$$

Figure 3.1 is a Venn-Euler diagram representing these sets and their elements.

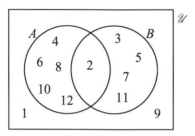

FIGURE 3.1: A Venn-Euler diagram.

Venn-Euler diagrams are very good at representing the relationships between sets. If $A \subseteq B$, then the region representing A will be enclosed inside the region representing B, so that every element in the region representing A is also inside that representing B. This is illustrated in figure 3.2 where the set A is represented by the shaded region.

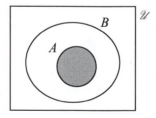

FIGURE 3.2: Venn-Euler diagrams representing subset.

Let A and B be sets.

The **intersection** of A and B, denoted $A \cap B$, is the set of all elements that belong to both A and B:

$$A \cap B = \{x : x \in A \text{ and } x \in B\}.$$

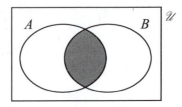

FIGURE 3.3: Venn-Euler diagram showing $A \cap B$.

Figure 3.3 is a Venn-Euler diagram representing the intersection; the shaded region of the diagram represents $A \cap B$.

The **union** of A and B, denoted $A \cup B$, is the set of all elements that belong to A or to B or to both:

$$A \cup B = \{x : x \in A \text{ or } x \in B\}.$$

Figure 3.4 is a Venn-Euler diagram representing the union; the shaded region of the diagram represents $A \cup B$.

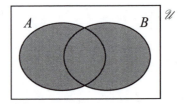

FIGURE 3.4: Venn-Euler diagram showing $A \cup B$.

There are obvious connections between intersection of sets and conjunction of propositions, and between union of sets and (inclusive) disjunction of propositions. Suppose that A and B are sets defined by propositional functions $P(x)$ and $Q(x)$, respectively:

$$A = \{x : P(x)\} \quad \text{and} \quad B = \{x : Q(x)\}.$$

Then

$$A \cap B = \{x : P(x) \wedge Q(x)\} \quad \text{and} \quad A \cup B = \{x : P(x) \vee Q(x)\}.$$

The definitions of intersection and union extend to more than two sets. Let $A_1, A_2, \ldots, A_n$ be sets.

Their **intersection** is:

$$\bigcap_{r=1}^{n} A_r = A_1 \cap A_2 \cap \cdots \cap A_n$$

$$= \{x : x \in A_1 \text{ and } x \in A_2 \text{ and } \ldots \text{ and } x \in A_n\}$$

$$= \{x : x \text{ belongs to } each \text{ set } A_r, \text{ for } r = 1, 2, \ldots, n\}.$$

Their **union** is:

$$\bigcup_{r=1}^{n} A_r = A_1 \cup A_2 \cup \cdots \cup A_n$$

$$= \{x : x \in A_1 \text{ or } x \in A_2 \text{ or } \ldots \text{ or } x \in A_n\}$$

$$= \{x : x \text{ belongs to } at \ least \ one \text{ set } A_r, \ r = 1, \ldots, n\}.$$

Two sets A and B are said to be **disjoint** if they have no elements in common; that is, if their intersection is the empty set, $A \cap B = \varnothing$. In a Venn-Euler diagram, we represent disjoint sets by drawing the regions representing them to be separated and non-overlapping. This is illustrated in figure 3.5.

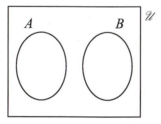

FIGURE 3.5: Venn-Euler diagram representing disjoint sets.

The **difference** of A and B, denoted $A - B$, is the set of all elements that belong to A but do not belong to B:

$$A - B = \{x : x \in A \text{ and } x \notin B\}.$$

The difference $A - B$ is sometimes denoted $A \setminus B$. Figure 3.6 is a Venn-Euler diagram representing the difference; the shaded region of the diagram represents $A - B$.

The **complement** of a set A, denoted $\bar{A}$, is the set of all elements that do not belong to A:

$$\bar{A} = \{x : x \notin A\} = \{x \in \mathscr{U} : x \notin A\}.$$

The complement of A is sometimes denoted A' or A^c. Note that $\bar{A} = \mathscr{U} - A$. Figure 3.7 is a Venn-Euler diagram representing the complement; the shaded region of the diagram represents $\bar{A}$.

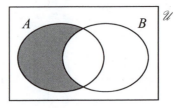

FIGURE 3.6: Venn-Euler diagram representing $A - B$.

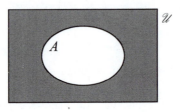

FIGURE 3.7: Venn-Euler diagram representing $\bar{A}$.

Examples 3.7

1. Let $\mathscr{U} = \{1, 2, 3, \ldots, 11, 12\}$,

   $A = \{n : n \text{ is a multiple of } 3\}$,

   $B = \{n : n \text{ is a factor of } 12\}$

   and $C = \{n : n \text{ is a multiple of } 4\}$.

List the elements of each of the following sets.

   (i) $A \cap (B \cup C)$
   (ii) $(A \cap B) \cup (A \cap C)$
   (iii) $A - (B \cap C)$
   (iv) $(A - B) \cup (A - C)$
   (v) $\overline{A \cup C}$
   (vi) $\bar{A} \cap \bar{C}$

Solution

It is helpful to have the elements of A, B, and C listed first:

   $A = \{3, 6, 9, 12\}$

   $B = \{1, 2, 3, 4, 6, 12\}$

   $C = \{4, 8, 12\}$.

   (i) $B \cup C$ comprises the elements that belong to B or to C or to both, so $B \cup C = \{1, 2, 3, 4, 6, 8, 12\}$. The intersection of this set with A comprises those elements common to both A and $B \cup C$; this gives $A \cap (B \cup C) = \{3, 6, 12\}$.

(ii) Firstly, we evaluate $A \cap B = \{3, 6, 12\}$ and $A \cap C = \{12\}$. The union of these two sets is $(A \cap B) \cup (A \cap C) = \{3, 6, 12\}$.

(iii) Firstly, $B \cap C = \{4, 12\}$. To form $A - (B \cap C)$, we take the elements of $A = \{3, 6, 9, 12\}$ and remove those that are also in $B \cap C = \{4, 12\}$. This gives $A - (B \cap C) = \{3, 6, 9\}$.

(iv) First note that $A - B$ contains those elements of A that are not also in B; in fact, 9 is the only such element, so $A - B = \{9\}$. Similarly, $A - C$ contains those elements of A that are not also in C, so $A - C = \{3, 6, 9\}$. Forming the union of these two sets gives $(A - B) \cup (A - C) = \{3, 6, 9\}$.

(v) First we find $A \cup C = \{3, 4, 6, 8, 9, 12\}$. The complement of this set contains all those elements of $\mathcal{U}$ that do not belong to $A \cup C$. Hence $\overline{A \cup C} = \{1, 2, 5, 7, 10, 11\}$.

(vi) Since $\bar{A} = \{1, 2, 4, 5, 7, 8, 10, 11\}$ and $\bar{C} = \{1, 2, 3, 5, 6, 7, 9, 10, 11\}$, their intersection is $\bar{A} \cap \bar{C} = \{1, 2, 5, 7, 10, 11\}$.

Note that, in this example, the sets are equal in pairs. For these sets, we have

$$A \cap (B \cup C) = (A \cap B) \cup (A \cap C),$$
$$A - (B \cap C) = (A - B) \cup (A - C),$$
and $\quad \overline{A \cup C} = \bar{A} \cap \bar{C}$

It is natural to ask whether these are properties of these particular sets or whether they are true for all sets. We will explore this for one of the pairs of sets in the next example.

2. The equality $A \cap (B \cup C) = (A \cap B) \cup (A \cap C)$ does, in fact, hold for all sets, and we can use Venn-Euler diagrams to illustrate this. In figure 3.8, we have identified the regions in a Venn-Euler diagram representing the two sets $A \cap (B \cup C)$ and $(A \cap B) \cup (A \cap C)$.

In the left-hand diagram, we have separately shaded the sets A and $B \cup C$ using diagonal lines (of different types). Then the intersection $A \cap (B \cup C)$ has both diagonal line shadings, which we have emphasised as the grey-shaded region.

In the right-hand diagram, we have separately shaded the sets $A \cap B$ and $A \cap C$ using diagonal lines (again, of different types). Then the union of these sets $(A \cap B) \cup (A \cap C)$ has either or both of the diagonal line shadings and again we have emphasised as the grey-shaded region.

The grey-shaded region in each diagram is the same, which illustrates that $A \cap (B \cup C) = (A \cap B) \cup (A \cap C)$. We should emphasise that the diagrams in figure 3.8 do not *prove* that $A \cap (B \cup C) = (A \cap B) \cup (A \cap C)$ for all sets A, B, and C.

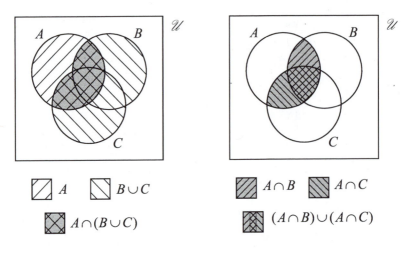

FIGURE 3.8: Illustrating $A \cap (B \cup C) = (A \cap B) \cup (A \cap C)$.

We have previously noted the connection between intersection of sets and conjunction of propositions and between unions of sets and disjunctions. This connection between sets and logic extends further. The complement of a set A , $\bar{A}$, contains all those elements of the universe not in A, so complement corresponds to negation. To be a little more precise, let $A = \{x : P(x)\}$ be a set defined by a predicate P. Then the complement of A is defined by $\bar{A} = \{x : \neg P(x)\}$.

In section 2.3, we used true to denote a tautology (a proposition that is always true) and false to denote a contradiction (a proposition that is always false). If we imagine these as defining sets then $\{x : \text{true}\}$ gives the universal set $\mathscr{U}$ since true is satisfied for all elements $x \in \mathscr{U}$. Similarly, $\{x : \text{false}\}$ gives the empty set $\varnothing$ since false is satisfied for no elements x.

In table 2.1 on page 39, we gave a list of standard logical equivalences. Each of these has its counterpart as an equality involving sets. For example, the identity illustrated in figure 3.8,

$$A \cap (B \cup C) = (A \cap B) \cup (A \cap C),$$

is the counterpart of a distributive law for propositions

$$P \wedge (Q \vee R) = (P \wedge Q) \vee (P \wedge R).$$

The set theory identities corresponding to the logical equivalences in table 2.1 are listed in the table below.

Set Theory Laws

Idempotent Laws	$A \cap A = A$
	$A \cup A = A$
Commutative Laws	$A \cap B = B \cap A$
	$A \cup B = B \cup A$
Associative Laws	$(A \cap B) \cap C = A \cap (B \cap C)$
	$(A \cup B) \cup C = A \cup (B \cup C)$
Absorption Laws	$A \cap (A \cup B) = A$
	$A \cup (A \cap B) = A$
Distributive Laws	$A \cap (B \cup C) = (A \cap B) \cup (A \cap C)$
	$A \cup (B \cap C) = (A \cup B) \cap (A \cup C)$
Involution Law	$\bar{\bar{A}} = A$
De Morgan's Laws	$\overline{A \cup B} = \bar{A} \cap \bar{B}$
	$\overline{A \cap B} = \bar{A} \cup \bar{B}$
Identity Laws	$A \cup \varnothing = A$
	$A \cap \mathscr{U} = A$
	$A \cup \mathscr{U} = \mathscr{U}$
	$A \cap \bar{A} = \varnothing$
Complement Laws	$A \cup \bar{A} = \mathscr{U}$
	$A \cap \bar{A} = \varnothing$
	$\bar{\varnothing} = \mathscr{U}$
	$\bar{\mathscr{U}} = \varnothing$

TABLE 3.1: Set theory laws.

Power set

Let A be any set. Then we can define a set comprising all the subsets of A. This is called the **power set of A**, denoted $\mathbb{P}(A)$.

Examples 3.8

1. The following is a list of sets A and their power sets $\mathbb{P}(A)$ for some small sets.

   $A = \varnothing$ $\qquad\qquad$ $\mathbb{P}(A) = \{\varnothing\}$

   $A = \{a\}$ $\qquad\qquad$ $\mathbb{P}(A) = \{\varnothing, \{a\}\}$

   $A = \{a, b\}$ $\qquad\quad$ $\mathbb{P}(A) = \{\varnothing, \{a\}, \{b\}, \{a, b\}\}$

   $A = \{a, b, c\}$ $\quad$ $\mathbb{P}(A) = \{\varnothing, \{a\}, \{b\}, \{c\}, \{a, b\}, \{a, c\}, \{b, c\}, \{a, b, c\}\}$

   $A = \{a, b, c, d\}$ $\quad$ $\mathbb{P}(A) = \{\varnothing, \{a\}, \{b\}, \{c\}, \{d\}, \{a, b\}, \{a, c\}, \{a, d\},$
   $\qquad\qquad\qquad\qquad\qquad \{b, c\}, \{b, d\}, \{c, d\}, \{a, b, c\}, \{a, b, d\},$
   $\qquad\qquad\qquad\qquad\qquad \{a, c, d\}, \{b, c, d\}, \{a, b, c, d\}\}$

   Note that, for every set A, the empty set $\varnothing$ and the set A itself both belong to the power set $\mathbb{P}(A)$.

   Also note that the power set of the empty set is *not* empty; $\mathbb{P}(\varnothing) = \{\varnothing\}$ which is a set containing a single element, namely the empty set. Using the analogy introduced in example 3.3.6, $\mathbb{P}(\varnothing) = \{\varnothing\}$ is a paper bag that contains another paper bag (albeit an empty one); this is different, of course, from an empty paper bag.

2. Let $A = \{a, b, c\}$ and $B = \{a, b\}$. Determine whether each of the following is true or false and give a brief justification.

   (i) $B \in \mathbb{P}(A)$

   (ii) $A \in \mathbb{P}(A)$

   (iii) $A \subseteq \mathbb{P}(A)$

   (iv) $\{\{a, c\}, B\} \subseteq \mathbb{P}(A)$

   (v) $\{a, c\} \in \mathbb{P}(A) \cap \mathbb{P}(B)$

   (vi) $\varnothing \subseteq \mathbb{P}(A)$

   Solution

   (i) $B \in \mathbb{P}(A)$ is true since $B = \{a, b\}$ is one of the elements listed in $\mathbb{P}(\{a, b, c\})$ in Example 1.

   (ii) $A \in \mathbb{P}(A)$ is true. For any set A, we have $A \subseteq A$, which means that A is an element of the power set $\mathbb{P}(A)$.

   (iii) This is false. Recall that $X \subseteq Y$ means each *element* of X is also an element of Y. In this case, the elements of A are letters (a, b, c), whereas the elements of $\mathbb{P}(A)$ are sets. Hence $A \not\subseteq \mathbb{P}(A)$.

   (iv) The statement is true. Each of the sets $\{a, c\}$ and $B = \{a, b\}$ is a subset of A and hence is an element of $\mathbb{P}(A)$. Therefore $\{\{a, c\}, B\} \subseteq \mathbb{P}(A)$.

(v) This is false. The set $\{a, c\}$ is not a subset of $B = \{a, b\}$, and so does not belong to $\mathbb{P}(B)$. Hence $\{a, c\} \notin \mathbb{P}(A) \cap \mathbb{P}(B)$.

(vi) The statement $\varnothing \subseteq \mathbb{P}(A)$ is true because the empty set is a subset of every set.

Exercises 3.2

1. For each of the following, draw a single Venn-Euler diagram and shade the region representing the given set.

   (i) $(A - B) \cup (B - A)$

   (ii) $(A \cup B) - (A \cap B)$

   (iii) $(A \cap B) \cup C$

   (iv) $\overline{A} \cap (B \cup C)$

   (v) $(A - B) \cup C$

2. Let $\mathscr{U} = \{n : n \in \mathbb{Z} \text{ and } 1 \leq n \leq 10\}$, $A = \{n : n \text{ is even}\}$, $B = \{n : n \text{ is prime}\}$, and $C = \{1, 4, 9\}$.

   Define each of the following sets by listing their elements.

   (i) $A \cap B$

   (ii) $A \cup B$

   (iii) $A - B$

   (iv) $B \cap C$

   (v) $\overline{A} \cap B$

   (vi) $A \cap (B \cup C)$

   (vii) $\overline{A \cup C}$

   (viii) $(A - C) - B$

3. Consider the sets A, B, C, D, and E represented by the following Venn-Euler diagram. (The sets C and E are represented by shaded regions.) For each of the following pairs of sets X and Y, state whether $X \subseteq Y$, $Y \subseteq X$, $X \cap Y = \varnothing$ (that is, X and Y are disjoint) or none of these.

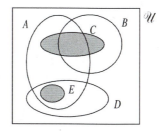

(i) $X = A \cup B, \quad Y = C$

(ii) $X = A \cap B, \quad Y = D$

(iii) $X = A \cap B, \quad Y = C$

(iv) $X = E, \quad Y = A \cap D$

(v) $X = B \cap C, \quad Y = C \cup D$

(vi) $X = A \cap E, \quad Y = D \cup E$

(vii) $X = C \cup E, \quad Y = A \cup D$

(viii) $X = C - B, \quad Y = D \cup E$

(ix) $X = A \cup D, \quad Y = B \cap E$

(x) $X = A - E, \quad Y = A - D$

4. (i) Show that if $A, B,$ and C are sets such that $C \subseteq A$ and $C \subseteq B$, then $C \subseteq A \cap B$.

   (ii) Show that if $A, B,$ and C are sets such that $A \subseteq C$ and $B \subseteq C$, then $A \cup B \subseteq C$.

5. The **symmetric difference** $A * B$ of two sets A and B is defined by

$$A * B = (A - B) \cup (B - A).$$

   (i) Let $A = \{2, 4, 6, 8, 10\}$, $B = \{4, 8, 12, 16\}$, and $C = \{1, 4, 16, 64\}$. List the elements of each of the following sets.

   (a) $A * B$

   (b) $A * C$

   (c) $A * (B \cap C)$

   (d) $A * (B \cup C)$

   (e) $(A * B) \cap (A * C)$

   (f) $(A * B) \cup (A * C)$

   (ii) For each of the following, draw a single Venn-Euler diagram and shade the region representing the given set.

   (a) $A * (B \cap C)$

   (b) $A * (B \cup C)$

   (c) $(A * B) \cap (A * C)$

   (d) $(A * B) \cup (A * C)$

6. List the elements of $\mathbb{P}(A)$ in each of the following cases.

   (i) $A = \{1, 2, 3\}$

   (ii) $A = \{1, \{2, 3\}\}$

(iii) $A = \{\{1\}, \{2, 3\}\}$

(iv) $A = \{\{1, 2, 3\}\}$

(v) $A = \mathbb{P}(\{1, 2\})$

7. Let $A = \{a, b\}$, and $B = \{b, c, d\}$.

Evaluate $\mathbb{P}(A)$, $\mathbb{P}(B)$, $\mathbb{P}(A \cap B)$, and $\mathbb{P}(A) \cap \mathbb{P}(B)$.

What do you notice about the relationship between $\mathbb{P}(A \cap B)$ and $\mathbb{P}(A) \cap \mathbb{P}(B)$?

Will this relationship always hold? If so, can you explain why; if not, can you explain why not?

3.4 The Cartesian product

The order in which the elements of a finite set are listed is immaterial; in particular, $\{x, y\} = \{y, x\}$. In some circumstances, however, order is significant. For instance, in coordinate geometry the points with coordinates $(1, 2)$ and $(2, 1)$, respectively, are distinct. We therefore wish to define, in the context of sets, something akin to the coordinates of points used in analytical geometry.

An **ordered pair** (x, y) of elements satisfies:

$$(x, y) = (a, b) \text{ if and only if } x = a \text{ and } y = b.$$

With this definition it is clear that (x, y) and (y, x) are different (unless $x = y$), so the order is significant. It could be argued, with justification, that we have not really *defined* the ordered pair, but merely listed a *property* which we desire of it. In fact, (x, y) can be defined in terms of (unordered) sets, but this is not particularly instructive. What is important about ordered pairs is the property above.

Let A and B be sets. Their **Cartesian product** [3] denoted $A \times B$ is the set of all ordered pairs where the first element is drawn from the set A and the second element is drawn from the set B:

$$A \times B = \{(x, y) : x \in A \text{ and } y \in B\}.$$

When $A = B$ we write A^2 for the Cartesian product $A \times A$. Thus, for example, the Cartesian plane is $\mathbb{R}^2 = \mathbb{R} \times \mathbb{R} = \{(x, y) : x \in \mathbb{R} \text{ and } y \in \mathbb{R}\}$.

---

[3] Named after the French mathematician and philosopher René Descartes (1596–1650), the founder of analytical geometry.

Since Venn-Euler diagrams have proved to be a useful representation of sets, we would like to have a diagrammatic representation of the Cartesian product $A \times B$ that relates it to the sets A and B. The idea for the diagrammatic representation comes from the coordinate plane $\mathbb{R}^2$ with perpendicular axes. For arbitrary sets, we represent A and B as horizontal and vertical line segments (rather than regions in the plane). Then $A \times B$ is represented as the rectangular region lying above A and to the right of B — see figure 3.9. An element $(a, b) \in A \times B$ is represented as the point lying above a and to the right of b.

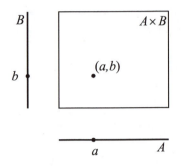

FIGURE 3.9: Diagram representing the Cartesian product $A \times B$.

Examples 3.9

1. Let $A = \{1, 2, 3, 4\}$ and $B = \{a, b, c\}$. Then

$$A \times B = \{(1, a), (1, b), (1, c), (2, a), (2, b), (2, c),$$
$$(3, a), (3, b), (3, c), (4, a), (4, b), (4, c)\}.$$

2. Let $A = \{1, 2, 3\}$, $B = \{2, 3, 4\}$ and $X = \{a, b, c\}$

   We will investigate the interaction between intersection and the Cartesian product.

   Since

$$A \times X = \{(1, a), (1, b), (1, c), (2, a), (2, b), (2, c), (3, a), (3, b), (3, c)\}$$

   and

$$B \times X = \{(2, a), (2, b), (2, c), (3, a), (3, b), (3, c), (4, a), (4, b), (4, c)\},$$

   it follows that their intersection is

$$(A \times X) \cap (B \times X) = \{(2, a), (2, b), (2, c), (3, a), (3, b), (3, c)\}.$$

Now $A \cap B = \{2, 3\}$. Hence

$$(A \cap B) \times X = \{(2, a), (2, b), (2, c), (3, a), (3, b), (3, c)\}.$$

Therefore, for these sets,

$$(A \times X) \cap (B \times X) = (A \cap B) \times X.$$

An obvious question is whether this equation is satisfied for all sets A, B and X or whether it was a particular property of these sets.

In fact, the equation holds for all sets, and this is illustrated in figure 3.10. The figure shows both $A \times X$ and $B \times X$ with different diagonal line shading. Their intersection, $(A \times X) \cap (B \times X)$, has both line shadings and this is emphasised by the gray shading. However, this grey-shaded region is the region lying above $A \cap B$, so it also represents $(A \cap B) \times X$.

The diagram in figure 3.10 is not a proof of this identity, but it does give a powerful visualisation of it.

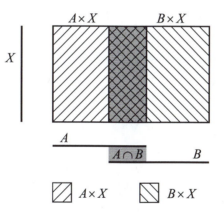

FIGURE 3.10: Diagram representing the interaction between Cartesian product and intersection.

We can extend the definition of Cartesian product to n sets. Ordered pairs readily generalise to ordered triples (x, y, z), ordered quadruples (w, x, y, z), or more generally, **ordered n-tuples** $(x_1, x_2, \ldots, x_n)$ with the property that

$$(x_1, x_2, \ldots, x_n) = (a_1, a_2, \ldots, a_n) \quad \Leftrightarrow \quad x_1 = a_1, x_2 = a_2, \ldots, x_n = a_n$$
$$\Leftrightarrow \quad x_r = a_r, \text{ for } r = 1, 2 \ldots, n.$$

Let $A_1, A_2, \ldots A_n$ be sets. Their **Cartesian product** is the set of all ordered

n-tuples where the first element lies in the first set A_1, the second element lies in the second set A_2, and so on:

$$A_1 \times A_2 \times \ldots \times A_n = \{(a_1, a_2, \ldots, a_n) : a_1 \in A_1, a_2 \in A_2, \ldots, a_n \in A_n\}.$$

Example 3.10

Let $A = \{1, 2, 3\}$, $B = \{a, b, c\}$ and $C = \{\alpha, \beta\}$. Then

$$\begin{aligned}
A \times B \times C = \{&(1, a, \alpha), (1, a, \beta), (1, b, \alpha), (1, b, \beta), (1, c, \alpha), (1, c, \beta), \\
&(2, a, \alpha), (2, a, \beta), (2, b, \alpha), (2, b, \beta), (2, c, \alpha), (2, c, \beta), \\
&(3, a, \alpha), (3, a, \beta), (3, b, \alpha), (3, b, \beta), (3, c, \alpha), (3, c, \beta)\}.
\end{aligned}$$

Exercises 3.3

1. Let $A = \{1, 2, 3\}$, $B = \{2, 3, 4\}$, $X = \{a, b, c\}$, and $Y = \{c, d, e\}$. Evaluate each of the following sets by listing their elements.

    (i) $A \times (X \cap Y)$

    (ii) $(A \cap B) \times (X \cap Y)$

    (iii) $(A \times X) \cap (B \times Y)$

    (iv) $(A \times Y) \cap (B \times X)$

    (v) $(A - B) \times X$

    (vi) $(A \times X) - (B \times X)$

    (vii) $(A \cup B) \times (X \cap Y)$

    (viii) $(A \cap X) \times Y$

2. This question explores how the Cartesian product interacts with the subset relation.

    (i) Choose sets A, B, X, and Y satisfying $A \subseteq X$ and $B \subseteq Y$. Evaluate $A \times B$ and $X \times Y$. What do you notice?

    (ii) Draw a diagram, similar to that in figure 3.10, to show that, if $A \subseteq X$ and $B \subseteq Y$, then $A \times B \subseteq X \times Y$.

    (iii) Is *every* subset of $X \times Y$ of the form $A \times B$ for some $A \subseteq X$ and $B \subseteq Y$? Explain your answer.

    (iv) Suppose sets A, B, X, and Y are such that $A \times B \subseteq X \times Y$. Does it *necessarily* follow that $A \subseteq X$ and $B \subseteq Y$?

3. This question considers the following identity:

$$(A \cap B) \times (X \cap Y) = (A \times X) \cap (B \times Y).$$

(i) Verify this identity for the sets $A = \{1,2,3,4\}, B = \{2,4,6\}, X = \{x,y\}$, and $Y = \{y,z\}$.

(ii) Draw two diagrams similar to that in figure 3.10, one showing

$$(A \cap B) \times (X \cap Y)$$

and one showing

$$(A \times X) \cap (B \times Y),$$

to illustrate the identity.

4. The aim now is to find and illustrate an identity, similar to that in question 3, for the set $(A \cup B) \times (X \cup Y)$.

(i) Using the same sets as in question 3 (i), $A = \{1,2,3,4\}, B = \{2,4,6\}, X = \{x,y\}$, and $Y = \{y,z\}$, show that

$$(A \cup B) \times (X \cup Y) \neq (A \times X) \cup (B \times Y).$$

Note that you *only* need to show the two sets are different; it should not be necessary to list *all* the elements of both sets.

(ii) By drawing similar diagrams to those in question 3 (ii), conjecture a general expression for

$$(A \cup B) \times (X \cup Y)$$

as a union of Cartesian products.

3.5 Functions and composite functions

In this section we introduce another of the central concepts of modern mathematics, that of a function or mapping. Although functions have been used in mathematics for several centuries, it is only comparatively recently that a rigorous and generally accepted definition of the concept has emerged. We will actually give two definitions of a function. The first definition, below, is a little informal but captures the idea of a function. Later, we will give a more formal definition based on Cartesian products of sets that makes a little more precise one aspect of the informal definition. We will use the terms 'function' and 'mapping' interchangeably.

Definition 3.1 (Informal definition of a function)
Let A and B be sets. A **function** or **mapping** $f : A \to B$ is a rule which associates, to each element $a \in A$, a unique element $b = f(a) \in B$.

This is a very general definition — A and B are any sets and f is any rule with the appropriate property. It is quite common to visualise the function rule as being encapsulated in a 'function machine'. This is a 'black box', illustrated in figure 3.11, which has the property that if an element $a \in A$ is fed into the machine, it produces as output the associated element $f(a) \in B$.

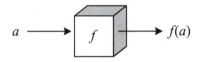

FIGURE 3.11: A function machine.

An alternative visualisation, which shows f linking the two sets A and B, is given in figure 3.12.

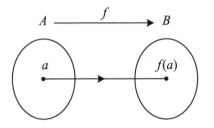

FIGURE 3.12: Diagram of a function showing the sets involved.

Both representations of a function are useful in different contexts, although figure 3.12 is probably more satisfactory as is shows the three different components of the function: the sets A and B and the rule associating elements of A with elements of B.

Let $f : A \rightarrow B$ be a function. Then the set A is called the **domain** of f and the set B is called the **codomain** of f. For $a \in A$, the element $b = f(a) \in B$ is called the **image** of a and we also write $a \mapsto b$ in this case. The set of all images of elements of A is called the **image** of f, $\operatorname{im} f$:

$$\operatorname{im} f = \{b \in B : b = f(a) \text{ for some } a \in A\} = \{f(a) : a \in A\}.$$

The image of f is illustrated in figure 3.13.

Examples 3.11

1. Let $A = \{a, b, c, d, e\}$ and $B = \{1, 2, 3, 4, 5, 6\}$. A function $f : A \rightarrow B$ is defined by
$$a \mapsto 2, \ b \mapsto 4, \ c \mapsto 2, \ d \mapsto 5, \ e \mapsto 1.$$

This function is illustrated in figure 3.14.

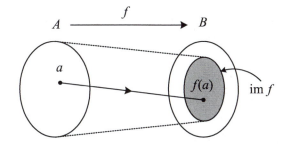

FIGURE 3.13: Diagram showing the image of a function.

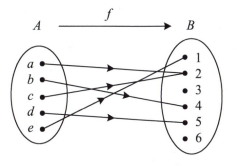

FIGURE 3.14: Illustrating the function in example 3.11.1.

Note that it is possible to have two different elements of A with the same image in B; in this example we have both $f(a) = 2$ and $f(c) = 2$. Also, it is possible to have elements of B that are not the image of any element of A; in this example, 3 is not the image of any $a \in A$.

The image of the function is $\operatorname{im} f = \{1, 2, 4, 5\}$.

2. Frequently, authors of textbooks will use phrases such as 'the function $f(x) = (x + 1)^2$' or 'the function $g(x) = \sin x / x$'. However, the expression itself does not define a function because the domain and codomain have not been specified. There is usually an implicit assumption that a numerical variable x ranges over the real numbers or a subset of it.

Thus the first expression defines a function

$$f : \mathbb{R} \to \mathbb{R}, \ f(x) = (x + 1)^2.$$

However the second function does not define a function $\mathbb{R} \to \mathbb{R}$ because the expression $\sin x / x$ is not defined when $x = 0$. Usually, in such a situation, we would wish to define the function to have as large a domain as possible. Hence the second expression defines a function

$$g : \mathbb{R} - \{0\} \to \mathbb{R}, \ g(x) = \frac{\sin x}{x}.$$

The domain $\mathbb{R} - \{0\}$ is the set of non-zero real numbers which is frequently denoted

$$\mathbb{R}^* = \{x \in \mathbb{R} : x \neq 0\}.$$

3. Find the image of each of the following functions.

   (i) $f : \mathbb{R} \to \mathbb{R}, \ f(x) = x^2$.

   (ii) $g : \mathbb{R} \to \mathbb{R}, \ g(x) = \dfrac{2x}{x^2 + 1}$.

Solution

   (i) For the square function, we have $f(x) = x^2 \geq 0$ for all $x \in \mathbb{R}$. Also, for any $y \geq 0$ taking $x = \sqrt{y} \in \mathbb{R}$, we have $f(x) = f\left(\sqrt{y}\right) = \left(\sqrt{y}\right)^2 = y$. Hence the image of f is

$$\operatorname{im} f = \{y \in \mathbb{R} : y \geq 0\}.$$

   The image of f is illustrated in figure 3.15.

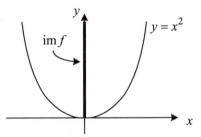

FIGURE 3.15: The image of the square function $\mathbb{R} \to \mathbb{R}, x \mapsto x^2$.

   (ii) Finding the image of $g : \mathbb{R} \to \mathbb{R}, \ g(x) = \dfrac{2x}{x^2 + 1}$ is more difficult since we cannot immediately 'see' what the image is.

   Now $y \in \operatorname{im} g$ if and only if

$$y = \frac{2x}{x^2 + 1} \quad \text{for some } x \in \mathbb{R}.$$

   Now this is equivalent to

$$yx^2 + y = 2x$$

   or

$$yx^2 - 2x + y = 0.$$

Regarding this as a quadratic equation in x and using the quadratic formula, we have, provided $y \neq 0$,

$$x = \frac{2 \pm \sqrt{4 - 4y^2}}{2y}.$$

In order that this has a real solution we require $y \neq 0$ and

$$4 - 4y^2 \geq 0.$$

Hence

$$y^2 \leq 1 \quad (\text{and } y \neq 0),$$

which means

$$-1 \leq y \leq 1 \quad (\text{and } y \neq 0).$$

Therefore, provided $-1 \leq y \leq 1$ and $y \neq 0$, there exists a real number x such that $y = g(x)$. The value $y = 0$ is a special case, but clearly $g(0) = 0$, so $0 \in \text{im } g$.

Therefore im g is the interval

$$\text{im } g = [-1, 1] = \{y \in \mathbb{R} : -1 \leq y \leq 1\}.$$

A sketch of the graph of g showing im g is given in figure 3.16.

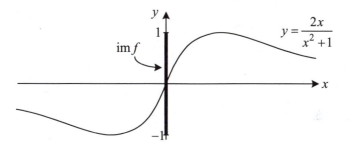

FIGURE 3.16: The image of the function $g : \mathbb{R} \to \mathbb{R}$, $g(x) = \dfrac{2x}{x^2 + 1}$.

The definition of a function $f : A \to B$ given in definition 3.1 raises an important question: what constitutes a 'rule'? Sometimes this will be a formula, like the functions defined in example 3.11.2 above. However the 'rule' may simply be describing explicitly which element is associated to which as in example 3.11.1 above. In this case, we can describe the 'rule' compactly just by listing, as pairs, the elements and their images:

$$(a, 2), \ (b, 4), \ (c, 2), \ (d, 5), \ (e, 1).$$

This list defines a subset of the Cartesian product $A \times B = \{(a, b) : a \in A, b \in B\}$ comprising all the pairs $(a, f(a))$

$$\{(a, f(a)) : a \in A\}.$$

This is the key to the formal definition of a function. Instead of trying to define what we mean by a rule (that associates elements of A with elements of B), instead we simply define the set of those pairs $(a, f(a))$ using any of the ways described earlier for defining sets.

Definition 3.2 (Formal definition of a function)
Let A and B be sets. A **function** or **mapping** $f : A \rightarrow B$ is a subset of the Cartesian product, $f \subseteq A \times B$, that satisfies:

for each $a \in A$ there exists a unique $b \in B$ such that $(a, b) \in f$.

Examples 3.12

1. The function $f : A \rightarrow B$ in example 3.11.1 above is defined by the set $f = \{(a, 2), (b, 4), (c, 2), (d, 5), (e, 1)\} \subseteq A \times B$.

2. Similarly, the two functions in example 3.11.2 are defined by the sets

$$f = \{(x, (x + 1)^2) : x \in \mathbb{R}\} \subseteq \mathbb{R}^2$$

$$\text{and} \quad g = \left\{\left(x, \frac{\sin x}{x}\right) : x \in \mathbb{R} - \{0\}\right\} \subseteq \mathbb{R} - \{0\} \times \mathbb{R}.$$

3. The square function was defined informally in example 3.11.3 (i) as $f : \mathbb{R} \rightarrow \mathbb{R}$, $f(x) = x^2$. We may define this formally as

$$f = \{(x, y) \in \mathbb{R}^2 : y = x^2\}.$$

Suppose we attempt to define the square root function g in a similar way as the set

$$g = \{(x, y) \in \mathbb{R} \times \mathbb{R} : x = y^2\}.$$

There is a problem with this, however. Figure 3.17 shows this set as a subset of $\mathbb{R}^2$. The set g does not satisfy the condition

for each $x \in \mathbb{R}$ there exists a unique $y \in \mathbb{R}$ such that $(x, y) \in g$

in definition 3.5 in two respects.

Firstly, for some $x \in \mathbb{R}$, there does not exist any $y \in \mathbb{R}$ such that $(x, y) \in g$. For example, there is no element $(-1, y) \in g$. Secondly, for some $x \in \mathbb{R}$, there exist two values of $y \in \mathbb{R}$ such that $(x, y) \in g$. For example, both $(4, 2) \in g$ and $(4, -2) \in g$.

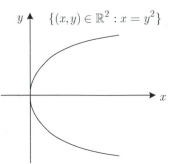

FIGURE 3.17: Attempting to define the square root function.

In order to overcome these problems and properly define a square root function, we need to restrict the domain and codomain. If we define

$$\mathbb{R}_{\geq 0} = \{x \in \mathbb{R} : x \geq 0\}.$$

then we may define this square root function informally as

$$g : \mathbb{R}_{\geq 0} \to \mathbb{R}_{\geq 0}, \ g(x) = \sqrt{x}$$

or formally as

$$g = \{(x, y) \in \mathbb{R}_{\geq 0} \times \mathbb{R}_{\geq 0} : x = y^2\} = \{(x, y) \in \mathbb{R}_{\geq 0} \times \mathbb{R}_{\geq 0} : y = \sqrt{x}\}.$$

Note that $\sqrt{x}$ means 'the non-negative square root of x'.

Composition of functions

Let $f : A \to B$ and $g : B \to C$ be two functions. If $x \in A$, then $y = f(x)$ belongs to B so we can 'apply' the function g to y to get $z = g(y) = g(f(x))$ which belongs to C. In symbols:

$$x \in A \ \Rightarrow \ y = f(x) \in B \ \Rightarrow \ z = g(y) = g(f(x)) \in C.$$

This association $x \mapsto g(f(x))$ defines a function $A \to C$, called the **composite** of f and g. The composite function is denoted $g \circ f$:

$$g \circ f : A \to C, (g \circ f)(x) = g(f(x)).$$

The composite function $g \circ f$ is illustrated in figure 3.18.

If we think of the functions f and g being represented by 'function machines', then the composite $g \circ f$ has a function machine that is obtained by connecting the output of f to the input of g. This is represented in figure 3.19.

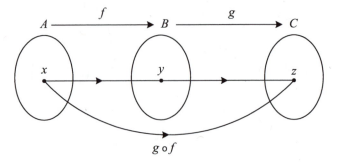

FIGURE 3.18: Illustrating the composite function.

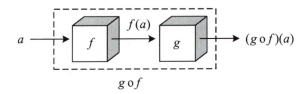

FIGURE 3.19: Composite of function machines.

Examples 3.13

1. Let $f : \mathbb{R} \to \mathbb{R}$, $f(x) = 4x - 1$ and $g : \mathbb{R} \to \mathbb{R}$, $g(x) = x^2 + 1$ be two functions. Evaluate:

   (i) $g \circ f(2)$

   (ii) $f \circ g(2)$

   (iii) $g \circ f(x)$

   (iv) $f \circ g(x)$

Solution

   (i) Since $g \circ f(2) = g(f(2))$, we first need to evaluate $f(2)$ and then 'apply' g to the result. We have $f(2) = 4 \times 2 - 1 = 7$, so

   $$g \circ f(2) = g(f(2)) = g(7) = 7^2 + 1 = 50.$$

   (ii) This time we evaluate $g(2) = 2^2 + 1 = 5$ and then 'apply' f to the result:
   $$f \circ g(2) = f(g(2)) = f(5) = 4 \times 5 - 1 = 19.$$

   (iii) Repeating the process in part (i) but with a general x gives

   $$g \circ f(x) = g(f(x)) = g(4x - 1) = (4x - 1)^2 + 1 = 16x^2 - 8x + 2.$$

(iv) Similarly, applying the function f to $g(x) = x^2 + 1$ gives

$$f \circ g(x) = f(g(x)) = f(x^2 + 1) = 4(x^2 + 1) - 1 = 4x^2 + 3.$$

2. Let $f : \mathbb{R} \to \mathbb{R}$, $f(x) = 2x + 1$. Define 'multiple' composite functions as follows:

$$f^{[2]} = f \circ f, \; f^{[3]} = f \circ (f \circ f) = f \circ f^{[2]}, f^{[4]} = f \circ f^{[3]} \dots.$$

Conjecture an expression for $f^{[n]}(x)$.

Solution

We have

$$f^{[2]}(x) = f \circ f(x) = f(2x + 1) = 2(2x + 1) + 1 = 4x + 3,$$
$$f^{[3]}(x) = f \circ f^{[2]}(x) = f(4x + 3) = 2(4x + 3) + 1 = 8x + 7,$$
$$f^{[4]}(x) = f \circ f^{[3]}(x) = f(8x + 7) = 2(8x + 7) + 1 = 16x + 15,$$
$$f^{[5]}(x) = f \circ f^{[4]}(x) = f(16x + 15) = 2(16x + 15) + 1 = 32x + 31.$$

From these examples we conjecture that

$$f^{[n]}(x) = 2^n x + (2^n - 1).$$

In chapter 8 we will develop a method of proof that will allow us to prove this conjecture.

Exercises 3.4

1. Three functions f, g, and h are defined as follows.

$$f : \mathbb{R} \to \mathbb{R}, \quad f(x) = |x - 1|$$
$$g : \mathbb{Z} \to \mathbb{R}, \quad g(x) = \frac{2x}{x^2 - 3}$$
$$h : \mathbb{R} \to \mathbb{R}, \quad h(x) = 2x + 1$$

For each of the following, either evaluate the expression or explain why it is not defined.

(i) $f\left(\frac{1}{2}\right)$

(ii) $g\left(\frac{1}{2}\right)$

(iii) $(g \circ h)\left(\frac{1}{2}\right)$

(iv) $(f \circ f)(-2)$

(v) $(g \circ g)(3)$

(vi) $(h \circ h)(x)$

(vii) $(f \circ h)(x)$

(viii) $(f \circ g)(x)$

2. Three functions f, g, and h are defined as follows.

$$f : \mathbb{R} \to \mathbb{R}, \quad f(x) = 2x + 1$$

$$g : \mathbb{R} \to \mathbb{R}, \quad g(x) = \frac{1}{x^2 + 1}$$

$$h : \mathbb{R} \to \mathbb{R}, \quad h(x) = \sqrt{x^2 + 1}$$

Find expressions for each of the following.

(i) $(f \circ g)(x)$ (iv) $(f \circ f)(x))$

(ii) $(g \circ f)(x)$ (v) $((f \circ g) \circ h)(x)$

(iii) $(g \circ h)(x)$ (vi) $(f \circ (g \circ h))(x)$

3. Let $f : A \to B$ and $g : C \to D$ be two functions.

What is the most general situation in which the composite function $g \circ f$ is defined? Draw a diagram to illustrate this situation.

4. Find im f, the image of f, for each of the following functions.

(i) $A = \mathbb{P}(\{a, b, c, d\})$, the power set of $\{a, b, c, d\}$,
$f : A \to \mathbb{Z}$, $f(C) = |C|$.

(ii) $f : \mathbb{Z} \to \mathbb{Z}$, $f(n) = n^2$.

(iii) $A = \{\text{countries of the world}\}$, $B = \{\text{cities of the world}\}$
$f : A \to B$, $f(X) = $ the capital city of X.

(iv) $A = \mathbb{P}(\{a, b, c, d\})$, $f : A \to A$, $f(X) = X \cap \{a\}$.

(v) $A = \mathbb{P}(\{a, b, c, d\})$, $f : A \to A$, $f(X) = X \cup \{a\}$.

5. Determine the image of each of the following functions.

(i) $f : \mathbb{R} \to \mathbb{R}$, $x \mapsto x^2 + 2$

(ii) $f : \mathbb{R} \to \mathbb{R}$, $x \mapsto (x + 2)^2$

(iii) $f : \mathbb{R} \to \mathbb{R}$, $x \mapsto 1/(x^2 + 2)$

(iv) $f : \mathbb{R} \to \mathbb{R}$, $x \mapsto 4x/(x^2 + 5)$

(v) $f : \mathbb{R} \to \mathbb{R}$, $x \mapsto \sqrt{x^2 + 1}$

6. Let $f : A \to B$ and $g : B \to C$ be two functions. Give the formal definition of the composite function $g \circ f : A \to C$ as a subset of $A \times C$.

3.6 Properties of functions

A function $f : A \to B$ is **injective** or **one-one** if different elements have different images; in other words if we *never* have the situation given in figure 3.20.

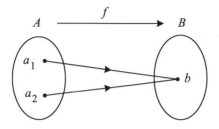

FIGURE 3.20: Different elements of the domain with the same image.

The statement that different elements have different images is:

$$\text{for all } a_1, a_2 \in A, \quad a_1 \neq a_2 \Rightarrow f(a_1) \neq f(a_2).$$

Recall from section 2.3, page 40, that a conditional statement $P \Rightarrow Q$ is logically equivalent to its contrapositive $\neg Q \Rightarrow \neg P$. For any $a_1, a_2 \in A$ the two statements

$$a_1 \neq a_2 \Rightarrow f(a_1) \neq f(a_2) \quad \text{and} \quad f(a_1) = f(a_2) \Rightarrow a_1 = a_2$$

are contrapositives of one another and hence are logically equivalent. In mathematics, it is generally easier to work with *equals* rather than *not equals*, so we will usually use the contrapositive of the conditional statement:

$$\text{for all } a_1, a_2 \in A, \quad f(a_1) = f(a_2) \Rightarrow a_1 = a_2.$$

Definition 3.3
Let $f : A \to B$ be a function. Then f is **injective** (or **one-one**) if:

$$\text{for all } a_1, a_2 \in A, \quad f(a_1) = f(a_2) \Rightarrow a_1 = a_2.$$

$\square$

We say that a function $f : A \to B$ is **surjective** or **onto** if every element of B is the image of some element of A; more colloquially, f is surjective if every element of B is 'hit' by an arrow coming from A. This is equivalent to saying that the image of f is 'all of' the codomain b, $\operatorname{im} f = B$. This is illustrated in the diagram in figure 3.21.

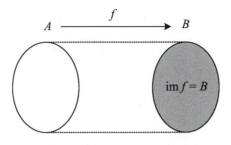

FIGURE 3.21: A surjective function.

Definition 3.4

Let $f : A \to B$ be a function. Then f is **surjective** (or **onto**) if:

for every $b \in B$, there exists $a \in A$ such that $f(a) = b$.

□

An injective function is also called an **injection** and a surjective function is also called a **surjection**.

Examples 3.14

1. Consider the function $f : \{a, b, c, d, e\} \to \{1, 2, 3, 4, 5, 6\}$ defined by

$$a \mapsto 2, \ b \mapsto 4, \ c \mapsto 2, \ d \mapsto 5, \ e \mapsto 1.$$

   This function is illustrated in figure 3.14.

   The function is not injective because $f(a) = 2 = f(c)$ (but $a \neq c$). Also f is not surjective because 3 in the codomain is not the image of any element of the domain, $3 \notin \text{im} f$. Note that $6 \notin \text{im} f$ either, but only one element not in $\text{im} f$ is needed to show that the function is not surjective.

2. Consider the square function $f : \mathbb{R} \to \mathbb{R}$, $f(x) = x^2$. The graph of f is given in figure 3.15.

   Clearly it is possible for different elements of the domain to have the same image; for example, $f(2) = 4 = f(-2)$. Therefore f is not injective.

   Also, since $x^2 \geq 0$ for all $x \in \mathbb{R}$, no negative real number is in the image of f; for example, $-1 \neq \text{im} f$. Therefore f is not surjective. In example 3.11.3 (i), we showed that $\text{im} f = \mathbb{R}_{\geq 0} = \{y \in \mathbb{R} : y \geq 0\}$. In this case, we don't need to determine $\text{im} f$ completely to show that f is not surjective; all we require is an element in the codomain $\mathbb{R}$ that is not in the image of the function $\text{im} f$.

Now consider the function $g : \mathbb{R}_{\geq 0} \to \mathbb{R}_{\geq 0}$, $g(x) = x^2$. In other words, g has the same 'rule' as f but its domain and codomain are different.

If $x \geq 0$ and $y \geq 0$ are such that $x^2 = y^2$, then x and y are equal. Hence g is injective. Similarly, if $y \geq 0$, then $y \in \operatorname{im} g$ (as shown in example 3.11.3 (i)). So g is also surjective.

Comparing the functions f and g, we can see that the definitions of the domain and codomain are crucially important in terms of the properties of the functions. The functions f and g both are defined by the same rule, although one function is both injective and surjective but the other is neither.

3. In the previous example, we saw that f was neither injective nor surjective and g was both injective and surjective. To emphasise that the properties of injectivity and surjectivity are independent of one another, define functions that are:

   (i) injective but not surjective

   (ii) surjective but not injective.

Solution

Simple solutions are provided using small finite sets for the domain and codomain which we do first in each case. A little more challenging is to find examples $A \to B$ where A and B are both subsets of $\mathbb{R}$ and we do this secondly in each case.

   (i) To construct a function $A \to B$ that is injective but not surjective we require that different elements of A have different images and also that there are elements of B that are not the image of any element of A.

   Let $A = \{a, b, c\}$ and $B = \{1, 2, 3, 4\}$ and define $f : A \to B$ by

   $$f(a) = 1, f(b) = 2, f(c) = 3.$$

   Clearly, different elements of A have different images, so f is injective. Also $4 \in B$ is not the image of any element of A so $4 \notin \operatorname{im} f$. Hence f is not surjective.

   For an example where the domain and codomain are subsets of $\mathbb{R}$, let

   $$g : \mathbb{R} \to \mathbb{R}, \ g(x) = e^x.$$

   If $x \neq y$, then $e^x \neq e^y$ so g is injective. Also $e^x > 0$ for all $x \in \mathbb{R}$ so, for example, $-1 \notin \operatorname{im} g$. Hence g is not surjective.

   (ii) To construct a function $A \to B$ that is surjective but not injective, we require that each element of B is the image of some element of

A and also that (at least) a pair of elements of A have the same image.

Let $A = \{a, b, c, d\}$ and $B = \{1, 2, 3\}$ and define $f : A \to B$ by

$$f(a) = 1, f(b) = 2, f(c) = 3, f(d) = 3.$$

Clearly, every element in B is the image of some element in A, so g is surjective. Also $g(c) = 3 = g(d)$ but $c \neq d$, so g is not injective.

For an example where the domain and codomain are subsets of $\mathbb{R}$, we can use the square function. Let

$$g : \mathbb{R} \to \mathbb{R}_{\geq 0}, \ g(x) = x^2.$$

Every $y \geq 0$ is the image of some $x \in \mathbb{R}$ so g is surjective, Since $g(2) = 4 = g(-2)$ but $2 \neq -2$, for example, it follows that g is not injective.

Reflecting on the examples we have considered of functions $f : A \to B$ where A and B are subsets of $\mathbb{R}$, we now describe how we can tell from its graph whether a function is injective and/or surjective.

Let $f : A \to B$ be a function where A and B are subsets of $\mathbb{R}$. Suppose, first, that f is *not* injective. Then there are distinct elements a_1 and a_2 in A such that $f(a_1) = f(a_2) = b$, say. This means that the horizontal line at height b meets the graph at points corresponding to $x = a_1$ and $x = a_2$ on the x-axis. This situation is illustrated in figure 3.22. If, on the other hand, f is injective, then this situation never occurs. In other words, a horizontal line through any point in B on the y-axis will not meet the graph in *more* than one point.

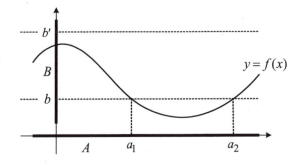

FIGURE 3.22: Horizontal line test for functions $\mathbb{R} \to \mathbb{R}$.

Suppose now that f is *not* surjective. Then there is an element $b' \in B$ which is not the image of any element of A. In terms of the graph, this means that the horizontal line through b' does not meet the graph at any point. Again, figure 3.22 illustrates this case. If, on the other hand, f is surjective, then this situation never occurs. In other words, a horizontal line through any point $b \in B$ on the y-axis will meet the graph in *at least* one point.

These considerations are summarised in the following theorem.

Theorem 3.2 (Horizontal line test)
Let $f : A \to B$ be a function, where A and B are subsets of $\mathbb{R}$. Then:

(i) f is injective if and only if every horizontal line through a point of B on the y-axis meets the graph of f *at most* once;

(ii) f is surjective if and only if every horizontal line through a point of B on the y-axis meets the graph of f *at least* once.

Bijective functions and inverse functions

We now turn our attention to a particularly important class of functions — those that are both injective and surjective.

Definition 3.5
A function $f : A \to B$ is **bijective**, or is a **bijection**, if it is both injective and surjective.

Given a function $f : A \to B$, 'reversing the arrows' does not necessarily define a function $B \to A$. Consider, for example, the function defined in example 3.11.1:

$$f : \{a, b, c, d, e\} \to \{1, 2, 3, 4, 5, 6\}, \ a \mapsto 2, \ b \mapsto 4, \ c \mapsto 2, \ d \mapsto 5, \ e \mapsto 1.$$

The arrow diagram for f was given in figure 3.14. Reversing the arrows gives the diagram shown in figure 3.23. This diagram is not that of a function $B \to A$ for two reasons. Firstly, the elements 3 and 6 are not associated with element of A. Secondly, the element 2 in B is associated with two elements, a and c in A. Thus the arrow diagram in figure 3.23 fails both aspects of the definition of a function given in definition 3.1.

One of the reasons why bijective functions are important is that if $f : A \to B$ is a bijection, then neither of these problems occurs when reversing the arrows. Hence, if $f : A \to B$ is a bijection, then 'reversing the arrows' does define a function $B \to A$. The resulting function $B \to A$ is called the **inverse function** of f and is denoted f^{-1}. By reversing the arrows we mean that if f gives an arrow from a to b, then f^{-1} gives an arrow from b to a. Symbolically, if $f(a) = b$, then $f^{-1}(b) = a$. We summarise these considerations in the following definition.

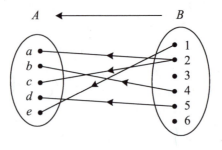

FIGURE 3.23: Reversing the arrows in example 3.11.1.

Definition 3.6

Let $f : A \to B$ be a bijective function. The function $f^{-1} : B \to A$ defined by

$$f^{-1}(b) = a \text{ if and only if } b = f(a)$$

is called the **inverse function** of f.

The definition of the inverse function of a bijection is illustrated in figure 3.24.

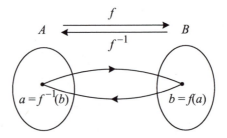

FIGURE 3.24: Defining the inverse function.

Examples 3.15

1. In example 3.14.2, we showed that the square function

$$g : \mathbb{R}_{\geq 0} \to \mathbb{R}_{\geq 0}, \ g(x) = x^2$$

is bijective. If $y = x^2 = g(x)$, then $x = \sqrt{y} = g^{-1}(y)$. Hence the inverse function is

$$g^{-1} : \mathbb{R}_{\geq 0} \to \mathbb{R}_{\geq 0}, \ g^{-1}(y) = \sqrt{y}.$$

Of course the variable name is not significant when defining a function, so we could have defined g^{-1} by $g^{-1}(x) = \sqrt{x}$.

2. Let $f : \mathbb{R} \to \mathbb{R}$ be defined by $f(x) = 3x - 1$.

The graph of f is a straight line — see figure 3.25. Therefore, by the horizontal line test, theorem 3.2, f is both injective and surjective.

To find the inverse function, we let $y = f(x)$ and then rearrange to obtain $x = f^{-1}(y)$:

$$y = f(x) \Rightarrow y = 3x-1 \Rightarrow 3x = y+1 \Rightarrow x = \tfrac{1}{3}(y+1) \Rightarrow f^{-1}(y) = \tfrac{1}{3}(y+1).$$

Hence the inverse function, defined in full, is

$$f^{-1} : \mathbb{R} \to \mathbb{R}, \ f^{-1}(y) = \tfrac{1}{3}(y + 1).$$

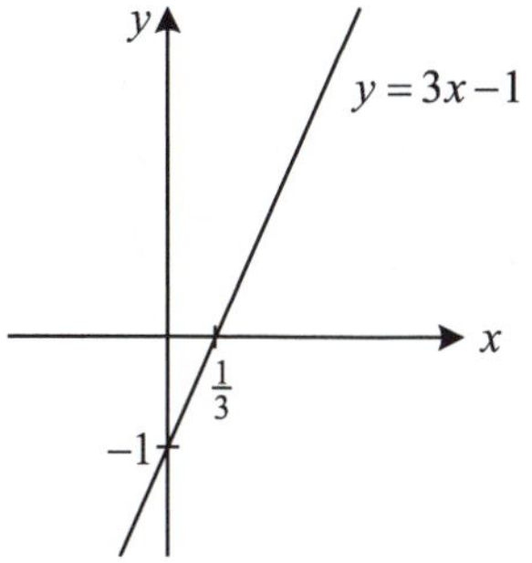

FIGURE 3.25: The graph of $f : \mathbb{R} \to \mathbb{R}$, $f(x) = 3x - 1$.

We have seen that if a function $f : A \to B$ is not bijective, then the process of reversing the arrows does not define a function $B \to A$. Let us examine this in a little more detail by considering separately the impact of the failure of a function to be injective and surjective when trying to define an inverse function.

If $f : A \to B$ is not injective, then there exists at least two distinct elements $a_1, a_2 \in A$ with the same image. This is illustrated in figure 3.26. Reversing the arrows does not define a function $B \to A$ because there is an element in B that is associated with more than one element of A. This 'problem' is not easily rectified. It would involve identifying some elements of the domain A to be removed, although there is not an obvious way to do this. If $f(a_1) = f(a_2)$, we would need to remove either a_1 or a_2 or both from the domain; the situation would be more complicated if there were several elements of A all with the same image. Thus a failure of injectivity may be regarded as a 'killer blow' to defining an inverse function.

Now consider the case $f : A \to B$ is injective but not surjective. This is illustrated in figure 3.27. In this case, reversing the arrows does not define a

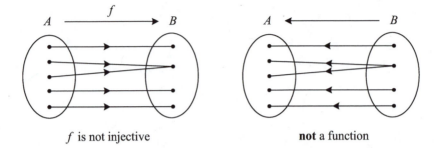

FIGURE 3.26: Reversing arrows when the function is not injective.

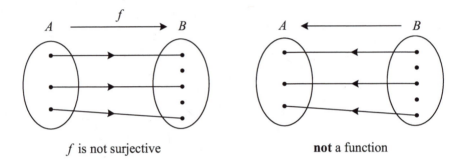

FIGURE 3.27: Reversing arrows when the function is not surjective.

function $B \to A$ because there are elements of B that are not associated with any element of A.

This problem is more easily fixed than a failure of injectivity. If we 'throw away' the elements of B that are not hit by arrows, then reversing the arrows does define a function. Throwing away the elements not hit by arrows just leaves the image of the function im f. So, if $f : A \to B$ is injective but not surjective, then changing the codomain to im f gives a bijective function $f : A \to \text{im} f$. This bijective function does have an inverse $f^{-1} : \text{im} f \to A$ defined by

$$f^{-1}(b) = a \text{ if and only if } f(a) = b \text{ for all } b \in \text{im} f.$$

This inverse function is illustrated in figure 3.28.

Exercises 3.5

1. For each of the following functions, determine whether or not the function is: (a) injective; (b) surjective.

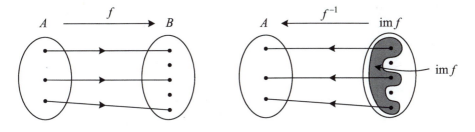

FIGURE 3.28: Defining an inverse when the function is not surjective.

(i) $f : \mathbb{R} \to \mathbb{R}$, $f(x) = \frac{1}{2}x - 7$.

(ii) $\mathbb{E} = \{2n : n \in \mathbb{Z}\} = \{\text{even integers}\}$
$f : \mathbb{Z} \to \mathbb{E}$, $f(n) = 2n + 4$.

(iii) $\mathbb{R}_{\geq 0} = \{x \in \mathbb{R} : x \geq 0\}$
$f : \mathbb{R}_{\geq 0} \to \mathbb{R}_{\geq 0}$, $f(x) = |x - 1|$.

(iv) $A = \{\text{countries of the world}\}$, $B = \{\text{cities of the world}\}$
$f : A \to B$, $f(X) = \text{capital city of } X$.

(v) $A = \{\text{countries of the world }\}$, $B = \{\text{cities of the world}\}$
$f : B \to A$, $f(X) = \text{country containing } X$.

(vi) $f : \mathbb{Z} \to \mathbb{Z}$, $f(n) = \begin{cases} \dfrac{n}{2} & \text{if } n \text{ is even} \\[2mm] \dfrac{n-1}{2} & \text{if } n \text{ is odd.} \end{cases}$

(vii) $f : \mathbb{Z} \to \mathbb{Z}$, $f(n) = \begin{cases} n+1 & \text{if } n \geq 0 \\ n & \text{if } n < 0. \end{cases}$

(viii) $A = \mathbb{P}(\{1, 2, 3, 4\})$
$f : A \to A$, $f(X) = X \cap \{1, 2\}$.

(ix) $A = \mathbb{P}(\{1, 2, 3, 4\})$
$f : A \to A$, $f(X) = X \cup \{1, 2\}$.

(x) A is any non-empty set
$f : A \to \mathbb{P}(A)$, $f(a) = \{a\}$.

2. Each of the following is the graph of a function $A \to B$ where A and B are subsets of $\mathbb{R}$.

Determine whether or not each function is: (a) injective; (b) surjective.

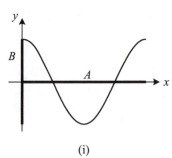

(i)

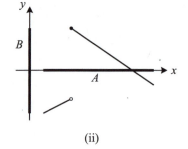

(ii)

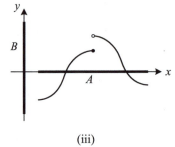

(iii)

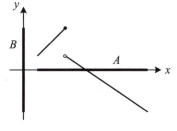

(iv)

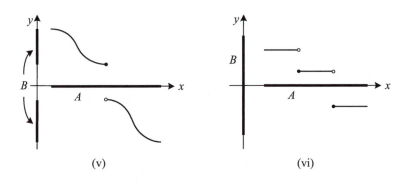

(v)

(vi)

3. Show that each of the following functions is a bijection and find its inverse.

   (i) $f : \mathbb{R} \to \mathbb{R}$, $f(x) = \dfrac{5x + 3}{8}$.

   (ii) $f : \mathbb{R} - \{-1\} \to \mathbb{R} - \{3\}$, $f(x) = \dfrac{3x}{x + 1}$.

   (iii) $f : [1, 3] \to [-2, 2]$, $f(x) = 2x - 4$.

   (iv) $f : \mathbb{R}^+ \to (0, 2)$, $f(x) = \dfrac{4}{x + 2}$.

   (v) $f : \mathbb{R}^2 \to \mathbb{R}^2$, $f(x, y) = (y, x)$.

   (vi) $f : \mathbb{R}^2 \to \mathbb{R}^2$, $f(x, y) = (2x - 1, 5y + 3)$.

(vii) $f : \mathbb{Z}^+ \to \mathbb{Z}, \ f(n) = \begin{cases} \dfrac{n}{2} & \text{if } n \text{ is even} \\[2mm] \dfrac{1-n}{2} & \text{if } n \text{ is odd.} \end{cases}$

4. Let $f : A \to B$ be a function that is neither injective nor surjective.

   (i) Show that it is possible to define a surjective function $f : A \to B'$ by replacing B with a suitable subset B'.

   (ii) Show that it is possible to define an injective function $f : A' \to B$ by replacing A with a suitable subset A'.

Chapter 4

The Structure of Mathematical Proofs

4.1   Introduction

In this chapter we introduce the key features of mathematical proof. It is very hard, some would say impossible, to define precisely what is meant by a mathematical proof as mathematicians themselves write them. The notion of proof in certain logical systems may be defined quite precisely, but this does not really describe what mathematicians do when constructing their proofs. The following professional mathematician's definition of proof, given by Slomson [11], serves as a useful starting point.

> A mathematical proof is a correct and convincing mathematical argument. Whether an argument is correct is a matter of logic: the conclusion must be a logical consequence of the premises. What counts as convincing will vary from person to person, and has changed with time.

In the following sections we will explain some of the general features that make a 'correct and convincing mathematical argument'. We will begin by returning to the work we began in section 1.5 and looking at some example proofs in rather more detail than is usual. From this we will extract some general features that are common to most mathematical proofs and so develop an informal framework for constructing proofs. In particular, we will explore the deductive method that is at the heart of mathematical reasoning. We will also consider how we approach universally quantified propositional functions of the form $\forall x \bullet P(x)$ where the universe for x is infinite. Returning to Slomson's requirement that a proof must be convincing, we will also discuss what knowledge the proof writer may assume the proof reader possesses. In section 4.4, we describe the method of direct proof which underpins most mathematical proofs. Our discussion in these sections will be geared towards developing an understanding of the principal features of mathematical proofs as they are usually written. In the final section, we explore in more depth Slomson's criterion for correctness: 'the conclusion must be a logical consequence of the premises'. We will develop a more formal framework for understanding what is a proof, more closely based on the notion of formal proof that we considered in section 2.5. Any mathematical proof will have a context and an intended audience, which our formal framework will need to take into account.

4.2   Some proofs dissected

In this section we look in some detail at some proofs of elementary properties of the positive integers. Our consideration of these proofs will be rather more detailed than is usual for a mathematics text. For each example, we will provide both a proof and a commentary on the proof. In the following section we will use these proofs to extract some general features and principals that will apply to many proofs.

Building on the proof, given in section 1.5 (page 11), that the square of an odd integer is odd, we begin by proving two similar results about evenness, oddness and divisibility.

Theorem 4.1
The sum and difference of two odd integers are even.

Before we begin to prove this, we need to understand precisely what the result is saying. Firstly, note that there is implicit universal quantification in the proposition:

> The sum and difference of *any* two odd integers are
> *always* even.

We may express this more formally using universal quantification as:

> For all integers m and n, if m and n are odd, then
> both $m + n$ and $m - n$ are even.

Proof. Suppose that m and n are odd integers.

Then there exist integers a and b such that $m = 2a + 1$ and $n = 2b + 1$.

Hence $m + n = (2a + 1) + (2b + 1) = 2a + 2b + 2 = 2(a + b + 1)$ where $a + b + 1$ is an integer. Therefore $m + n$ is even.

Similarly, $m - n = (2a + 1) - (2b + 1) = 2a - 2b = 2(a - b)$ where $a - b$ is an integer. Therefore $m - n$ is even.

$\square$

Commentary

The original statement of the result, 'the sum and difference of two odd integers are even', did not obviously give the logical structure of the proposition. We first needed to rewrite it so that the universal quantification became evident, 'for all integers m and n, if m and n are odd, then both $m+n$ and $m-n$ are even'. If we define the following predicates over the universe of integers

$$O : \quad \ldots is\ odd$$
$$E : \quad \ldots is\ even,$$

then we could symbolise the proposition as

$$\forall m \, \forall n \bullet (O(m) \wedge O(n)) \Rightarrow (E(m+n) \wedge E(m-n)).$$

In the proof, we assumed that m and n are odd, $O(m) \wedge O(n)$, and from this deduced that $m+n$ is even and $m-n$ is even, $E(m+n) \wedge E(m-n)$.

In fact, we proved $m+n$ is even separately from the proof that $m-n$ is even. Essentially, we gave two separate proofs: one for $(O(m) \wedge O(n)) \Rightarrow E(m+n)$ and one for $(O(m) \wedge O(n)) \Rightarrow E(m-n)$. However, from example 2.4.2 (page 36), we know that

$$[P \Rightarrow (Q \wedge R)] \equiv [(P \Rightarrow Q) \wedge (P \Rightarrow R)].$$

This means that, to prove a statement of the form $P \Rightarrow (Q \wedge R)$, it is sufficient to prove both $P \Rightarrow Q$ and $P \Rightarrow R$, which is precisely what we have done in this case.

Theorem 4.2

For all integers m and n, if m and n are odd, then $m^2 - n^2$ is divisible by 4.

Perhaps the most obvious proof of this statement would be to follow a similar strategy to that used in the previous example. This would produce the following proof (which starts in exactly the same way as the proof in the previous example).

Proof 1. Suppose that m and n are odd integers.

Then there exist integers a and b such that $m = 2a + 1$ and $n = 2b + 1$.

$$\begin{aligned}
\text{Therefore} \quad m^2 - n^2 &= (2a+1)^2 - (2b+1)^2 \\
&= (4a^2 + 4a + 1) - (4b^2 + 4b + 1) \\
&= 4a^2 + 4a - 4b^2 - 4b \\
&= 4(a^2 + a - b^2 - b) \quad \text{where } a^2 + a - b^2 - b \in \mathbb{Z}.
\end{aligned}$$

Hence $m^2 - n^2$ is divisible by 4. $\qquad\qquad\square$

This proof is perfectly correct and, we hope, easy to follow. There is an alternative proof, however, that builds on the result of example 4.1 above. We give this alternative proof below.

Proof 2. Suppose that m and n are odd integers.

Then both $m + n$ and $m - n$ are even, so there exist integers a and b such that $m + n = 2a$ and $m - n = 2b$. Therefore

$$m^2 - n^2 = (m - n)(m + n) = (2a)(2b) = 4ab \text{ where } ab \in \mathbb{Z}.$$

Hence $m^2 - n^2$ is divisible by 4.

$\square$

Commentary

There are several observations to make regarding these proofs. Perhaps the most obvious is that this example indicates that proofs are not unique. There will frequently be several *essentially* different ways of proving a result; that is, proofs which differ in approach or substance, and not just in the style of presentation. Which of several proofs is to be preferred will be a matter of judgement and will depend on various factors. For example, some people will prefer shorter, more sophisticated proofs that may be harder to understand; others may prefer proofs that are perhaps longer but do not rely on so much knowledge on the part of the reader.

The two proofs given above illustrate this. The first proof is more elementary in the sense that it assumes nothing other than the meaning of even and odd and some algebraic manipulation. The second proof is shorter and, one might argue, more elegant but relies on the results proved in the previous example and so is assuming greater 'background knowledge' on behalf of the reader. In fact, this occurs all the time — the proofs of some results rely on other propositions that need to have been proved earlier. We shall explore this idea — that proofs build on proofs that in turn build on other proofs etc. — in more detail in section 4.5 below.

The prime factorisation theorem says that every integer greater than 1 can be factored into prime numbers, and is part of the so-called 'Fundamental Theorem of Arithmetic'. The full Fundamental Theorem goes on to state that the factorisation for a given integer is unique apart from the ordering of the prime factors. However, the proof of the uniqueness part is more sophisticated and can wait until later (see chapter 7).

Theorem 4.3 (The Prime Factorisation Theorem)
Every integer greater than 1 can be expressed as a product of prime numbers.

Before embarking on a proof of the theorem, we need to understand precisely what a prime number is. A simple definition is that a **prime number** is an integer greater than 1 which is not divisible by any positive integer except 1 and itself. Thus 5 is a prime number since it is not divisible by any positive integer except 1 and 5, whereas 6 is not a prime number since it is divisible by 2, for instance, which is different from both 1 and 6 itself. (Notice that, according to the definition, 1 is not a prime number.)

Now that we understand the term 'prime number', we could embark on a search for a proof of the theorem. It is usually best though to ensure that we first understand thoroughly what the theorem is really saying. Often the most effective way of achieving this is to look at some examples.

Consider the integer 24. We can write $24 = 2 \times 12$, which expresses 24 as the product of a prime number (2) and a non-prime number (12). Since 12 is not prime, we can now look for its factors. We continue in this way as follows:

$$24 = 2 \times 12$$
$$= 2 \times 3 \times 4$$
$$= 2 \times 3 \times 2 \times 2.$$

We have now expressed 24 as a product of prime numbers: $2 \times 3 \times 2 \times 2$. Of course, there are other such expressions, for example $2 \times 2 \times 2 \times 3$, but this just contains the same prime numbers written in a different order.

We may carry out this process on any positive integer. For example, carrying out the process on $1234\,567\,890$ takes rather longer and is trickier but eventually produces the following expression:

$$1234\,567\,890 = 10 \times 123\,456\,789$$
$$= 2 \times 5 \times 3 \times 41\,152\,263$$
$$= 2 \times 5 \times 3 \times 3 \times 13\,717\,421$$
$$= 2 \times 3 \times 3 \times 5 \times 3607 \times 3803.$$

In this case, the last step is the hardest: factorising $13\,717\,421 = 3607 \times 3803$ and knowing that these factors are prime so that no further factorisation is possible. The theorem says that we can obtain such an expression for any integer bigger than 1.

Now that we have some intuitive 'feel' for the theorem, we can begin the search for a proof. In fact, the basis of the proof is already contained in the examples above. There, for example, we found the prime factors of 24 by first finding two factors (2 and 12) and then finding factors of these where possible, and so on. Since this process can be applied to any integer greater than 1, we can construct a proof of the general result. (A shorter, more sophisticated proof of this result will be given later — see chapter 8.) We can now give an informal proof of the Prime Factorisation Theorem.

Proof. Let n be any integer greater than 1. If n is prime then there is nothing to prove as n itself is already expressed as a 'product' of primes, albeit in a rather trivial way.

If n is not prime, then there exist factors n_1 and n_2, each greater than 1, such that $n = n_1 \times n_2$.

Now consider n_1 and n_2 in turn. If n_1 is composite (that is, not prime), then it can be expressed as a product of two integers each greater than 1, say $n_1 = m_1 \times m_2$. Similarly, either n_2 is prime or it can be expressed as a product $n_2 = m_3 \times m_4$ where m_3 and m_4 are greater than 1. At this stage we have expressed n in one of the following four ways:

$$n = n_1 \times n_2 \qquad \text{(if both } n_1 \text{ and } n_2 \text{ are prime),}$$
$$n = m_1 \times m_2 \times n_2 \qquad \text{(if } n_1 \text{ is composite and } n_2 \text{ is prime),}$$
$$n = n_1 \times m_3 \times m_4 \qquad \text{(if } n_1 \text{ is prime and } n_2 \text{ is composite),}$$
$$n = m_1 \times m_2 \times m_3 \times m_4 \quad \text{(if } n_1 \text{ and } n_2 \text{ are both composite).}$$

Next consider each m_i in turn and continue the process. At every step in the process, each factor is either prime or is split into two smaller factors. Therefore, this 'subdivision' process must eventually stop. When the process stops, the result is an expression of the form

$$n = p_1 \times p_2 \times \ldots \times p_k$$

where each p_i is prime. Therefore we have shown that n can be expressed as a product of primes.

$\square$

Commentary

Our treatment of this theorem and its proof is more detailed than is usual for a mathematics text. There are several reasons for this. One is our desire to give some indication at least of how the proof might be discovered, rather than just presenting the proof itself. If we are to learn how to construct proofs, it is clearly desirable to gain some insight into how a proof evolves from underlying ideas. It will not be sufficient just to study completed proofs. Another reason is to indicate the importance of precisely defined terms, such as 'prime number'.

Perhaps the most important lesson to learn from this example is that any mathematical proof is an exercise in communication. A correct but incomprehensible proof is of little use to anyone (and we hope our proof does not fall into that category). In writing proofs, clarity and comprehensibility, as well as correctness, are important goals.

Exercises 4.1

1. Prove each of the following.
    (i) The sum of any two consecutive integers is odd.
    (ii) For all integers n, if n is prime, then $n^2 + 5$ is not prime.
    (iii) For all integers n, if $n - 2$ is divisible by 4, then $n^2 - 4$ is divisible by 16.

2. (i) Prove that the sum of any three consecutive positive integers is divisible by 3.
    (ii) Prove that the sum of any five consecutive positive integers is divisible by 5.

3. (i) Prove that the product of any three consecutive positive integers is divisible by 6.
    (ii) Prove that the product of any five consecutive positive integers is divisible by 24.

4. (i) Prove that the sum of the squares of any two consecutive positive integers is odd.
    (ii) Prove that the sum of the squares of any three consecutive positive integers is one less than a multiple of 3.

4.3   An informal framework for proofs

In this section we give an informal description of one of our principal proof techniques, the method of direct proof. By looking in a bit more detail at simple examples of proofs, we identify some general features that will appear in many other proofs. In other words, we will extract some of the essential structure of our examples that will provide a general framework for many proofs.

To begin with, we look a little more closely at the structure of one half of the proof in theorem 4.1. We repeat the result and its proof below.

Theorem 4.1
The sum of two odd integers is even.

Proof. Suppose that m and n are odd integers.
Then there exist integers a and b such that $m = 2a + 1$ and $n = 2b + 1$.
Hence $m + n = (2a + 1) + (2b + 1) = 2a + 2b + 2 = 2(a + b + 1)$ where $a + b + 1$ is an integer.
Therefore $m + n$ is even. $\qquad\square$

Recall that the statement being proved is a universally quantified propositional function 'for all integers m and n, if m and n are odd, then $m + n$ is even'. In symbols,

$$\forall m \, \forall n \bullet (O(m) \wedge O(n)) \Rightarrow E(m + n)$$

where the universe is the integers, O denotes the predicate '... *is odd*', and E denotes the predicate '... *is even*'.

There are two important features of our proof.

- To prove the universally quantified statement, we actually proved $(O(m) \wedge O(n)) \Rightarrow E(m + n)$ for *arbitrary* integers m and n. By 'arbitrary' we mean that m and n were *not* assumed to have any special characteristics (such as being equal to 347, being greater than 17 or being prime or whatever) other than the property of 'being an odd integer'. In other words, m and n could have been any odd integers.

  Although the statement was universally quantified, we did not prove it for *every* odd integer m and *every* odd integer n individually. Clearly, this is impossible as there are infinitely many odd integers. Instead, we proved this for arbitrary 'representative' odd integers m and n.

  This distinction between universally quantified variables '$\forall m$' and '$\forall n$' on the one hand and arbitrary integers m and n on the other, is quite subtle. In part, the distinction is masked by the fact that we are using the same letter to denote both the universally quantified variable and its arbitrary representative. Some people prefer to make this distinction more explicit by using different letters, as follows.

  Theorem 4.1
  For all integers m and n, if m and n are odd, then $m + n$ is even.

  Proof. Suppose that k and ℓ are odd integers.
  Then there exist integers a and b such that $k = 2a + 1$ and $\ell = 2b + 1$.
  Hence $k + \ell = (2a + 1) + (2b + 1) = 2a + 2b + 2 = 2(a + b + 1)$ where $a + b + 1$ is an integer.
  Therefore $k + \ell$ is even. $\square$

- The second feature is that, to prove the conditional statement

  $$\text{if } m \text{ and } n \text{ are odd, then } m + n \text{ is even}$$

  (for our arbitrary integers m and n), we *assumed* m and n were odd and from this we made a *sequence of deductions* concluding with the statement $m + n$ is even. Symbolically, to prove

  $$(O(m) \wedge O(n)) \Rightarrow E(m + n)$$

we assumed the antecedent $O(m) \wedge O(n)$ and, from this, made a sequence of deductions to establish the consequent $E(m + n)$.

These two features of our proof are examples of general principles which we now describe.

Principle of Universal Generalisation

To prove a universally quantified statement of the form $\forall x \bullet P(x)$, it is sufficient to prove $P(k)$ where k is an *arbitrary* element of the universe for x.

This principle allows us to prove statements over infinite universes. If a propositional function is universally quantified over a finite universe, it is possible in principle to prove it by verifying the statement individually for each element of the universe. Of course, if the universe is finite but very large, this approach may not be feasible.

For example, suppose we wish to prove

if m and n are odd then $m + n$ is even

over the universe for m and n of positive integers less than 6. We could do this as follows.

Proof. Let:

$n = 1$	$m = 1$:	then $m + n = 2$ which is even;
$n = 1$	$m = 3$:	then $m + n = 4$ which is even;
$n = 1$	$m = 5$:	then $m + n = 6$ which is even;
$n = 3$	$m = 1$:	then $m + n = 4$ which is even;
$n = 3$	$m = 3$:	then $m + n = 6$ which is even;
$n = 3$	$m = 5$:	then $m + n = 8$ which is even;
$n = 5$	$m = 1$:	then $m + n = 6$ which is even;
$n = 5$	$m = 3$:	then $m + n = 8$ which is even;
$n = 5$	$m = 5$:	then $m + n = 10$ which is even.

This verifies the result for each pair (m, n) where m and m are odd positive integers less than 6, so the result follows. $\square$

When the universe is infinite, this approach is clearly impossible and the Principle of Universal Generalisation is required. The crucial point to note about the arbitrary element k of the universe is not whether it is labelled n or k, but rather that it is not assumed to possess any attribute that is not possessed by every element of the universe. In other words, it is a 'general' element of the universe.

The second feature of our proof that is a general principle is that we began with an initial proposition P and then made a sequence of deductions culminating in a final proposition Q. We call this the 'deductive method'. Essentially, this says that mathematical proofs use deductive reasoning as described in section 1.4 rather than inductive reasoning, described in section 1.3, which is the basis of most human knowledge and belief.

Deductive Method

A proof starts with some initial statement or statements and proceeds by making a sequence of deductions that follow logically until a final statement is obtained.

As we shall see, different types of proof have different initial and final statements. The important aspect of the deductive method is that a proof comprises a sequence of statements that follow logically from earlier statements in the proof. Later in this chapter, we will examine in more detail what we mean by 'follows logically'; see section 4.5.

There is one further important observation that we wish to make regarding our proof of theorem 4.1. Although the proof itself is straightforward, it assumes a certain familiarity with the integers and their properties. In particular, the following properties are among those assumed, either explicitly or implicitly.

- *The oddness property.*

  An integer n is odd if and only if it can be expressed as $n = 2a + 1$ for some integer a.

- *The evenness property.*

  An integer n is even if and only if it can be expressed as $n = 2a$ for some integer a.

- *Associativity and commutativity of addition.*

  By equating $(2a + 1) + (2b + 1)$ with $2a + 2b + 2$, we are assuming that:

  - we can ignore the brackets and regroup terms as necessary so that we can rewrite $(2a + 1) + (2b + 1)$ as $2a + (1 + 2b) + 1$
  - we can equate $1 + 2b$ with $2b + 1$.

  Of these assumptions, the first relies on the *associative property* of addition

$$\text{for all } a, b, c \in \mathbb{Z}, \ a + (b + c) = (a + b) + c,$$

and the second relies on the *commutative property* of addition

$$\text{for all } a, b \in \mathbb{Z}, \ a + b = b + a.$$

- *Distribution law of multiplication over addition.*

  By equating $2a + 2b + 2$ with $2(a + b + 1)$ we are also assuming that 'factorisation works'. This relies on the *distributive property* of multiplication over addition:

  $$\text{for all } a, b, c \in \mathbb{Z}, \ a(b + c) = ab + ac.$$

Of course, we would not normally consider a simple algebraic re-arrangement $(2a+1)+(2b+1) = 2a+2b+2 = 2(a+b+1)$ in such minute detail. Indeed, we would normally complete the re-arrangement automatically without giving a second thought to the algebraic processes and steps required. This is because we are very familiar with the basic properties of addition and multiplication of integers, so much so that they feel like 'second nature' to us. However, for the current discussion, the important point is not the details of *which* properties of the integers we are using, but rather the fact that there are *some* properties that the proof uses.

This is quite typical. In almost any proof, there will be properties about the system under consideration that will be assumed implicitly but not referred to explicitly. The previous discussion highlighted some properties of the 'system' of integers (and their addition and multiplication) that were implicit in the proof in theorem 4.1. We refer to all of these properties as our 'background knowledge' of the system under consideration.

Background knowledge

In any proof, the properties of the system under consideration that are assumed implicitly but not referred to explicitly, will be referred to as our **background knowledge** of the system.

By its very nature, this background knowledge is somewhat vague and ill-defined and it will vary from context to context. For example, two specialists in a particular branch of mathematics will have a very extensive background knowledge of their field that will not be shared by a layperson or novice in the field. Although it is ill-defined, it is nevertheless important. A proof will only be understandable by a reader if they share the background knowledge that the proof writer assumed when writing the proof.

Anyone writing a proof will need to consider how much background knowledge they may assume or wish to assume. This will depend on who is the intended audience for the proof as well as the preferences of the proof writer. In section 4.2, we gave two different proofs of theorem 4.2.

> *For all integers m and n, if m and n are odd, then $m^2 - n^2$ is divisible by 4.*

The first proof assumed less prior knowledge than the second proof, but was longer and, perhaps, less elegant. This example illustrates in a small way that there is often a trade-off between 'elementary' proofs, those that assume less background knowledge, and more 'sophisticated' proofs. Since elementary proofs have less to draw on, they can sometimes be longer and more complex in their reasoning. Set against this, there is a potential danger with more sophisticated proofs in that the reader may not possess all the knowledge that is assumed of them.

We conclude this section with a brief consideration of how the previous discussion relates to the following proof in set theory. In this example, we first prove the following result about sets.

Theorem 4.4
For all sets A and B, $A \cap B \subseteq A \cup B$.

Proof. Let A and B be sets.

Let $x \in A \cap B$. Then $x \in A$ and $x \in B$. Hence it follows that $x \in A$ or $x \in B$, so $x \in A \cup B$.

We have shown that: $x \in A \cap B \Rightarrow x \in A \cup B$. Therefore $A \cap B \subseteq A \cup B$. $\square$

How does this example relate to our previous discussion? The first point to note is the use of the Principle of Universal Generalisation. The statement to be proved is universally quantified ('for all sets A and B ...') and the proof starts by letting A and B be arbitrary sets. In other words, although it is not mentioned explicitly, the proof implicitly used the Principle of Universal Generalisation.

Secondly, the background knowledge used in the proof draws on section 3.3. In particular, it explicitly uses the meaning of

- intersection: $x \in X \cap Y$ if and only if $x \in X$ and $x \in Y$;
- union: $x \in X \cup Y$ if and only if $x \in X$ or $x \in Y$;
- subset: $X \subseteq Y$ if and only if, for all x, $x \in X \Rightarrow x \in Y$.

Thirdly, the proof follows the deductive method. The initial statements are 'A is a set', 'B is a set', and '$x \in A \cap B$'. Deductions are then made firstly to obtain the statement $x \in A \cup B$ and then, using the background knowledge of the meaning of subset, the final statement $A \cap B \subseteq A \cup B$ is deduced.

In summary, the framework for proofs that we have developed has the following components and each of our proofs has exhibited these.

- **Principle of Universal Generalisation**: to prove a universally quantified statement $\forall x \bullet P(x)$, the statement $P(k)$ is proved for an arbitrary element of the universe k.

- **Use of background knowledge**: in any proof, certain properties of the system under consideration will be assumed, implicitly or explicitly.

- **Deductive method**: a proof starts with some initial statement or statements and proceeds by making a sequence of deductions that follow logically until a final statement is obtained.

Exercises 4.2

1. For each of the proofs in exercise 4.1:

   (a) identify how the proof uses the Principle of Universal Generalisation;

   (b) identify explicitly some of the background knowledge that is used in the proof;

   (c) describe how the proof uses the deductive method — what are the initial statements and what is the final statements?

4.4 Direct proof

In the previous section we described a proof as following the deductive method: start with an initial statement or statements and make a sequence of deductions until a final statement is obtained. However, we did not describe what comprises the initial and final statements. In this section, we consider this in more detail and describe the most common structure of proof, known as 'direct proof'. In fact, each of the proofs given earlier in this chapter have been direct proofs.

We begin by looking again at theorem 4.1, the first theorem and proof in section 4.3. The following summarises the overall structure of the proof but omits the details of the actual deductions made.

Theorem 4.1

For all integers m and n, if m and n are odd, then $m + n$ is even.

Proof. Suppose that m and n are odd integers.

$$\vdots$$

Therefore $m + n$ is even.   □

The structure of the proof is to assume that m and n are arbitrary odd integers — these are the initial statements — and to deduce that $m+n$ is even — this is the final statement. With the universe being the integers and with predicates O for 'odd' and E for 'even', the statement being proved is the conditional

$$(O(m) \wedge O(n)) \Rightarrow E(m+n).$$

The structure of the proof was to *assume* the antecedent $O(m) \wedge O(n)$ and *deduce* the consequent $E(m+n)$. This general structure is called the 'method of direct proof', which we now define.

Method of Direct Proof (for $P \Rightarrow Q$)

To prove a proposition of the form $P \Rightarrow Q$, it is sufficient to *assume* P and, from this, *deduce* Q.

It is easy to see why this method works. Recall the truth table for the conditional $P \Rightarrow Q$, given on page 21 and repeated below.

P	Q	$P \Rightarrow Q$
T	T	T
T	F	F
F	T	T
F	F	T

The only situation when $P \Rightarrow Q$ is false is when P is true and Q is false, so if we wish to establish the truth of $P \Rightarrow Q$, it is this case that we must rule out. In other words, we must show that when P is true, Q cannot be false. We do not need to consider the situation when P is false because $P \Rightarrow Q$ is true in this case, regardless of the truth or falsity of Q. Assuming P to be true at the beginning of the proof has the effect of restricting attention to the first two rows of the truth table. Deducing that Q is true in this case, rules out the second row as a possibility and hence shows that the conditional $P \Rightarrow Q$ is true.

In section 4.2, we gave two different proofs of the following theorem.

Theorem 4.2

For all integers m and n, if m and n are odd, then $m^2 - n^2$ is divisible by 4.

The statement being proved is a conditional $P \Rightarrow Q$ where P is 'm and n are odd' and Q is '$m^2 - n^2$ is divisible by 4'. Both proofs followed the same structure, described as follows.

Suppose that m and n are odd integers.
$$\vdots$$
Therefore $m^2 - n^2$ is divisible by 4.

Each proof assumed P and from this deduced Q; in other words, each proof was a direct proof. The two proofs differed in the sequence of deductions used to obtain Q from P, but not in their overall structure.

We have also seen two other proofs, in theorems 4.3 and 4.4, where the statements being proved are not naturally expressed as a conditional, $P \Rightarrow Q$.

- Prime Factorisation Theorem 4.3: every integer greater than 1 can be expressed as a product of prime numbers.

- Theorem 4.4: for all sets A and B, $A \cap B \subseteq A \cup B$.

The method of direct proof of a conditional, assume P and then deduce Q, does not really apply in these cases as the proposition being proved is not (obviously) of the form $P \Rightarrow Q$. If we examine those proofs, we see that they start by considering arbitrary elements of the universe and then assuming some piece of background knowledge about the appropriate system: the integers or sets. In the case of the Prime Factorisation Theorem, the proof considers an arbitrary integer greater than 1 and starts with the property that every such integer is either prime or can be factorised in some way. For the set theory result, the background knowledge that gets the proof going is the knowledge that, to prove that one set is a subset of another, we must show that every element of the first set is an element of the second set. So each of these proofs have the same overall structure: assume some background knowledge and proceed to make deductions until the final desired conclusion is reached. We still regard these as direct proofs because we proceed directly to the appropriate conclusion.

In each of these cases, the statement of the theorem is of the general form $\forall x \bullet Q(x)$ or $\forall x \, \forall y \bullet Q(x, y)$. The proofs also apply the Principle of Universal Generalisation which reduces the statement to be proved to a proposition Q involving an arbitrary element or elements of the appropriate universe. In each case, the proof begins with some background knowledge and proceeds by making deductions until the statement Q is obtained. So each of the proofs followed the following method of direct proof but started with some aspect of background knowledge rather than the antecedent P of a conditional $P \Rightarrow Q$. We can summarise this as follows.

Method of Direct Proof (for Q)

To prove the proposition Q, it is sufficient to *assume* appropriate background knowledge and, from this, *deduce Q*.

The distinction we have made between theorems expressed as conditional statements $P \Rightarrow Q$ and those that are expressed as a non-conditional proposition Q is somewhat less clear cut than we have indicated. Suppose we have a proposition of the form $\forall x \bullet Q(x)$ where x belongs to some universe $\mathscr{U}$. By making the universe explicit, we could express this as a conditional $\forall x \bullet x \in \mathscr{U} \Rightarrow Q(x)$. For example, theorem 4.4 could have been expressed as a conditional as follows: 'if A and B are sets, then $A \cap B \subseteq A \cup B$'.

We now look at some further examples of direct proofs of the two different types; that is, direct proofs of $P \Rightarrow Q$ and direct proofs of Q.

Example 4.1

In this example, we consider the interaction between composition of functions, described in section 3.5, with the properties of injectivity and surjectivity, defined in section 3.6. Two obvious questions to ask are:

Is the composite of two injective functions an injective function?

Is the composite of two surjective functions a surjective function?

Let f and g be two functions such that the composite $g \circ f$ is defined — figure 3.18 on page 114 illustrates the setup. Recall that, informally, a function is injective if different elements always have different images. Suppose this is true for both f and g and consider the composite $g \circ f$, which has rule 'apply f then apply g'. Starting with different elements, applying f gives different elements since f is injective; then applying g still gives different elements since g is injective. This suggests that the composite of two injections is also an injection.

Now suppose that f and g are surjective. Then each element of their respective codomain is the image of some element in the domain; informally, every element of the codomain is 'hit by an arrow' coming from some element of the domain. Again consider $g \circ f$. The codomain of $g \circ f$ is the same as the codomain of g. Since g is surjective, every element of this codomain is hit by a 'g-arrow'. But the starting point of this g-arrow lies in the codomain of f, so it is hit by an 'f-arrow' as f is surjective. Putting the two arrows together, we see that every element of the codomain of $g \circ f$ is hit by a composite 'f followed by g arrow'. This suggests that the composite of two surjections is also an surjection.

The following theorem summarises the preceding discussion.

Theorem 4.5

Let $f : A \to B$ and $g : B \to C$ be two functions with composite function $g \circ f : A \to C$.

(i) If f and g are both injective then so, too, is $g \circ f$.

(ii) If f and g are both surjective then so, too, is $g \circ f$.

Proof.

(i) Suppose that $f : A \to B$ and $g : B \to C$ are both injective functions.

Let $a_1, a_2 \in A$ and let $b_1 = f(a_1)$ and $b_2 = f(a_2)$. To show that $g \circ f$ is injective, we need to show that, if $(g \circ f)(a_1) = (g \circ f)(a_2)$, then $a_1 = a_2$. Now

$$(g \circ f)(a_1) = (g \circ f)(a_2)$$
$$\Rightarrow \quad g(f(a_1)) = g(f(a_2))$$
$$\Rightarrow \quad g(b_1) = g(b_2) \qquad \text{(since } f(a_1) = b_1, f(a_2) = b_2\text{)}$$
$$\Rightarrow \quad b_1 = b_2 \qquad \text{(since } g \text{ is injective)}$$
$$\Rightarrow \quad f(a_1) = f(a_2) \qquad \text{(since } f(a_1) = b_1, f(a_2) = b_2\text{)}$$
$$\Rightarrow \quad a_1 = a_2 \qquad \text{(since } f \text{ is injective)}.$$

Hence $g \circ f$ is injective.

(ii) Suppose that $f : A \to B$ and $g : B \to C$ are both surjective functions.

Let $c \in C$. To show that $g \circ f$ is surjective, we need to show that $(g \circ f)(a) = c$ for some $a \in A$.

Since g is surjective, there exists $b \in B$ such that $g(b) = c$. Also, f is surjective, there exists $a \in A$ such that $f(a) = b$. Therefore, there exists $a \in A$ such that

$$(g \circ f)(a) = g(f(a)) = g(b) = c.$$

Hence $g \circ f$ is surjective.

$\square$

Commentary

Each of the two parts of the proof is a direct proof of a conditional $P \Rightarrow Q$. We started each proof by assuming P: f and g are both injective or both surjective functions. Then each proof made a number of deductions to obtain Q: $g \circ f$ is injective or surjective.

Note also that each proof uses implicitly the Principal of Universal Generalisation: each part proves the result for arbitrary injective or surjective functions, as appropriate. Indeed, as we have seen previously, even the universal quantification in the statement of the theorem is implicit. The structure of the results is as follows, although the universal quantification was not explicit in the statement of the theorem:

> for all functions f and g, if f and g are injective/surjective,
> then $g \circ f$ is injective/surjective.

Finally, the principal background knowledge used in each case is the meaning of the terms 'injective' and 'surjective', both in terms of what we may assume about f and g and also the property that must be established for $g \circ f$.

Examples 4.2 (Properties of the modulus function)
The **modulus** $|x|$ of a real number $x \in \mathbb{R}$ is defined by

$$|x| = \begin{cases} x & \text{if } x \geq 0 \\ -x & \text{if } x < 0. \end{cases}$$

This defines a function $|\cdot| : \mathbb{R} \to \mathbb{R}$, $x \mapsto |x|$ which has graph given in figure 4.1.

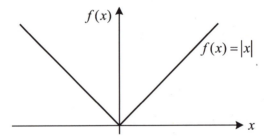

FIGURE 4.1: The modulus function.

The modulus function is important in mathematics. For example, it is an important part of definitions (and hence proofs) of limits in real analysis; we will consider some examples of these proofs in section 5.4. In the following examples, we will explore some of the properties of the modulus function.

1. The first property we will prove is the following inequality, known as the **triangle inequality** for the modulus function.

   Theorem 4.6
   For all $x, y \in \mathbb{R}$, $|x + y| \leq |x| + |y|$.

   Proof. Without loss of generality, we may assume that $x \leq y$.

Since the values of $|x|$ and $|y|$ depend on whether x and y are positive or negative, we will split the proof into four cases:

Case 1: $0 \leq x \leq y$

Case 2: $x < 0 \leq y$ and $x + y \geq 0$

Case 3: $x < 0 \leq y$ and $x + y < 0$

Case 4: $x \leq y < 0$

Case 1. In this case, $|x| = x$, $|y| = y$ and, since $x + y \geq 0$, we also have $|x + y| = x + y$. Therefore

$$|x + y| = x + y = |x| + |y|,$$

so the result holds in this case.

Case 2. In this case $|x| = -x$, which means that $x < 0 < -x$. Also $|y| = y$ and $|x + y| = x + y$. Therefore

$$|x + y| = x + y < -x + y = |x| + |y|,$$

so the result holds in this case.

Case 3. Again we have $|x| = -x$. Also $|y| = y$, which means that $-y < 0 < y$. Since $x + y$ is negative, we have $|x + y| = -(x + y)$. Therefore

$$|x + y| = -(x + y) = -x - y < -x + y = |x| + |y|,$$

so the result holds in this case.

Case 4. In this case we have $|x| = -x$ and $|y| = -y$. Since $x + y < 0$ also, we have $|x + y| = -(x + y)$. Therefore

$$|x + y| = -(x + y) = -x - y = |x| + |y|,$$

so the result holds in this case.

We have shown that the result holds in all four cases. Hence, for all $x, y \in \mathbb{R}$, $|x + y| \leq x + y$.   □

Commentary

The opening phrase, 'without loss of generality we may assume that ...', is quite common in proofs. What this means is that there are several cases — two in our example — where the reasoning for each case is essentially the same. Therefore we will present the reasoning in only one of the cases. In our example, the two cases are $x \leq y$ and $y \leq x$. Since the statement we are proving, $|x + y| \leq |x| + |y|$, is symmetric in x and

y, we could obtain the proof in the case $y \leq x$ from the proof when $x \leq y$ simply by swapping the roles of x and y. Hence there is no loss of generality by making the additional assumption $x \leq y$.

Since the definition of the modulus function involves two cases ($x \geq 0$ and $x < 0$), the proof naturally splits into four cases. These comprise three 'main' cases, one of which has two 'sub-cases', as illustrated in figure 4.2. To complete the proof, we need to show that the required result holds in each of the four cases. The argument in each case is straightforward and just requires careful consideration of which terms are positive and which are negative.

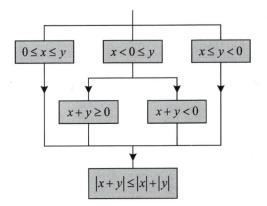

FIGURE 4.2: Illustrating the four cases in the proof of $|x + y| \leq |x| + |y|$.

When it is sensible to split a proof into cases is often a matter of judgement. On the one hand, the individual cases themselves are usually more straightforward to deal with because we will have made some additional assumptions for each of the cases. Balanced against this is the additional work required to prove the result for each case separately.

2. In this example, we will explore the relationship between the values of the modulus of various expressions. There are (at least) two reasons for doing so. Firstly, a good facility for arguments of this kind is essential for an understanding of those parts of real analysis that consider convergence of sequences, limits of functions, continuity and differentiability of functions, and so forth. From our perspective, however, a second reason for considering these examples, is that they illustrate the interplay between universal and existential quantification that we considered in section 2.4.

(i) The first result we consider is the following.

There exists a real number $\delta > 0$ such that

$$|x + 1| < \delta \implies |3x^2 - x - 4| < 1.$$

The first point to note here is that, although the existential quantification is explicit, there is a hidden universal quantification: it is implicit that the implication $|x+1| < \delta \Rightarrow |3x^2 - x - 4| < 1$ holds for all real numbers x.

We can think of the proposition as asserting that we may control the size of $|3x^2 - x - 4|$, in the sense of ensuring it is less than 1, provided we are permitted to control the size of $|x+1|$, in the sense of ensuring that it is less than some specified δ. So, we begin by considering the expression that we are trying to control,

$$|3x^2 - x - 4| = |(x+1)(3x-4)| = |x+1|\,|3x-4|.$$

Here we have first factorised the expression and then used the property, $|ab| = |a||b|$, whose proof we leave as an exercise; see exercise 4.3.1. Of these two terms, we have control over $|x+1|$ as we can ensure that it is less than some $\delta > 0$, but we need to consider $|3x - 4|$ more carefully. The 'trick' is to rewrite $3x - 4$ in terms of $x+1$ and then apply the triangle inequality proved in the previous example, as follows:

$$
\begin{aligned}
|3x - 4| &= |3(x+1) - 7| \\
&\leq |3(x+1)| + |-7| \quad \text{(by the triangle inequality)} \\
&= 3|x+1| + 7. \quad \text{(simplifying)}
\end{aligned}
$$

Recall that we are trying to establish the existence of $\delta > 0$ such that

$$|x+1| < \delta \Rightarrow |3x^2 - x - 4| < 1.$$

Suppose, for a moment, that $\delta = 1$. Then, for $|x+1| < 1$, we have

$$|3x - 4| \leq 3|x+1| + 7 < 3 + 7 = 10.$$

From this we have,

$$|3x^2 - x - 4| = |x+1|\,|3x-4| < 10|x+1| \text{ (since } |3x-4| \leq 10),$$

which will be less than 1 provided $|x+1| < 1/10$. So if we now take $\delta = 1/10$, then both the required inequalities, $|3x-4| \leq 10$ and $|x+1| < 1/10$, will hold when $|x+1| < \delta$. We can now piece this together and write as a coherent proof of the result.

There exists a real number $\delta > 0$ such that

$$|x+1| < \delta \Rightarrow |3x^2 - x - 4| < 1.$$

Proof. Let $\delta = \frac{1}{10}$. For all x satisfying $|x + 1| < \delta$, we have

$$
\begin{aligned}
\left|3x^2 - x - 4\right| &= |(x+1)(3x-4)| \\
&= |x+1|\,|3x-4| \\
&= |x+1|\,|3(x+1) - 7| \\
&\leq |x+1|(3|x+1| + 7) \quad \text{(triangle inequality)} \\
&< 10|x+1| \quad \text{(since } |x+1| < \delta < 1\text{)} \\
&\leq 10 \times \tfrac{1}{10} = 1 \quad \text{(since } |x+1| < \delta = \tfrac{1}{10}\text{).}
\end{aligned}
$$

Hence we have shown that, when $\delta = \frac{1}{10}$, $|x + 1| < \delta \Rightarrow |3x^2 - x - 4| < 1$.

$\square$

Commentary

The structure of the statement being proved is an existentially quantified proposition: $\exists \delta \bullet P(\delta)$. (We also note that $P(\delta)$ involved a universal quantification over x, but that does not alter the overall form of the statement as an existentially quantified proposition.) We proved the proposition by finding an explicit value for δ ($\delta = 1/10$) which 'worked'. The bulk of the proof is concerned with showing that the chosen value of δ has the required property. Later, in chapter 7, we will see occasions where existence of an object is established without an explicit value being found.

This example also illustrates that the process of finding a proof is sometimes quite different from how the proof is eventually presented. In this case, the realisation that $\delta = 1/10$ is sufficient comes at the end of what we might call the 'discovery process'. However, in the proof itself, the choice of $\delta = 1/10$ comes right at the beginning.

It is also worth noting that the choice of $\delta = 1/10$ is not unique. Recall that we established

$$\left|3x^2 - x - 4\right| \leq |x+1|(3|x+1| + 7).$$

Suppose, in the bracket, we chose $|x+1| \leq 1/3$. Then $3|x+1| + 7 \leq 8$, so that $|3x^2 - x - 4| \leq 8|x+1|$ and we can then ensure that $|3x^2 - x - 4| < 1$ by ensuring, in addition, that $|x+1| \leq 1/8$. To ensure both inequalities hold, we require $\delta < 1/3$ and $\delta < 1/8$. Therefore, an alternative choice for δ which makes the proof work is $\delta = 1/8$.

(ii) In this example we consider the following result.

For all $\varepsilon > 0$, there exists $\delta > 0$ such that

$$|x + 2| < \delta \ \Rightarrow \ |x^2 - 4| < \varepsilon.$$

This is similar to the previous example, but with another level of quantification. In the previous example we were required to show that $|3x^2 - x - 4|$ was less than some *specific* value, namely 1, by choosing $|x+1| < \delta$. Now we have to show that $|x^2 - 4|$ is less than any *arbitrary* $\varepsilon > 0$ again by choosing $|x + 2| < \delta$.

Proof. Given $\varepsilon > 0$, let $\delta = \min\left\{1, \dfrac{\varepsilon}{5}\right\}$.

Then $\delta > 0$ and $|x + 2| < \delta$ implies both that $|x + 2| < 1$ and that $|x + 2| < \dfrac{\varepsilon}{5}$. Now

$$
\begin{aligned}
|x^2 - 4| &= |(x + 2)(x - 2)| \\
&= |x + 2|\,|(x + 2) - 4| \\
&\leq |x + 2|(|x + 2| + 4) \quad \text{(triangle inequality)} \\
&< 5|x + 2| \quad \text{(since } |x + 2| < 1) \\
&\leq 5 \times \frac{\varepsilon}{5} = \varepsilon \quad \text{(since } |x + 2| < \frac{\varepsilon}{5}).
\end{aligned}
$$

Hence we have shown that, for the given choice of δ, $|x+2| < \delta \ \Rightarrow$ $|x^2 - 4| < \varepsilon$. $\qquad\qquad\square$

Commentary

This time we did not show the discovery process. Hence the choice of $\delta = \min\{1, \varepsilon/5\}$ at the start of the proof seems to come out of nowhere.

What we actually did was to work through the first part of the reasoning:

$$
\begin{aligned}
|x^2 - 4| &= |(x + 2)(x - 2)| \\
&= |x + 2|\,|(x + 2) - 4| \\
&\leq |x + 2|(|x + 2| + 4) \quad \text{(triangle inequality)}.
\end{aligned}
$$

We then realised that, provided $|x+2| < 1$, the term in the bracket would be less than 5. Hence we could then ensure the whole expression was less than ε provided, *in addition*, that $|x + 2| < \varepsilon/5$. To ensure that both of these steps work, we then recognised that we could choose $\delta = \min\{1, \varepsilon/5\}$. In other words, this choice of δ comes after a careful discovery process even though, in the proof itself, this is not shown.

It is also worth remarking on the structure of the statement in this example. As we noted at the start, it is similar to the previous example but with an extra quantification: $\forall \varepsilon \, \exists \delta \bullet P(\varepsilon, \delta)$ where the universe of discourse for both ε and δ is the positive real numbers. As noted in section 2.4 (see examples 2.12, questions 2 and 3), the order of quantification is significant. In this case, because the universal quantification of ε comes before the existential quantification of δ, the value of δ may depend on ε. In other words, for each value of ε there must be an appropriate value of δ, but different values of ε may have different values of δ.

A case study: metric spaces

In the next few pages, we introduce a topic that, we suspect, will be unfamiliar to many readers. Our aim here is to show how a new piece of abstract mathematics is built from a few basic assumptions together with what we might call a 'standard toolkit of mathematics'. The basic assumptions will usually be chosen to capture key properties that are common to many examples, possibly in diverse areas, that will provide a common framework worthy of study. We do not wish to specify precisely what is contained in the 'standard toolkit of mathematics', but it will include elements of propositional and predicate logic, some set theory, the simple properties of integers, rational numbers and real numbers, the properties of 'equals' and inequalities, and the idea of functions and their basic properties. For the purposes of this discussion, we will assume the material covered in chapters 2 and 3. Those readers who wish to, may omit this discussion and go straight to section 4.5.

We aim to define a 'metric space' that will capture different circumstances in which we have some space where we can measure distance. In other words, we may think of a metric space as providing a general framework for 'space where we can measure distances'. An important observation is that the notion of 'distance' — or 'metric' — is not universal but will vary according to context as the following examples illustrate.

Examples 4.3

1. As a familiar example, the straight line distance between points $P(x_1, y_1)$ and $Q(x_2, y_2)$ in the plane is given by

$$d(P, Q) = \sqrt{(x_1 - x_2)^2 + (y_1 - y_2)^2}.$$

Mathematically, this is often referred to as the **Euclidean distance** or **Euclidean metric**. When applied to points on a map, for example, this is sometimes referred to as the 'distance the crow flies' – see figure 4.3. The formula for $d(P, Q)$ follows by applying Pythagoras' theorem to the right-angled triangle in the figure.

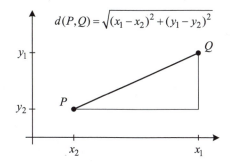

$$d(P,Q) = \sqrt{(x_1 - x_2)^2 + (y_1 - y_2)^2}$$

FIGURE 4.3: The Euclidean metric.

2. Residents of Manhattan in New York, for example, would not find the 'crow flies' distance very useful simply because this is not way they are able to travel around the city. They may refer instead to the distance between locations in the city as being 'three blocks west and two blocks north', for example. The mathematical version of this is what we shall call the **Manhattan distance** or **Manhattan metric** d_M between points $P(x_1, y_1)$ and $Q(x_2, y_2)$ in the plane defined by

$$d_M(P,Q) = |x_1 - x_2| + |y_1 - y_2|.$$

The d_M distance is the sum of the horizontal and vertical distances between the points — see figure 4.4.

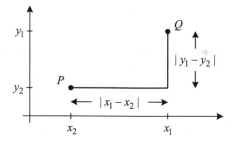

FIGURE 4.4: The Manhattan metric.

3. It is well known that the paths aeroplanes take when flying between cities are arcs of great circles. (A great circle on a sphere is the intersection of the sphere with a plane that passes through the centre of the sphere.) Aeroplanes take these paths because they are the shortest distances between points on the sphere. We call this great circle distance the **sphere distance** or **sphere metric** d_S between the points P and Q — see figure 4.5.

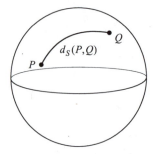

FIGURE 4.5: The spherical metric.

4. Our final example is not geometric in nature. A **binary word** is simply a string of 0s and 1s, such as 0111000110101110. Binary words are used to store and communicate data electronically. The distance between two (equal length) binary words is the number of positions, or bits, in which the words differ. We refer to this as the **binary word distance** or **binary word metric**. For example, if $\mathbf{x} = 10110101$ and $\mathbf{y} = 11010111$ are two 8-bit binary words, then their distance is $d(\mathbf{x}, \mathbf{y}) = 3$ because they differ in the second, third, and seventh bits.

Each of the previous examples gives a sensible interpretation of 'distance' between 'points' where, in the last example, we need to interpret points as binary words. If we are to study 'spaces where we can measure distances' in a general context, then we need to extract the key properties that a function must possess in order to be a (sensible) distance function. Let's suppose we have a set of 'points' X; these could be points in the plane or on a sphere or something less geometric such as binary words. The following are reasonable properties that we would want any distance function, or metric, to possess.

- The distance function d should take a pair of points (x, y) as 'input' and return a result $d(x, y)$ that is a real number.

- The distance $d(x, y)$ should be non-negative, $d(x, y) \geq 0$.

- The distance $d(x, y)$ should only be zero when $x = y$.

- It should not matter whether we measure distance from x to y or from y to x: $d(x, y) = d(y, x)$.

- The last property is a little less obvious but arises from thinking about the lengths of sides of triangles. If we have three points x, y, and z, then the (direct) distance from x to z should be no greater than the distance that goes via y: $d(x, z) \leq d(x, y) + d(y, z)$. This is illustrated in figure 4.6.

It turns out that these properties are just those needed to define a sensible distance function for which (a) we can build a useful theory and (b) there are many examples in different contexts. The formal definition is the following.

Definition 4.1
Let X be a non-empty set. A **metric** on X is a function $d : X \times X \to \mathbb{R}$ satisfying the following three conditions.

(M1) For all $x, y \in X$, $d(x, y) \geq 0$ and $d(x, y) = 0$ if and only if $x = y$.

(M2) For all $x, y \in X$, $d(x, y) = d(y, x)$.

(M3) For all $x, y, z \in X$, $d(x, z) \leq d(x, y) + d(y, z)$.

A set X together with a metric is called a **metric space**.

The condition (M3) is called the **triangle inequality** for the metric d.

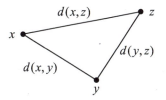

FIGURE 4.6: The triangle inequality.

Although it is not immediately obvious, each of the distance functions in examples 4.3 defines a metric. In each case, it is clear that we have defined the appropriate function $d : X \times X \to \mathbb{R}$. To verify that the function defines a metric we need to prove that it satisfies the three conditions (M1), (M2), and (M3) in definition 4.1. We will establish this for the Manhattan metric in the next theorem.

Theorem 4.7
The function

$$d : \mathbb{R}^2 \times \mathbb{R}^2 \to \mathbb{R}, \ d((x_1, y_1), (x_2, y_2)) = |x_1 - x_2| + |y_1 - y_2|$$

defines a metric on $\mathbb{R}^2$, called the Manhattan metric.

Proof. We show that d satisfies the three properties in definition 4.1.

(M1) Let $(x_1, y_1), (x_2, y_2) \in \mathbb{R}^2$.

     Then $d((x_1, y_1), (x_2, y_2)) = |x_1 - x_2| + |y_1 - y_2| \geq 0$ since the modulus function is non-negative.

     Also $\ \ d((x_1, y_1), (x_2, y_2)) = 0 \ \Leftrightarrow \ |x_1 - x_2| + |y_1 - y_2| = 0$

$$\Leftrightarrow \ |x_1 - x_2| = 0 \text{ and } |y_1 - y_2| = 0$$

$$\Leftrightarrow \ x_1 = x_2 \text{ and } y_1 = y_2$$

$$\Leftrightarrow \ (x_1, y_1) = (x_2, y_2).$$

     Hence d satisfies (M1).

(M2) To establish (M2), we will need the following property of modulus

$$|a - b| = |-(a - b)| = |b - a|. \tag{*}$$

     Let $(x_1, y_1), (x_2, y_2) \in \mathbb{R}^2$.

     Then $\ \ d((x_1, y_1), (x_2, y_2)) = |x_1 - x_2| + |y_1 - y_2|$

$$= |x_2 - x_1| + |y_2 - y_1| \qquad \text{(using (*))}$$

$$= d((x_2, y_2), (x_1, y_1)).$$

     Hence d satisfies (M2).

(M3) To establish (M3), we will need the corresponding property for the modulus function,

$$|a - c| \leq |a - b| + |b - c|. \tag{**}$$

     This follows from the triangle inequality for the modulus function, example 4.2.4.6, page 146, by letting $x = a - b$ and $y = b - c$.

     Let $(x_1, y_1), (x_2, y_2), (x_3, y_3) \in \mathbb{R}^2$.

     Then $\ \ d((x_1, y_1), (x_2, y_2)) + d((x_2, y_2), (x_3, y_3))$

$$= (|x_1 - x_2| + |y_1 - y_2|) + (|x_2 - x_3| + |y_2 - y_3|)$$

$$= (|x_1 - x_2| + |x_2 - x_3|) + (|y_1 - y_2| + |y_2 - y_3|)$$

$$\geq |x_1 - x_3| + |y_1 - y_3| \qquad \text{(using (**))}$$

$$= d((x_1, y_1), (x_3, y_3)).$$

     Hence d satisfies (M3).

Since d satisfies the three conditions (M1), (M2), and (M3), it is a metric on $\mathbb{R}^2$. $\square$

In order to give a flavour of how an abstract theory of metric spaces might be developed that is both independent of individual examples but applying to all metric spaces, we consider the generalizations of discs and spheres in 2-dimensional and 3-dimensional spaces.

A **disc** in $\mathbb{R}^2$ with centre $\mathbf{a} \in \mathbb{R}^2$ and radius $r \in \mathbb{R}^+$ is the set of those points $\mathbf{x}$

whose distance from the centre is less than or equal to the radius. We denote the disc with centre $\mathbf{a}$ and radius r as

$$D(\mathbf{a}, r) = \{\mathbf{x} \in \mathbb{R}^2 : d(\mathbf{x}, \mathbf{a}) \leq r\}.$$

In three-dimensional space, the corresponding concept is the **sphere** $S(\mathbf{a}, r)$ with $\mathbf{a} \in \mathbb{R}^3$ and radius $r \in \mathbb{R}^+$ given by

$$S(\mathbf{a}, r) = \{\mathbf{x} \in \mathbb{R}^3 : d(\mathbf{x}, \mathbf{a}) \leq r\}.$$

Clearly these definitions rely only on the notion of distance and therefore they generalise to any metric space. The corresponding concept in a metric space is that of a 'closed ball', which we now define.

Definition 4.2
Let X be metric space with metric d. Let $a \in X$ and $r \geq 0$ be a real number. The (**closed**) **ball**, centre a and radius r in X is

$$B(a, r) = \{x \in X : d(x, a) \leq r\}.$$

Example 4.4
Using the Manhattan metric for $\mathbb{R}^2$, closed balls are diamond-shaped rather than disc-shaped as in the Euclidean metric. For example, consider the unit ball (that is, unit radius) centred at the origin $\mathbf{0} = (0, 0)$,

$$B(\mathbf{0}, 1) = \{\mathbf{x} \in \mathbb{R}^2 : d(\mathbf{x}, \mathbf{0}) \leq 1\} = \{(x_1, x_2) \in \mathbb{R}^2 : |x_1| + |x_2| \leq 1\}.$$

The ball, shown in figure 4.7, is bounded by four lines:

$$
\begin{aligned}
x_1 + x_2 &= 1 \quad \text{in the first quadrant where } x_1 \geq 0, x_2 \geq 0 \\
-x_1 + x_2 &= 1 \quad \text{in the second quadrant where } x_1 \leq 0, x_2 \geq 0 \\
-x_1 - x_2 &= 1 \quad \text{in the third quadrant where } x_1 \leq 0, x_2 \leq 0 \\
x_1 - x_2 &= 1 \quad \text{in the fourth quadrant where } x_1 \leq 0, x_2 \geq 0.
\end{aligned}
$$

Examples 4.5
In these examples, we see how we can use what is familiar — discs in $\mathbb{R}^2$ coming from the Euclidean metric — to motivate results in the general context. However, we also sound a note of caution. Although the familiar context can motivate the more general case, we also need to be aware that any proofs must only rely on the general context and must not assume anything that may be true in the familiar context but not true more generally.

1. We first use our intuition about when discs in $\mathbb{R}^2$ are disjoint or subsets of one another to state and prove corresponding results for general metric spaces. Motivated by figure 4.8, it should be clear that, for two discs $D(\mathbf{a}, r)$ and $D(\mathbf{b}, s)$ in $\mathbb{R}^2$,

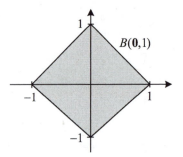

FIGURE 4.7: A closed ball in the Manhattan metric.

(i) if $d(\mathbf{a}, \mathbf{b}) > r + s$, then the discs are disjoint, $D(\mathbf{a}, r) \cap D(\mathbf{b}, s) = \varnothing$;

(ii) if $d(\mathbf{a}, \mathbf{b}) + r < s$, then $D(\mathbf{a}, r) \subseteq D(\mathbf{b}, s)$.

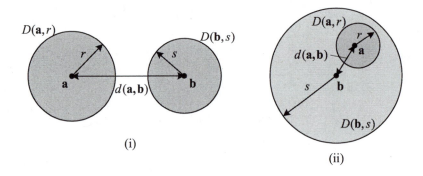

FIGURE 4.8: Relationships between discs in $\mathbb{R}^2$.

Motivated by the situation in $\mathbb{R}^2$ with the Euclidean metric, we now state and prove the corresponding results for a general metric space.

Theorem 4.8

Let X be a metric space with metric d, and let $B(a, r), B(b, s)$ be two closed balls in X.

(i) If $d(a, b) > r + s$, then the balls are disjoint, $B(a, r) \cap B(b, s) = \varnothing$.

(ii) If $d(a, b) + r \leq s$, then $B(a, r) \subseteq B(b, s)$.

Proof.

(i) Let $B(a, r), B(b, s)$ be two closed balls in X and suppose $d(a, b) > r + s$.

To prove that the balls are disjoint, we will show that if an element

$x \in X$ belongs to one of the balls, then it does not belong to the other.

So suppose that $x \in B(a, r)$. This means that $d(x, a) \leq r$. Note that, by (M2), we will use $d(x, a)$ and $d(a, x)$ interchangeably.

By the triangle inequality (M3) we have $d(a, x) + d(x, b) \geq d(a, b)$. Therefore $d(x, b) \geq d(a, b) - d(a, x)$.

Since $d(a, b) > r + s$ and $d(a, x) \leq r$, we have

$$d(x, b) \geq d(a, b) - d(a, x) > (r + s) - r = s$$

so $x \notin B(b, s)$.

Hence the balls are disjoint, $B(a, r) \cap B(b, s) = \varnothing$.

(ii) Let $B(a, r)$ and $B(b, s)$ be two closed balls in X and suppose that $d(a, b) + r \leq s$.

To prove that $B(a, r)$ is a subset of $B(b, s)$ we need to show that, if $x \in B(a, r)$, then $x \in B(b, s)$.

So suppose $x \in B(a, r)$. Then $d(x, a) \leq r$.

Note that, since $d(a, b) + r \leq s$ we have $d(a, b) \leq s - r$.

Again, the triangle inequality (M3) gives $d(x, b) + d(b, a) \leq d(x, a)$. Therefore

$$d(x, b) \leq d(x, a) - d(a, b) \leq r - (r - s) = s$$

since $d(a, b) \leq s - r$.

Therefore $x \in B(b, s)$. We have proved that:

$$x \in B(a, r) \implies x \in B(b, s).$$

Hence $B(a, r) \subseteq B(b, s)$.

□

2. In example 1 (i) above, we showed that if the distance between ball centres is greater than the sum of the radii, then the balls are disjoint. For the Euclidean metric in $\mathbb{R}^2$, the converse is also true: if the balls (discs, in this case) are disjoint, then the distance between their centres must be greater than the sum of the radii.

Surprisingly, perhaps, this is *not* true generally for metric spaces. To show this, we give an example of a metric space and two balls which are disjoint but where the distance between their centres is less than the sum of their radii.

Let X be any non-empty set. We need X to contain at least two elements because we need to define two different balls. We leave it as an exercise (see exercise 4.1.4) to show that the function

$$d : X \times X \to \mathbb{R}, \ d(x,y) = \begin{cases} 0 & \text{if } x = y \\ 1 & \text{if } x \neq y \end{cases}$$

defines a metric on X. This is called the **discrete metric** on X.

Now let $X = \mathbb{R}$ with the discrete metric and consider a ball $B\left(a, \frac{3}{4}\right)$ with center $a \in \mathbb{R}$ and radius $\frac{3}{4}$. If $x \neq a$ then, by definition, $d(x,a) = 1 > \frac{3}{4}$, so $x \notin B\left(a, \frac{3}{4}\right)$. Therefore the only element of X that lies in the ball $B\left(a, \frac{3}{4}\right)$ is a itself (as $d(a, a) = 0$, of course). Hence

$$B\left(a, \tfrac{3}{4}\right) = \{a\}.$$

Now consider the two balls $B\left(0, \frac{3}{4}\right) = \{0\}$ and $B\left(1, \frac{3}{4}\right) = \{1\}$. Clearly the balls have no elements in common, $B\left(0, \frac{3}{4}\right) \cap B\left(1, \frac{3}{4}\right) = \varnothing$. However, the distance between their centres $d(0, 1) = 1$ is less than the sum of radii $\frac{3}{4} + \frac{3}{4} = \frac{3}{2}$. Therefore the converse of theorem 4.8 (i) is false for the discrete metric: if $B(a, r) \cap B(b, s) = \varnothing$ it does not *necessarily* follow that $d(a, b) > r + s$.

Exercises 4.3

1. Prove each of the following properties of the modulus function.

   (i) For all $x \in \mathbb{R}$, $x \leq |x|$.

   (ii) For all $x, y \in \mathbb{R}$, $|xy| = |x||y|$.

   (iii) For all $x, y \in \mathbb{R}$, $||x| - |y|| \leq |x - y|$.
   Hint: begin by applying the triangle inequality for $|\cdot|$ (page 146) to the expression $|(x - y) + y|$.

2. (i) Prove that there exists a real number $\delta > 0$ such that $|x - 1| < \delta \Rightarrow |x^3 - 1| < 2$.

   (ii) Prove that, for all real numbers $\varepsilon > 0$, there exists a real number $\delta > 0$ such that $|x - 1| < \delta \Rightarrow |2x^2 + 3x - 5| < \varepsilon$.

   (iii) Prove that, for all real numbers $\varepsilon > 0$, there exists a real number $\delta > 0$ such that $|x + 2| < \delta \Rightarrow |x^3 + x^2 + 4| < \varepsilon$.

   (iv) Prove that, for all real numbers $\varepsilon > 0$, there exists a real number $\delta > 0$ such that $|x - 1| < \delta \Rightarrow \left|\dfrac{x - 1}{x}\right| < \varepsilon$.

3. Verify that the binary word metric, defined in example 4.3.4, does indeed define a metric on the set of all binary words of some fixed length.

4. Let X be any non-empty set and let d be defined by

$$d : X \times X \to \mathbb{R}, \ d(x, y) = \begin{cases} 0 & \text{if } x = y \\ 1 & \text{if } x \neq y. \end{cases}$$

(i) Show that d is a metric on X. This is called the **discrete metric** on X and (X, d) is called a **discrete metric space**.

(ii) Describe the closed balls in the discrete metric space (X, d).

(iii) In example 4.5.2, we showed that the converse of theorem 4.8 (i) is false for the discrete metric.

Investigate whether theorem 4.8 (ii) is true or false for the discrete metric.

4.5 A more formal framework

In this section, we provide a more formal discussion of the nature of proof. Those readers who are mainly interested in various techniques of proof, rather than an underlying framework for proof, may omit this section.

To understand more fully what mathematicians mean by proof, we first need to look a little more closely at the nature of mathematics itself. From a formal standpoint, mathematics operates according to something known as the 'axiomatic method'. This was first introduced by Euclid over two thousand years ago and has subsequently evolved, particularly during the last two hundred years, into the current modus operandi of mathematics. We begin this section by giving a brief, if somewhat incomplete, description of the axiomatic method which governs the development of any mathematical theory.

In outline, a branch of mathematics starts with a set of premises and proceeds by making deductions from these assumptions using the methods of logic described in chapter 2. The premises are called 'axioms', the statements deduced from them are the theorems, and the sequences of deductions themselves are the proofs of the theorems. The case study on metric spaces given in section 4.4 was intended to illustrate in microcosm the development of a mathematical theory in this way. In the remainder of this section, we shall expand upon and make more precise this overview of the formal description of the mathematics, referring to the metric spaces example as appropriate.

Of course, mathematics is a discipline engaged in by human beings and, in practice, it does not develop in quite such a precise and orderly manner as we have just indicated. In the remaining chapters of this book, we shall consider in more detail how mathematicians really go about exploring their mathematical landscapes.

We begin by considering the Prime Factorisation Theorem 4.3: *every integer greater than 1 can be expressed as a product of prime numbers*. To comprehend the statement, we needed precise meanings for the terms 'divisible', 'prime number', and so on. This shows that any mathematical theory will need precisely stated definitions. However, it is not possible to define all the terms used in a given mathematical theory. A little thought should indicate why this is so. Consider the definition of a prime number given in section 4.2: 'a prime number is an integer greater than 1 which is not divisible by any positive integer except 1 and itself'. This relates the term 'prime number' to more basic concepts such as 'integer', 'positive', the number '1' and 'divisible'. Any definition is like this — it relates the term being defined to other terms. Some or all of these other terms may then be defined using yet more terms, and so on. Clearly, this process of definition must stop somewhere or we would spend all our time defining yet more and more terms and never get round to doing any mathematics proper. Therefore, some terms must be left undefined. This discussion indicates what should be the first ingredients in any axiomatic mathematical theory. They are the **undefined** or **primitive** terms.

In the same way that we cannot define every term, so we cannot prove every result. For example, in our proof of the Prime Factorisation Theorem, we used (implicitly as well as explicitly) various properties of the integers and prime numbers. For the most part, we assumed that these properties were familiar and did not need referring to explicitly. If we were required to prove these properties, the proofs would need to be based on some other statements about the integers, and so on. Again, to avoid an infinite regression, we are forced to have some statements which will not be proved.[1] These are the **axioms** of the theory.

In the example of a metric space given in definition 4.1, the axioms are the properties listed as (M1), (M2), and (M3). In this case, the undefined terms are harder to identify since the notion of a metric space builds on other concepts such as 'set', 'function', and 'real number'.

As we have mentioned, Euclid is generally recognised as the first person to state axioms explicitly in around 300 BC. Just five axioms were the basis for his famous development of geometry. To Euclid, however, axioms did not require proof because they were basic statements about the real physical world

---

[1] The Greek philosopher Aristotle (384–322 BC) was well aware of this. In his Metaphysics, Aristotle wrote, 'Now it is impossible that there should be demonstration of absolutely everything, for there would be an infinite regress, so that even then there would be no proof.' Indeed, he went on to say of those who took the contrary view, 'Such a man, as such, is no better than a vegetable.' See [3] for further details.

which he took to be *self-evidently true*. (The Greek word *axioma* — αξιωμα — means 'that which is thought fitting'.) Although mathematicians no longer view axioms in this way, the Euclidean perspective still lingers in our culture. In non-mathematical discourse or writing we may come across the phrase 'it is axiomatic that . . . ' meaning that what follows is not open to question. To see why mathematicians were forced to abandon the Euclidean view of axioms, we indulge in a brief digression to describe the birth of non-Euclidean geometry.

One of Euclid's axioms, the parallel axiom, states that, for every line ℓ and every point P not lying on ℓ, there exists a line m containing P which is parallel to ℓ in the sense that the two lines never meet; see figure 4.9.

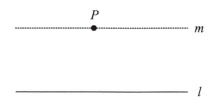

FIGURE 4.9: Euclid's parallel axiom.

To claim that this statement is self-evidently true is problematic; the problem lies in the word 'never' in the statement that the lines ℓ and m never meet. This means that, *no matter how far they are extended*, the lines will not meet. Since it is not possible to extend the lines forever, to claim that the parallel axiom is *self-evidently* true seems at best to be overstating the case. There was enough doubt about this axiom for the mathematical community to spend some two thousand years attempting to show that it could be deduced from Euclid's other axioms. If this could have been achieved, then there would have been no need to include the proposition as an axiom because it would have been a theorem. Eventually, however, it was discovered that the axiom could not be deduced from the remaining Euclidean axioms.

In the first half of the nineteenth century, two young mathematicians, the Hungarian János Bolyai and the Russian Nikolai Lobachevsky,[2] independently of one another found a geometry in which the parallel axiom is false. This new geometry shares the remaining Euclidean axioms and its discovery or invention (depending on your point of view) showed finally that the parallel axiom could not be deduced from the remaining Euclidean axioms. The reason for this is quite simple. If the parallel axiom were deducible from the remaining axioms, then it would be a theorem and so it would not be possible to construct a consistent geometry where the parallel axiom property was false. Since

---

[2] It is highly probable that the great German mathematician Carl Friedrich Gauss shared this discovery. Although he published relatively little, Gauss was the foremost mathematician of his time — many would say the greatest ever — and he was acutely aware of the controversy which would inevitably result from the discovery of non-Euclidean geometry.

Bolyai and Lobachevsky found a geometry in which the parallel axiom was contradicted, it follows that it is not possible to deduce the axiom from the other Euclidean axioms. This geometry of Bolyai and Lobachevsky is usually referred to simply as 'non-Euclidean geometry' even though there are many geometries — for example, the geometry of a sphere — that are different from Euclid's geometry.

The existence of two geometries, one in which the parallel axiom is true and one in which it is false, has certain implications. In particular, it is not possible for both the parallel axiom and its negation to be true, self-evidently or otherwise. Mathematicians were therefore forced to re-think their views of the nature of axioms.

Today, we no longer regard axioms as self-evident truths, but simply as statements about the undefined terms which are taken as assumptions to serve as the basic building blocks of the theory. It is not necessary for axioms to reflect any perceived property of the 'real world', such as the geometry of points and lines in a plane. In principle, we are free to choose any consistent set of axioms as the starting point for a mathematical theory. The requirement of consistency, though, is vitally important. A set of axioms is **consistent** if it is not possible to deduce from the axioms some proposition P as well as its negation $\neg P$. If it were possible to infer both P and $\neg P$, then the axioms contain a hidden self-contradiction which make the system useless. Recall from section 2.5 that if an argument has inconsistent premises, then it is automatically valid no matter what the conclusion — see example 2.14.4 on page 74. When applied to axiom systems, this means that it is possible to deduce any proposition whatsoever from an inconsistent set of axioms. The modern perspective has replaced self-evidence by consistency as the paramount criterion for an axiom system.[3]

We have said that, in principle, any consistent set of axioms can serve as the framework for a mathematical theory. In practice, though, mathematicians do not choose their axiom systems arbitrarily. Some sets of axioms are tailor-made for a particular purpose and others are studied because they have interesting and far-reaching applications. The reasons for studying a particular axiom system lie outside the system itself and relate to possible interpretations of the system. For example, the reason why metric spaces are studied in the abstract (that is, via the axioms given in definition 4.1) is that there are many examples of systems where the axioms are satisfied. Any result we may prove about abstract metric spaces will then apply to every example. For instance, the results in theorem 4.8 apply to closed balls in all metric spaces.

---

[3] Although crucially important, consistency is somewhat elusive. In 1931, the Austrian logician Kurt Gödel showed that any set of axioms for the arithmetic of the positive integers could not formally be proved to be consistent. Since elementary arithmetic is fundamental to just about all of mathematics, this is a rather depressing state of affairs. Although we know that axiom systems must be consistent, we will frequently be unable to prove them to be so.

These properties of balls are used in practice when designing error-correcting codes, where the set of words within a certain distance of codewords (using the binary word metric described in example 4.3.4) is important in determining the error-correcting properties of the code.

We will take a slightly informal view of the notion of an 'axiom system'. In describing and axiom system, we will not require that everything is 'axiomatised'; in other words, we will allow 'standard' mathematical symbols and terminology to be used in describing the system without these having to be described by their own set of axioms. In particular, we will generally assume language of sets and functions as part of this background knowledge that we can draw on. In the definition of a metric space, for example, we used the notions of sets and set membership, the Cartesian product, functions, the real numbers etc. Roughly speaking, we will assume the logical framework introduced in chapter 2 and the theory of sets and functions introduced in chapter 3 as forming this 'standard' mathematical background that we will draw on freely in describing axiom systems.

Definition 4.3
An **axiom system** comprises a collection of undefined terms, which may be words, phrases or symbols, and a collection of statements, called **axioms**, involving the undefined terms (as well as 'standard' mathematical symbols and terminology).

A mathematical theory can now be described as the evolution of an axiom system by the use of deductive reasoning to prove theorems about the terms of the system. Definitions can be, and in practice always are, introduced to smooth the flow of the theory. They serve to simplify notation. In principle, definitions are unnecessary. In practice, we could never get very far if we had only the language of the undefined terms to use. For example, once we have introduced the definition of a prime number we can use this concise term freely without having to refer constantly to an 'integer greater than 1 which is not divisible by any positive integer other than 1 and itself'. The basic core of the theory is its theorems and their proofs which we consider in more detail below.

There is an analogy for an axiom system which may prove helpful here. We could loosely compare the development of an axiomatic mathematical theory with the construction of a building from, say, bricks and mortar. The raw materials — sand, cement, clay and so on — are like the undefined terms and symbols of the system. The first layer of bricks forming the foundations of the building represents the axioms. It is vitally important that this first layer of bricks is laid properly if any building constructed on top is not to collapse. This is analogous to the consistency requirement of the axioms — if the axioms are inconsistent, then any theory developed from them will 'collapse'. At this stage, of course, there is no building but only foundations together with raw materials and rules which will permit a building to be constructed. So it is with an axiom system — the system itself is just the basic framework from which a theory can be developed. A building rises from its foundations by brick being laid on top of brick using mortar to hold the structure in place. In the mathematical context, each individual brick could be likened to a theorem and the mortar holding it firmly in place is its proof.

When defining an axiom system, we often have an interpretation of the axioms in mind; that is, a situation or situations where the undefined terms are given meanings such that the axioms are true propositions. Before defining a metric space, we introduced four specific situations of a set together with a distance function. Each of these may be regarded as an interpretation of the abstract notion of a metric space. In theorem 4.7, we proved this for the Manhattan metric by showing that each of the axioms (M1), (M2), and (M3) was satisfied in this case. The following definition captures this idea.

Definition 4.4
An **interpretation** of an axiom system is a situation where the undefined terms of the system are given a meaning. An interpretation is called a **model** of the axiom system if the axioms, when interpreted according to the given meanings, are true propositions.

Note that it is in an interpretation of the system where meaning is introduced and we may think of models as 'correct' interpretations. Models are the raison d'être of axiom systems. The reason that axiom systems are useful is because they provide information about their models, which is where meaning resides and hence is where our interest lies. Indeed, it is the models which really determine which axioms systems are studied. We have said that, in principle, any consistent axiom system is just as valid or worthy of study as any other. In practice though, some axiom systems are more important and hence more deeply studied than others. The importance of any axiom system lies in its models and not in some intrinsic property of the system itself.

Axiom systems studied by mathematicians fall into one of two main categories which serve separate purposes. Some axiom systems, like those of a metric space, have many different models. In addition to metric spaces, examples which some readers may have encountered are the axiom systems for various kinds of algebraic objects such as groups, rings, fields, vector spaces, Boolean algebras, monoids, and the like. In each of these cases there are many examples of the particular algebraic structure. Each example is a model of the axiom system. An important advantage in studying the axiom system in such cases is that of economy of labour. If we can prove some theorem directly from the axioms, then it must be the case that the theorem becomes a true proposition in every model of the axiom system. Thus we will know that every example of the particular algebraic structure will possess whatever property is described by the theorem. For instance, using the group theory axioms (see definition 5.4

on page 201 below) it is not too difficult to prove that inverses are unique. (This proof is given in chapter 7; see theorem 7.15 on page 298.) From this we know that, in every example of a group, inverses are unique and we do not need to prove this fact for each and every group.

The second category of axiom systems commonly studied comprise those which have essentially only one model. In other words, all models are for all practical purposes the same. (The notion of two models being 'essentially the same' is one which can be made completely precise. The word mathematicians use for this notion is **isomorphic**, which is derived from Greek and means literally 'having the same shape or form.' The details of how 'being essentially the same' is given a precise meaning need not concern us here.) Usually in these cases, the model is a familiar structure such as the set of integers or the set of real numbers or the Euclidean geometry of two- or three-dimensional space. Here, the purpose of using axiom systems is rather different. The axioms represent a few basic properties of the given structure from which it is possible to deduce many other properties of the structure (the theorems of the system). For example, there is an axiom system with thirteen axioms, describing an algebraic object called a 'complete ordered field'. It can be shown that there is essentially only one example of a complete ordered field in the sense that all models of the axiom system are equivalent in a very precise way. 'The' example is the set $\mathbb{R}$ of real numbers together with the operations of addition and multiplication as well as the usual less-than-or-equal-to ordering, $\leq$, of real numbers. What this means is that all the usual properties of the real numbers can be deduced from just thirteen axioms. The advantage of the axiomatic approach is that we need assume only a limited number of properties, as the remainder can be rigorously deduced from these. It also means that, in a sense, the thirteen axioms define what we mean by the system of real numbers, i.e., the axioms characterise the real number system.

We now turn to a consideration of how a mathematical theory based on an axiom system develops by proving theorems using the rules of logic. We need first to define precisely what we mean by a theorem. Consider, for example, the theorem known to anyone who has studied elementary geometry: *the sum of the angles of any triangle is* 180°. In fact, this is a theorem of Euclidean geometry, but is not a theorem of the non-Euclidean geometry described by Bolyai and Lobachevsky. Non-Euclidean geometry has a different set of axioms from Euclidean geometry; Euclid's parallel axiom is replaced in non-Euclidean geometry by a different axiom concerning parallel lines. Hence the two axiom systems have different sets of theorems. In fact, the corresponding theorem in the non-Euclidean geometry described by Bolyai and Lobachevsky states: *all triangles have angle sum less than* 180°. The important point here is that what is or is not a theorem depends on the particular axiom system.

Informally, a theorem in an axiom system is a result that 'can be deduced from' the axioms and any background knowledge that is assumed. We now wish to consider this informal notion in a little more detail. In some systems, such as propositional logic, the deduction rules can be made very precise. Indeed, we have given a description of this in section 2.5. Below we describe, much less formally than for propositional logic, some of the deduction rules that we may apply when reasoning in an axiom system.

We will use K to denote our background knowledge. We do not wish to make precise exactly what is included in K as this will depend on the context. However, it will usually include things like the properties of equality (such as, if $a = b$ and $b = c$, then $a = c$), simple properties of sets (such as, if $A \subseteq B$ and $B \subseteq A$, then $A = B$), simple properties of the integers (such as $a + b = b + a$) and so forth.

In section 2.5, we used the notation $\{P_1, P_2, \ldots, P_n\} \vdash Q$ to denote that the proposition Q can be deduced from the propositions $P_1, P_2, \ldots, P_n$ using the deduction rules for propositional logic. We now describe something akin to this for axiom systems. Let Γ be a set of statements and let T be another statement within the current axiom system. We use $\Gamma \rightsquigarrow T$ to denote that T 'can be deduced from' the background knowledge K and the statements in Γ. This is not standard mathematical notation. We will read this 'squiggly' arrow $\rightsquigarrow$ as 'can be deduced from' or simply 'follows from'. Shortly we will outline some of the deduction rules that will apply in axiom systems. First we wish to introduce the notions of *theorem* and *proof*. A theorem is simply a proposition that can be deduced in the axiom system and a proof is a sequence of deductions that shows this.

Definition 4.5

Let $A = \{A_1, A_2, \ldots, A_k\}$ be a collection of axioms that comprises an axiom system and let K be the assumed background knowledge within which the axiom system sits.

(i) A **theorem** in the axiom system is a statement T that may be deduced from the axioms of the system, $A \rightsquigarrow T$.

(ii) A **proof** of a theorem T is a sequence $P_1, P_2, \ldots, P_n$ of statements such that, $P_n = T$ and, for $r = 1, \ldots, n$, either P_r is an axiom or it is deducible from the previous statements in the proof:

$$P_r \in A \quad \text{or} \quad A, \{P_1, \ldots, P_{r-1}\} \rightsquigarrow P_r.$$

Note that, according to this definition, the first step in a proof will always be an axiom because there are no prior statements from which it may be deduced. This is analogous to our formal deductions in proposition logic where the first propositions to appear in the deductions were always premises. Later, we will modify this notion slightly, to allow theorems that we have previously proved to appear as statements in the proof rather than having always to begin right back with the axioms.

In section 2.5 (page 67), we introduced deduction rules for propositional logic that allow us to establish logical implication of propositions. We now consider a similar collection of deduction rules for axiom systems. We may regard these rules as describing some of the properties of the 'follows from' relation $\rightsquigarrow$. The rules are described in an analogous way to the deduction rules for propositional logic. Each rule has two lines; on the first line we have a collection of 'follows from' statements and on the second line is a single 'follows from' statement. The rule is intended to capture the fact that, if we can establish each of the 'follows from' statements in the first line, then we may deduce the 'follows from' statement of the second line. As we shall see, these rules underpin some of the methods of proof that are considered elsewhere in this text.

Conjunction Rule

This rule says that if we can separately deduce both P and Q from the same set of statements Γ, then we may deduce their conjunction $P \wedge Q$.

$$\frac{\Gamma \rightsquigarrow P; \ \Gamma \rightsquigarrow Q}{\Gamma \rightsquigarrow P \wedge Q}$$

We used this rule in our proof in theorem 4.1 where, to prove

the sum and difference of two odd integers is even,

we first separated this out into two separate statements

the sum of two odd integers is even

and *the difference of two odd integers is even.*

We then proved each of these statements individually.

Conditional Rule

This rule says that if we can deduce Q from Γ and P, then we can deduce the conditional $P \Rightarrow Q$ from Γ.

$$\frac{\Gamma, P \rightsquigarrow Q}{\Gamma \rightsquigarrow (P \Rightarrow Q)}$$

This rule underpins the method of direct proof of a conditional $P \Rightarrow Q$. It says that, to deduce $P \Rightarrow Q$ from a collection of statements Γ, we can instead add P to our assumptions and from this deduce Q. This was precisely the method of direct proof of $P \Rightarrow Q$: assume P and deduce Q.

Modus Ponens Rule

The next rule is one that allows us to make step-by-step deductions in a proof. It states that if we can deduce P from Γ and we can separately deduce Q from Γ and P, then we can deduce Q from Γ.

$$\frac{\Gamma \rightsquigarrow P; \ \Gamma, P \rightsquigarrow Q}{\Gamma \rightsquigarrow Q}$$

We have called this the Modus Ponens rule although it does not quite mirror Modus Ponens for propositional logic.

$$\frac{P; \ P \Rightarrow Q}{Q}$$

However, using the Conditional rule, from $\Gamma, P \rightsquigarrow Q$ we may deduce $\Gamma \rightsquigarrow (P \Rightarrow Q)$. Were we to replace $\Gamma, P \rightsquigarrow Q$ with $\Gamma \rightsquigarrow (P \Rightarrow Q)$ in the statement of the rule, then it *would* look like the corresponding rule in propositional logic.

$$\frac{\Gamma \rightsquigarrow P; \ \Gamma \rightsquigarrow (P \Rightarrow Q)}{\Gamma \rightsquigarrow Q}$$

In exercise 4.4.1 we show that the 'converse' of the Conditional rule also holds. In other words, if we can deduce $P \Rightarrow Q$ from Γ, then we can deduce Q from Γ and P. Hence these two deductions are essentially equivalent.

Adding Assumptions Rule

This rule says that if we can deduce P from Γ, then we can also deduce P from Γ together with some additional statements Γ_1.

$$\frac{\Gamma \rightsquigarrow P}{\Gamma, \Gamma_1 \rightsquigarrow P}$$

Equivalence Rule

This rule says that if P and Q are logically equivalent, and if we can deduce P from Γ, then we can deduce Q from Γ.

$$\frac{\Gamma \rightsquigarrow P, \ P \equiv Q}{\Gamma \rightsquigarrow Q}$$

This rule is intuitively obvious because, if $P \equiv Q$, then P and Q are true under exactly the same circumstances. In fact, this rule underpins several methods of proof that will be considered in later chapters.

Cases Rule

This rule says that if we can deduce T separately from Γ and P and also from Γ and Q, then we can deduce T from Γ, and $P \vee Q$.

$$\frac{\Gamma, P \rightsquigarrow T; \ \Gamma, Q \rightsquigarrow T}{\Gamma, P \vee Q \rightsquigarrow T}$$

We may think of P and Q in the rule as covering two separate cases; for example, $x \geq 0$ and $x \leq 0$ in the case of real numbers. The rule then says that if we can deduce a theorem T in each of the two cases, then we can deduce T in the 'combined' case $P \vee Q$.

We used this (with four cases rather than two) in example 4.2.1, in the proof of the triangle inequality for the modulus function $|x + y| \leq |x| + |y|$. We actually gave separate mini-proofs for each of four separate cases (each of which also assumed $x \leq y$). Letting T stand for $|x+y| \leq |x|+|y|$, we separately established

1. $(0 \leq x \leq y) \rightsquigarrow T$
2. $(x < 0 \leq y \wedge x + y \geq 0) \rightsquigarrow T$
3. $(x < 0 \leq y \wedge x + y < 0) \rightsquigarrow T$
4. $(x \leq y < 0) \rightsquigarrow T$.

Note that some of the rules above mirror precisely deduction rules for propositional logic. For example, compare the Conjunction rules in propositional logic and axiom systems.

$$\frac{P, \ Q}{P \wedge Q} \qquad \frac{\Gamma \rightsquigarrow P; \ \Gamma \rightsquigarrow Q}{\Gamma \rightsquigarrow P \wedge Q}$$

Similarly, compare the Cases rule above with the Rule of case analysis 2 given in exercise 2.4.2 (iv).

$$\frac{P \Rightarrow R, \ Q \Rightarrow R}{(P \vee Q) \Rightarrow R} \qquad \frac{\Gamma, P \rightsquigarrow T; \ \Gamma, Q \rightsquigarrow T}{\Gamma, P \vee Q \rightsquigarrow T}$$

Rather than giving an 'axiom system version' for each of the rules of deduction in propositional logic individually, instead we give a generic rule that allows any of the propositional logic deduction rules to be mirrored in an axiom system.

Propositional Logic Rule

Suppose that there is a propositional logic deduction rule of the following form.

$$\frac{P_1, P_2, \ldots, P_n}{Q}$$

Then the following is a deduction rule in an axiom system.

$$\frac{\Gamma \rightsquigarrow P_1, \ \Gamma \rightsquigarrow P_2, \ \ldots, \ \Gamma \rightsquigarrow P_n}{\Gamma \rightsquigarrow Q}$$

We have described a proof of a theorem T in an axiom system A as a sequence of propositions $P_1, P_2, \ldots, P_n = T$ where each is either an axiom, $P_r \in A$, or can be deduced from earlier propositions in the proof $\{P_1, \ldots, P_{r-1}\} \rightsquigarrow$

P_r. However, a glance at any of the proofs in this chapter, will reveal that mathematical proofs are not set out as a sequence of numbered steps in the way we did for deductions in propositional logic. Similarly, they do not always start with axioms; indeed, rarely do proofs begin with axioms. Instead, proofs are written much less formally and the starting point is usually some piece of background knowledge or some theorem that has already been proved. We have allowed for this in our description of the deduction rules above by allowing a non-specified collection of statements Γ (together with background knowledge) in each of the deduction rules.

Given a collection of axioms A, the first theorems proved will, of necessity, depend directly on the axioms. In other words, for the first theorem(s) proved we will have $\Gamma = A$. However, thereafter, any theorems that we have already proved can be added to Γ. In this way we may regard Γ as being (selected from) the axioms A and all of the previously proved theorems.

We conclude this chapter with a brief consideration of how our deduction rules shape the structure of some proofs. Consider the Modus Ponens rule:

$$\frac{\Gamma \rightsquigarrow P; \ \Gamma, P \rightsquigarrow Q}{\Gamma \rightsquigarrow Q}$$

We may regard the notation $\Gamma \rightsquigarrow P$ as meaning that there is a sequence of steps that means we can deduce P from Γ.

$$\Gamma$$
$$\vdots \text{(sequence of steps 1)}$$
$$P$$

Similarly $\Gamma, P \rightsquigarrow Q$ means there is a second sequence of deduction steps beginning with Γ and P ending with Q.

$$\Gamma$$
$$P$$
$$\vdots \text{(sequence of steps 2)}$$
$$Q$$

We can then splice these two sequences of deduction steps together to establish $\Gamma \rightsquigarrow Q$.

$$\Gamma$$
$$\vdots \text{(sequence of steps 1)}$$
$$P$$
$$\vdots \text{(sequence of steps 2)}$$
$$Q$$

This really says that we may continue to make deductions that build on earlier parts of the proof. If we have deduced P in the first half of the proof, we may use it in any deductions we may make in the second half of the proof.

As we have noted, the Conditional rule underpins the method of direct proof of $P \Rightarrow Q$. Suppose that there is a sequence of deductions that shows $\Gamma, P \rightsquigarrow Q$. Then the Conditional rule gives $\Gamma \rightsquigarrow (P \Rightarrow Q)$. Table 4.1 illustrates the Conditional rule alongside how a direct proof of $P \Rightarrow Q$ might be structured.

Formal structure	**'Real' proof**
Γ	
P	Assume P
$\vdots$ (sequence of steps)	$\vdots$ (sequence of steps)
Q	Q
$P \Rightarrow Q$ (conditional rule)	Hence $P \Rightarrow Q$

TABLE 4.1: Comparing Conditional rule with direct proof of $P \Rightarrow Q$.

Finally, we consider the Cases rule

$$\frac{\Gamma, P \rightsquigarrow T; \ \ \Gamma, Q \rightsquigarrow T}{\Gamma, P \vee Q \rightsquigarrow T}$$

and relate this to a 'real' proof using cases. The notation $\Gamma, P \rightsquigarrow T$ means there is a sequence of deduction steps starting with Γ and P and ending with T.

$$\Gamma$$
$$P$$
$$\vdots \text{ (sequence of steps 1)}$$
$$T$$

Similarly, $\Gamma, Q \rightsquigarrow T$ means there is a sequence of deduction steps starting with Γ and Q and ending with T.

$$\Gamma$$
$$Q$$
$$\vdots \text{ (sequence of steps 2)}$$
$$T$$

Combining these together, one after the other, together with the Cases rule allows us to deduce $\Gamma, P \vee Q \rightsquigarrow T$. This is illustrated in the left-hand column of Table 4.2; the right-hand column shows how a 'real' proof with two cases might be structured.

Formal structure	'Real' proof
Γ	There are two cases.
P	Case 1: P
$\vdots$ (sequence of steps 1)	$\vdots$ (sequence of steps 1)
T	T
Q	Case 2: Q
$\vdots$ (sequence of steps 2)	$\vdots$ (sequence of steps 2)
T	T
$(P \vee Q) \Rightarrow T$ (proof by Cases rule).	Hence in both cases T.

TABLE 4.2: Comparing formal structure and real proof using cases.

Exercises 4.4

1. Show that the 'converse' of the Conditional deduction rule is also a deduction rule. In other words, show that if Q can be deduced from Γ and P, then $P \Rightarrow Q$ can be deduced from Γ.

$$\frac{\Gamma \rightsquigarrow (P \Rightarrow Q)}{\Gamma, P \rightsquigarrow Q}$$

2. Using the deduction rules in this section, establish the following Modus Tollens rule.

$$\frac{\Gamma \rightsquigarrow \neg Q; \ \Gamma, P \rightsquigarrow Q}{\Gamma \rightsquigarrow \neg P}$$

3. Construct a table, similar to Tables 4.1 and 4.2, that compares the formal version of a proof of $P \wedge Q$ using the Conjunction rule with the structure of a real proof of a proposition of the form $P \wedge Q$.

Chapter 5

Finding Proofs

5.1 Direct proof route maps

In section 4.3, we developed the notion of a direct proof of a conditional proposition $P \Rightarrow Q$. According to the method, described on page 142, we assume P and from this deduce Q. In practice, of course, the challenge is to find the chain of deductions that takes us from P to Q.

We will use the analogy of taking a car journey, say, from point A to point B when we are not completely familiar with the road network. Before setting out on the journey we need to plan the route from A to B. There are various techniques that we might deploy in deciding which route to take. Suppose we have a road atlas but the locations A and B are on different pages so that we cannot 'see' the whole journey in a single place. We might find some intermediate location, C say, and then plan two 'sub-journeys': one from A to C and the other from C to B. Each of these sub-journeys might also be subdivided further. In deciding which location C to select, we might reason along the following lines: 'if I can find a route to C, then I can see how I can complete the journey because I can see how to get from C to my destination B'. We will see that these techniques are analogous to those we may apply in finding proofs.

With the advent of modern technology, there are other aids that help us find a route from A to B. For example, we might have a satellite navigation system, either fitted to the car or as a mobile device. Although stories abound of 'sat nav' systems taking motorists on inappropriate routes, many people rely on them. There are also mapping services available online, such as Google maps, that will also find a route from A to B (at least when A and B are on the same land mass).

Returning to the notion of proof, we can describe the process of finding a direct proof of $P \Rightarrow Q$ as finding a 'route of deductions' from P to Q. We will know the starting point of the journey, P, and the destination, Q, but the route itself will be unknown. Worse still, we will not have at our disposal a sat nav or Google maps that will find a route 'automatically' and we will probably not even have a complete road atlas. To develop proof-finding skills, we need to develop the equivalent of our own 'sat nav skills' but in the world of logical deductions rather than that of a road network. We might call this developing our 'proof navigation' skills, or 'proof nav', skills. There is no algorithm or process that will always find a route of deductions from P to Q. Instead, we will need to develop strategies and techniques that we can deploy in different situations and, crucially, we will need to develop a lot of experience of finding proofs in different circumstances.

In the situation where we are trying to find a direct proof of a proposition Q (as opposed to a conditional proposition $P \Rightarrow Q$), there is a further complication: we know the destination Q but we are not given the starting point of the journey of deductions. For a direct proof of Q, described on page 144, we must start with *any* appropriate background knowledge and then proceed to deduce Q. So, in addition to finding a route that ends at Q, we must also select a suitable starting point for the journey. Often this can be difficult, especially when there is no obvious piece of background knowledge that will enable us to start the chain of deductions.

Example 5.1

Consider the following statement.

> *For all positive real numbers* x, $x + \dfrac{1}{x} \geq 2$.

The obvious questions to ask are: Is the statement true? If so, how do we prove it? To get a feel for the first question, we try some examples:

$$x = 1: \quad x + \frac{1}{x} = 1 + 1 = 2 \geq 2$$

$$x = 2: \quad x + \frac{1}{x} = 2 + \tfrac{1}{2} \geq 2$$

$$x = \tfrac{2}{3}: \quad x + \frac{1}{x} = \tfrac{2}{3} + \frac{1}{\tfrac{2}{3}} = \tfrac{2}{3} + \tfrac{3}{2} \geq 2$$

$$x = \pi: \quad x + \frac{1}{x} = \pi + \frac{1}{\pi} > \pi \geq 2.$$

In each of these (few) examples, the result is true. Since there does not seem to be anything special about the examples, we may start to have some confidence that the result is true and turn our attention to the second question.

As we described earlier, we need to find a route, comprising statements that follow logically from one another, from the starting point 'x is a positive real number' to the desired conclusion '$x + 1/x \geq 2$'. The problem here is the generality of the starting point. Some of the initial deductions we could make from $x > 0$ are, for example, $1/x > 0$, $2x > 0$ or $\sqrt{x} > 0$. These are illustrated in figure 5.1. However, it is not very clear that any of these is a useful starting point.

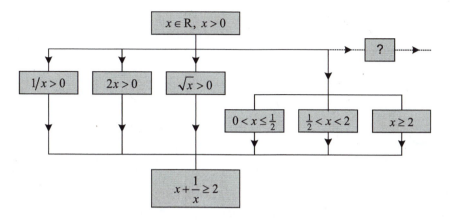

FIGURE 5.1: Some possible starting points in the proof.

Another possibility, also illustrated in figure 5.1, is that we could consider cases. If $x \geq 2$, then $x + 1/x \geq 2$. If $0 < x \leq \frac{1}{2}$, then $1/x \geq 2$ so again $x + 1/x \geq 2$. It might be tempting, therefore, to separate the proof into three cases:

$$0 < x \leq \tfrac{1}{2}, \quad \tfrac{1}{2} < x < 2, \quad \text{and} \quad x \geq 2.$$

As we have indicated, in the first and third cases it is easy to complete the proof. However, it is not so obvious how to proceed in the second case. Indeed, it is not clear that this middle case, $\frac{1}{2} < x < 2$, is any easier to deal with than the general case $x > 0$. Hence the consideration of cases is probably not very helpful either.

At this stage, we might be starting to become concerned as we cannot see a route of deductions from $x > 0$ to $x + 1/x \geq 2$. A useful technique in such cases is to 'work backwards' from the final result. In other words, think of a proposition R such that, *if we could prove R*, then we could deduce the final result. If we can do that, can we then find another proposition S such that, if we could prove S, then we could deduce R and hence complete the proof? Continuing in this way, we would hope to be able to work backwards far enough that we could see how to connect up our starting position, $x > 0$, with where we have reached working backwards.

In our example, as a first step, if we could show that $x^2 + 1 \geq 2x$, then since x is positive, we could divide through by x to obtain $x + 1/x \geq 2$. This is illustrated in figure 5.2. In fact, the way we make this 'backwards step' is probably to reason *in the wrong direction* by making the deduction

$$x + \frac{1}{x} \geq 2 \;\Rightarrow\; x^2 + 1 \geq 2x.$$

If we do this, we must always be careful to check that the reasoning 'goes the other' way as well. In our case, this works. The implication $x + 1/x \geq 2 \Rightarrow x^2 + 1 \geq 2x$ follows because we can multiply the inequality by the positive quantity x. The reverse implication $x^2 + 1 \geq 2x \Rightarrow x + 1/x \geq 2$ follows similarly: we can multiply the inequality by the positive quantity $1/x$.

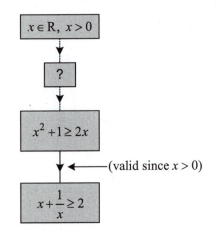

FIGURE 5.2: Working backwards from the final result.

We continue working backwards. Firstly, note that

$$x^2 + 1 \geq 2x \;\Rightarrow\; x^2 - 2x + 1 \geq 0$$

and this implication also reverses. Now the left-hand side $x^2 - 2x + 1 = (x-1)^2$ and we know that the square of any real number is greater than or equal to zero. In other words, we have found our starting deduction, the first step in the proof, $(x-1)^2 \geq 0$. By working backwards, we have obtained a statement that we know to be true. We can now work forwards to construct the proof itself as follows.

Proof. Let x be a positive real number. Then $x - 1$ is also a real number, so $(x-1)^2 \geq 0$. Hence we may proceed as follows.

$$
\begin{aligned}
(x-1)^2 \geq 0 \;&\Rightarrow\; x^2 - 2x + 1 \geq 0 \\
&\Rightarrow\; x^2 + 1 \geq 2x \\
&\Rightarrow\; x + \frac{1}{x} \geq 2 \quad \text{(dividing by } x > 0\text{)}
\end{aligned}
$$

$\square$

The actual proof itself is straightforward. It just uses a little algebra and the facts that (a) the square of a real number is non-negative and (b) multiplying or dividing an inequality by a positive quantity maintains the inequality. However, what is not at all obvious is that the first step in the proof should be the observation that $(x-1)^2 \geq 0$. We can only really see that this is a helpful first step using the process of working backwards from the desired conclusion.

Example 5.2

Investigate the truth or otherwise of the statement:

> *for all positive integers k, the sum of any k consecutive positive integers is divisible by k.*

If the statement is true, give a proof. If the statement is not true, describe the most general situation where it is true and provide a proof in this case.

Solution

We begin investigating this situation by considering some examples.

$k = 3$ We consider examples of sums of three consecutive integers:
$2 + 3 + 4 = 9$ which is divisible by 3,
$3 + 4 + 5 = 12$ which is divisible by 3,
$5 + 6 + 7 = 18$ which is divisible by 3,
$17 + 18 + 19 = 54$ which is divisible by 3.
Hence, for these examples, the result holds.

$k = 4$ Since $2 + 3 + 4 + 5 = 14$, which is not divisible by 4, the result is false in general for $k = 4$.

$k = 5$ We consider examples of sums of five consecutive integers:
$2 + 3 + 4 + 5 + 6 = 20$ which is divisible by 5,
$3 + 4 + 5 + 6 + 7 = 25$ which is divisible by 5,
$5 + 6 + 7 + 8 + 9 = 35$ which is divisible by 5,
$17 + 18 + 19 + 20 + 21 = 95$ which is divisible by 5.
Hence, for these examples, the result holds.

$k = 6$ Since $2 + 3 + 4 + 5 + 6 + 7 = 27$, which is not divisible by 6, the result is false in general for $k = 6$.

At this stage, we *know* that the result is not true for all positive integers k and we may *believe* that, when k is odd, the sum of any k consecutive positive integers is divisible by k. However, we need to be a little careful: the examples above suggest that the result holds for the odd prime numbers $k = 3$ and 5 but not for the even composite (non-prime) numbers $k = 4$ and 6. So perhaps the result depends on whether or not k is a prime rather than just any odd integer. To test this, we consider further examples: the only even prime number, 2, and a non-prime odd number 9.

$k = 2$ Since $2 + 3 = 5$, which is not divisible by 2, the result is false in general for $k = 2$.

$k = 9$ We consider examples of sums of nine consecutive integers:
$2 + 3 + 4 + 5 + 6 + 7 + 8 + 9 + 10 = 54$ which is divisible by 9,
$3 + 4 + 5 + 6 + 7 + 8 + 9 + 10 + 11 = 63$ which is divisible by 9,
$5 + 6 + 7 + 8 + 9 + 10 + 11 + 12 + 13 = 81$ which is divisible by 9,
$17 + 18 + 19 + 20 + 21 + 22 + 23 + 24 + 25 = 189$ which is divisible by 9.
Hence, for these examples, the result holds.

These examples are consistent with our conjecture that the result holds for odd positive integers only. We can now state our conjecture:

for all positive integers k, if k is odd, then the sum of any k consecutive positive integers is divisible by k.

We will give two proofs of this drawing on different background knowledge. The first proof builds directly on the examples we have considered where k is odd. This is very common: careful consideration of well-chosen examples can often lead to an argument that applies generally. Our first proof uses the observation, which the sharp-eyed reader will have spotted from our examples, that in any sum of an odd number of consecutive integers, there is a middle integer, and those either side of the middle integer may be combined into pairs with the same sum. For example, the middle integer of

$$17 + 18 + 19 + 20 + 21 + 22 + 23 + 24 + 25$$

is 21 and, either side of 21, working outwards, we have $20 + 22 = 42$, $19 + 23 = 42$, $18 + 24 = 42$, and $17 + 25 = 42$.

Proof 1. Let k be an odd positive integer.

The case $k = 1$ is trivial since every positive integer is divisible by 1, so we suppose $k \geq 3$.

Then $k = 2m + 1$ for some positive integer m.

Any sum S of $k = 2m + 1$ consecutive positive integers has a middle integer n, say, and m integers either side of this, thus:

$$S = (n - m) + \cdots + (n - 1) + n + (n + 1) + \cdots + (n + m).$$

Pairing off each $n - r$ less than n with the corresponding $n + r$ greater than

n gives

$$
\begin{aligned}
S &= (n - m) + \cdots + (n - 1) + n + (n + 1) + \cdots + (n + m) \\
&= [(n - m) + (n + m)] + \cdots + [(n - 1) + (n + 1)] + n \\
&= 2n + \cdots + 2n + n \\
&= 2mn + n \qquad \text{(since there are } m \text{ 'paired' terms)} \\
&= (2m + 1)n \\
&= kn.
\end{aligned}
$$

Therefore $S = kn$ is divisible by k. $\qquad\qquad\square$

Our second proof is both more straightforward and more sophisticated. It is more straightforward because it does not rely on spotting that a sum of $2m+1$ consecutive positive integers has a 'middle' integer together with integers on either side that may be paired off appropriately. It is more sophisticated since it assumes more background knowledge; in particular, is uses the following formula for the sum of the first r positive integers

$$
1 + 2 + \cdots + r = \tfrac{1}{2}r(r + 1). \qquad\qquad (*)
$$

Proof 2. First note that any sum of k consecutive integers can be written as follows, where n is the first (that is, smallest) integer in the sum:

$$
n + (n + 1) + \cdots + (n + k - 1).
$$

Let k be an odd positive integer. Then $k = 2m + 1$ for some positive integer m.

A sum S of $k = 2m + 1$ consecutive positive integers can be written as follows (by replacing k with $2m + 1$ in the expression above).

$$
\begin{aligned}
S &= n + (n + 1) \cdots + (n + 2m) \\
&= (n + n + \cdots + n) + (1 + 2 + \cdots + 2m) \qquad \text{(rearranging)} \\
&= kn + \tfrac{1}{2} \times 2m \times (2m + 1) \qquad\qquad \text{(using } (*) \text{ above)} \\
&= kn + m(2m + 1) \qquad\qquad\qquad \text{(simplifying)} \\
&= k(n + m) \qquad\qquad\qquad\qquad\quad \text{(since } k = 2m + 1\text{).}
\end{aligned}
$$

Therefore $S = k(n + m)$ is divisible by k. $\qquad\qquad\square$

Euler's phi function and the Smarandache function

In this section we explore proof-finding techniques using two functions defined on the set of positive integers. The definition of each function depends on the notion of divisibility: recall that, for $m, n \in \mathbb{Z}^+$, we say m **divides** n if $n = km$ for some integer k; in this case we also say that m is a **factor** of n. If $r \in \mathbb{Z}^+$ divides both m and n, then r is a common factor of m and n. The **greatest common divisor**, or **highest common factor**, of m and n is the largest integer that divides them both; we use $\gcd(m, n)$ to denote the greatest common divisor of m and n. Finally, we say that m and n are **coprime** if their greatest common divisor is 1, $\gcd(m, n) = 1$. For example, the greatest common divisor of 18 and 45 is 9, $\gcd(18, 45) = 9$; in particular 18 and 45 are not coprime. However, 18 and 35 are coprime as they have no common divisors greater than 1.

Definition 5.1
Euler's phi function, also known as Euler's **totient function**, $\phi : \mathbb{Z}^+ \to \mathbb{Z}^+$ is defined by

$$\phi(n) = \text{number of integers } 1 \leq a \leq n \text{ coprime with } n$$
$$= |\{a \in \mathbb{Z}^+ : 1 \leq a \leq n \text{ and } \gcd(a, n) = 1\}|.$$

Here are some examples:

$\phi(6) = 2$ since there are 2 integers coprime with 6, namely 1 and 5,
$\phi(7) = 6$ since there are 6 integers coprime with 7, namely $1, 2, 3, 4, 5, 6$,
$\phi(8) = 4$ since there are 4 integers coprime with 8, namely $1, 3, 5, 7$,
$\phi(9) = 6$ since there are 6 integers coprime with 9, namely $1, 2, 4, 5, 7, 8$.

The function was introduced by Euler in the eighteenth century and has been much studied since. Indeed, the properties of the function play a crucial role in the so-called RSA public key encryption system that is used, for example, to encrypt data sent via the Internet. We will consider two properties of the function, the first of which is very easy to prove.

Theorem 5.1
If p is prime, then $\phi(p) = p - 1$.

Before embarking on the proof, let us consider some examples. In the examples above, there was only one prime: $\phi(7) = 6$ because the integers $1, 2, 3, 4, 5$ and 6 are all coprime with 7. Similarly, all the integers $1, 2, 3, \ldots, 10$ are coprime with 11, so $\phi(11) = 10$. The general pattern is probably clear now: if p is prime, then all of the integers $1, 2, \ldots, p - 1$ are coprime with p. Hence we are ready to write the proof.

Proof. Let p be prime. Then the only (positive) factors of p are 1 and p.

Therefore, in the range, $1 \leq a \leq p$, all of the integers $1, 2, \ldots p - 1$ are coprime with p. There are $p - 1$ integers in this list so $\phi(p) = p - 1$. $\square$

Theorem 5.2

If p and q are distinct primes, then $\phi(pq) = (p-1)(q-1) = \phi(p)\phi(q)$.

The proof of theorem 5.1 became clear from considering one or two examples, so let's begin again by considering examples.

$p = 3, q = 5$ Here $pq = 15$. Consider the integers $1, 2, 3, \ldots, 15$.
The multiples of 3 — $3, 6, 9, 12, 15$ — are not coprime with 15. Similarly the multiples of 5 — $5, 10, 15$ — are not coprime with 15. All the remaining integers $1, 2, 4, 7, 8, 11, 13, 14$ are coprime with 15. There are 8 of these.
Hence $\phi(15) = 8 = 2 \times 4 = (3-1) \times (5-1)$, as required.

$p = 5, q = 7$ Here $pq = 35$. Consider the integers $1, 2, 3, \ldots, 35$.
The multiples of 5 — $5, 10, 15, 20, 25, 30, 35$ — are not coprime with 35. Similarly the multiples of 7 — $7, 14, 21, 28, 35$ — are not coprime with 35. All the remaining integers 1, 2, 3, 4, 6, 8, 9, 11, 12, 13, 16, 17, 18, 19, 22, 23, 24, 26, 27, 29, 31, 32, 33, 34 are coprime with 35.
There are 24 of these, so $\phi(35) = 24 = 4 \times 6 = (5-1) \times (7-1)$, as required.

These examples hold the clue to the proof — it is difficult to see a pattern in the integers that are coprime with 35, but it is easier to see which integers are *not* coprime with 35. In other words, instead of counting the integers that are coprime with pq, a better strategy will be to count the number of integers that are *not* coprime with pq. It is this strategy that allows us to complete the proof.

Proof. Let p and q be distinct primes and consider the list of integers $1, 2, 3, \ldots, pq$.

In the list, there are q integers that are multiples of p, namely $p, 2p, 3p, \ldots, qp$. Any pair of these has a common factor p, so that kp and pq are not coprime for $k = 1, 2, \ldots, q$. Similarly, there are p integers that are multiples of q, namely $q, 2q, 3q, \ldots, pq$. Again, any pair of these has a common factor q so that ℓq and pq are not coprime for $\ell = 1, 2, \ldots, p$. These two lists combined contain a total of $p + q - 1$ integers since pq appears in both lists and is the only integer belonging to both.

Since p and q are prime, all of the remaining integers are coprime with pq. Therefore, the total number of integers coprime with pq is

$$pq - (p + q - 1) = pq - p - q + 1 = (p-1)(q-1).$$

The last part of the inequality, $(p-1)(q-1) = \phi(p)\phi(q)$, follows from theorem 5.1, which gives $\phi(p) = p - 1$ and $\phi(q) = q - 1$. $\square$

We now turn our attention to another function, known as the Smarandache function, whose definition relies on the notion of divisibility. This was first considered by Édouard Lucas, a French mathematician, in 1883 and subsequently rediscovered by Florentin Smarandache, a Romanian mathematician just under a hundred years later in the late 1970s.

Definition 5.2
The **Smarandache function** $S : \mathbb{Z}^+ \to \mathbb{Z}^+$ is defined by

$$S(n) = \text{the smallest positive integer } m \text{ such that } n \text{ divides } m!$$

To help get a feel for the function, let's consider some examples.

$n = 3$ 3 does not divide $1! = 1$ or $2! = 2$ but 3 does divide $3! = 6$. Hence $S(3) = 3$.

$n = 4$ 4 does not divide $1! = 1$, $2! = 2$ or $3! = 6$ but 4 does divide $4! = 24$. Hence $S(4) = 4$.

$n = 5$ 5 does not divide $1! = 1$, $2! = 2$, $3! = 6$ or $4! = 24$. However 5 does divide $5! = 120$. Hence $S(5) = 5$.

$n = 6$ 6 does not divide $1! = 1$ or $2! = 2$, but 6 does divide $3! = 6$. Hence $S(6) = 3$.

When considering the first few examples above, we might have begun to suspect that the function just gives $S(n) = n$, but this is not the case as the example $S(6) = 3$ shows. Were we to have continued, we would have found $S(7) = 7, S(8) = 4, S(9) = 6, S(10) = 5$ and so on.

Our aim is to explore the function and prove some of its simple properties. The sequence of values $S(1), S(2), S(3), \ldots, S(n), \ldots$ is available as sequence number A002034 of the On-Line Encyclopedia of Integer Sequences.[1] A plot of the first 100 terms of the sequence is given in figure 5.3. From this it is clear both that the values of $S(n)$ exhibit some complex behaviour but also that there are some identifiable patterns.

In figure 5.3 it is clear that all the values lie on or below the line $y = x$ and there are a number of values that appear on the line $y = x$. In other words, $S(n) \leq n$ for all n and $S(n) = n$ for some values of n. From the limited number of examples above, we may guess that the values of n for which $S(n) = n$ are the prime numbers. This is our first result about the Smarandache function.

Theorem 5.3
For all $n \in \mathbb{Z}^+$, $S(n) \leq n$ and $S(n) = n$ if n is prime.

---

[1] Available at http://oeis.org/A002034.

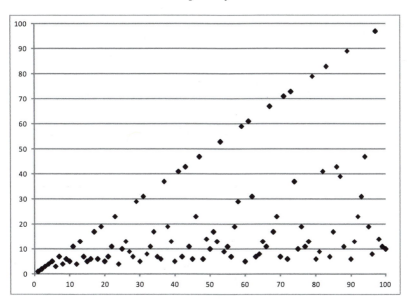

FIGURE 5.3: The Smarandache function.

This is really two theorems, or at least two separate statements about $S(n)$: for all n, $S(n) \leq n$ and for all n, if n is prime, then $S(n) = n$. Let us try to understand both parts by considering $S(11)$. We are seeking the first term in the sequence $1!, 2!, 3!, 4!, \ldots$ that is divisible by 11:

11 does not divide $1! = 1$

11 does not divide $2! = 1 \times 2$

11 does not divide $3! = 1 \times 2 \times 3$

11 does not divide $4! = 1 \times 2 \times 3 \times 4$

$\vdots$

11 does not divide $10! = 1 \times 2 \times 3 \times 4 \times \ldots \times 9 \times 10$

11 *does* divide $11! = 1 \times 2 \times 3 \times 4 \times \ldots \times 9 \times 10 \times \mathbf{11}$.

This example is probably sufficiently general to indicate how the proof of each part of the theorem can be constructed.

Proof. Let n be any positive integer.

Consider the sequence of factorials $1!, 2!, 3!, 4!, \ldots, n!, (n+1)!, (n+2)! \ldots$. Clearly n divides $n! = 1 \times 2 \times 3 \ldots \times (n-1) \times n$. Therefore the *smallest* $m!$ that is divisible by n must either be $n!$ itself or something smaller than $n!$. Hence $S(n) \leq n$.

Now suppose that n is equal to some prime p, say, and consider the sequence of factorials $1!, 2!, 3!, 4!, \ldots, p!$.

It is a property of primes that says if p divides a product ab, then p divides one of the factors, p divides a or p divides b or both. Since all of the factors of $1!, 2!, 3!, 4!, \ldots, (p-1)!$ are less than p, it follows that p does not divide any of these factorials.

Hence $p!$ is the smallest factorial that is divisible by p, so $S(p) = p$. □

For our second theorem about S, we mirror theorem 5.2 and evaluate $S(pq)$ where p and q are distinct primes. Again we begin by considering some examples.

$p = 3,\ q = 5$ Then $pq = 15$, so we need to find the smallest factorial that is divisible by 15.
Consider a factorial expression for $n \geq 5$:

$$1 \times 2 \times 3 \times 4 \times 5 \times \ldots \times n.$$

Clearly 5! contains both 3 and 5 as factors, so 15 divides 5!. However any factorial smaller that 5! will not be divisible by 5.
Hence $S(15) = 5$.

$p = 5,\ q = 11$ Then $pq = 77$, so we need to find the smallest factorial that is divisible by 55.
Consider a factorial expression for $n \geq 11$:

$$1 \times 2 \times 3 \times 4 \times 5 \times 6 \times \ldots \times 10 \times 11 \times \ldots \times n.$$

Again 11! contains both 5 and 11 as factors, so 77 divides 11!. However any factorial smaller that 11! will not be divisible by 11.
Hence $S(55) = 11$.

As before these examples are probably sufficiently general for us to be able to identify both the correct theorem for $S(pq)$ and its proof.

Theorem 5.4
If p and q are distinct primes, then $S(pq) = \max\{p, q\}$, the maximum of p and q.

Proof. Without loss of generality[2] we may suppose that $p < q$. Consider $q!$:

$$q! = 1 \times 2 \times 3 \times \ldots \times p \times \ldots \times q.$$

---

[2] We discussed the use of the phrase 'without loss of generality ...' in proofs in the commentary to the proof, in example 4.2.1, of the triangle inequality for the modulus function on page 147.

It is clear that both p and q are factors of $q!$ so that pq divides $q!$. By theorem 5.3, q will not be a factor of any factorial less than $q!$. Hence pq does not divide any factorial less than $q!$.

Therefore $q!$ is the smallest factorial that has pq as a factor, so $S(pq) = q = \max\{p, q\}$. $\qquad\square$

Exercises 5.1

1. (i) Let a be a real number. Prove that for all non-zero real numbers x,
    $$1 + \frac{a^2}{x^2} \geq \frac{2a}{x}.$$

    (ii) Prove that, if x and y are non-negative real numbers, then $x^2 \geq y^2 \Rightarrow x \geq y$.
    Hint: consider the expression $x^2 - y^2$.

    (iii) Prove that, for all real numbers x and y, $x^2 \geq y^2 \Rightarrow |x| \geq |y|$.

    (iv) Prove that, for all real numbers x and y, if $x \neq 0$ or $y \neq 0$, then $x^2 + xy + y^2 > 0$.

2. (i) Prove that, for all positive integers n, $n^3 - n$ is divisible by 3.

    (ii) Prove that, for all integers m and n, if $m + n$ is even, then mn is the difference of two squares; in other words, $mn = a^2 - b^2$ for some integers a and b.

    (iii) Prove that, for all integers $n \geq 2$, the sequence of $n - 1$ integers
    $$n! + 2, n! + 3, n! + 4, \ldots, n! + n$$
    are all composite.

3. Prove that, for all $a \geq 1$, $\sqrt{a + \sqrt{a}} + \sqrt{a - \sqrt{a}} \leq 2\sqrt{a}$.

4. Firstly, we introduce some notation. For positive integers m and n, let $m|n$ denote 'm divides n'.

   Prove each of the following properties of the divisibility relation on the set $\mathbb{Z}^+$.

    (i) For all $m, n, r \in \mathbb{Z}^+$, if $m|n$ and $n|r$, then $m|r$.

    (ii) For all $m, n, r, s \in \mathbb{Z}^+$, if $m|n$ and $r|s$, then $mr|ns$.

    (iii) For all $a, m, n, r, s \in \mathbb{Z}^+$, if $a|m$ and $a|n$, then $a|(rm + sn)$.

5. Prove each of the following about Euler's phi function.

(i) If p is prime, then $\phi(p^2) = p(p-1)$.

(ii) More generally, if p is prime and a is a positive integer, then $\phi(p^a) = p^a - p^{a-1}$.

(iii) If m and n are coprime, then $\phi(mn) = \phi(m)\phi(n)$.

Note that this result, which generalises theorem 5.2, is quite tricky to prove. One approach is to arrange the integers $1, 2, 3, \ldots, mn$ into an array with n rows and m columns as follows.

1	2	3	$\ldots$	m
$m+1$	$m+2$	$m+3$	$\ldots$	$2m$
$\vdots$	$\vdots$	$\vdots$	$\ldots$	$\vdots$
$(n-1)m+1$	$(n-1)m+2$	$(n-1)m+3$	$\ldots$	nm

Then count which columns contain (some) integers that are coprime with mn and, finally, for each such column, count how many entries are coprime with mn.

(iv) If $n = p_1^{\alpha_1} p_2^{\alpha_2} \ldots p_k^{\alpha_k}$ is the factorisation of n, where $p_1, p_2, \ldots, p_k$ are distinct primes and $\alpha_1, \alpha_2, \ldots, \alpha_n$ are positive integers, then

$$\phi(n) = n \left(1 - \frac{1}{p_1}\right) \left(1 - \frac{1}{p_2}\right) \ldots \left(1 - \frac{1}{p_k}\right).$$

In your proof, you may use any of the results above. Note that the existence of the expression $n = p_1^{\alpha_1} p_2^{\alpha_2} \ldots p_k^{\alpha_k}$ follows from the Prime Factorisation Theorem, page 132.

6. Let S be the Smarandache function.

   (i) Prove that, for all $n \in \mathbb{Z}^+$, $S(n!) = n$.

   (ii) Prove that, if p is prime then $S(p^2) = 2p$.

   (iii) Evaluate $S(2^k)$ for $k = 1, 2, \ldots 10$.

   (iv) Extend the argument given in the proof of theorem 5.4 to show that, if $n = p_1 p_2 \ldots p_k$ where $p_1, p_2, \ldots, p_k$ are primes and $p_1 < p_2 < \cdots < p_k$, then $S(n) = p_k$.

7. Let $\mathbf{A} = \begin{pmatrix} a & b \\ c & d \end{pmatrix}$ be a 2×2 real matrix such that $\mathbf{A}^2 = \mathbf{0}$ where $\mathbf{0} = \begin{pmatrix} 0 & 0 \\ 0 & 0 \end{pmatrix}$ is the 2×2 zero matrix.

   Prove that $d = -a$.

8. (i) Consider the proof of theorem 5.2. Try to identify (some of) the background knowledge that is assumed in the proof.

   (ii) Repeat part (i) for the proof of theorem 5.4.

5.2 Examples from sets and functions

In this section we apply some of the techniques of direct proof to explore further the properties of sets and functions. We assume familiarity with the material introduced in chapter 3. In particular, we consider the interaction between the notion of cardinality firstly with power sets and secondly with the injective and surjective properties of functions.

Cardinality and power sets

Let A be a finite set. Recall from chapter 3, page 86, that the cardinality of A, denoted $|A|$, is the number of elements in A. We may also consider the power set of A, denoted $\mathbb{P}(A)$, which is the set of all the subsets of A — see page 99. In this section we explore the interaction between these concepts and the notions of subset, intersection, etc.

In example 3.8.1, we gave the following list of sets and their power sets.

$$A = \varnothing \qquad \mathbb{P}(A) = \{\varnothing\}$$
$$A = \{a\} \qquad \mathbb{P}(A) = \{\varnothing, \{a\}\}$$
$$A = \{a,b\} \qquad \mathbb{P}(A) = \{\varnothing, \{a\}, \{b\}, \{a,b\}\}$$
$$A = \{a,b,c\} \qquad \mathbb{P}(A) = \{\varnothing, \{a\}, \{b\}, \{c\}, \{a,b\}, \{a,c\}, \{b,c\}, \{a,b,c\}\}$$
$$A = \{a,b,c,d\} \quad \mathbb{P}(A) = \{\varnothing, \{a\}, \{b\}, \{c\}, \{d\}, \{a,b\}, \{a,c\}, \{a,d\},$$
$$\{b,c\}, \{b,d\}, \{c,d\}, \{a,b,c\}, \{a,b,d\},$$
$$\{a,c,d\}, \{b,c,d\}, \{a,b,c,d\}\}$$

If we now list only the cardinalities of the sets rather than their elements, we obtain the following table.

$|A|$	$|\mathbb{P}(A)|$
0	1
1	2
2	4
3	8
4	16

From this table, the obvious conjecture to make is the following:

$$\text{if } |A| = n \text{ then } |\mathbb{P}(A)| = 2^n.$$

How might we prove this conjecture? A good starting point would be to consider in more detail one of the examples above. It is usually sensible to consider an example that is not too simple, so that it is not sufficiently general, and not too complicated, so that it becomes unmanageable. In this case we will consider the set with 4 elements, $A = \{a, b, c, d\}$.

We aim to count the elements of its power set $\mathbb{P}(A)$ in such a way that we can 'see' a general argument that will work for all sets. An obvious strategy is to split the power set into smaller pieces, according to the cardinality of its elements as follows.

Size of subset	Subsets	Count
empty set	$\varnothing$	1
1-element sets	$\{a\}, \{b\}, \{c\}, \{d\}$	4
2-element sets	$\{a, b\}, \{a, c\}, \{a, d\}, \{b, c\}, \{b, d\}, \{c, d\}$	6
3-element sets	$\{a, b, c\}, \{a, b, d\}, \{a, c, d\}, \{b, c, d\}$	4
4-element sets	$\{a, b, c, d\}$	1
		16

The next step is to consider how easy this will be to generalise to a set with n elements (for an arbitrary n). Suppose $|A| = n$. Then its power set obviously contains one empty set and one set with n elements, namely A itself. It is also easy to realise that there are n subsets of A containing a single element — there is a single element subset $\{a\}$ for each $a \in A$. However, now the reasoning gets a little more tricky. We need to be able to calculate the number of two-element subsets, the number of three-element subsets, and so on, for our set with n elements. This can be done but requires background knowledge of permutations and combinations. In particular, to form a subset of A containing r elements, say, we need to choose exactly r of its elements to go into the subset. Thus to complete the proof along these lines we would need to know, in general, how many ways there are of choosing r objects from n (distinct) objects. Some readers will no doubt know that this is the binomial coefficient

$$^{n}C_r = \binom{n}{r} = \frac{n!}{(n - r)!r!}.$$

To complete the proof, we also need to be able to add the numbers of subsets with r elements for each $r = 0, 1, \ldots, n$. This requires another piece of background knowledge, the Binomial Theorem. We leave this as an exercise — see exercise 5.2.2 — and instead pursue a different line of reasoning here.

Returning to the example $A = \{a, b, c, d\}$, instead of splitting the subsets into collections of equal cardinality, we consider which *elements* belong to the subsets. For example, the element a belongs to the following eight subsets:

$$\{a\}, \{a, b\}, \{a, c\}, \{a, d\}, \{a, b, c\}, \{a, b, d\}, \{a, c, d\}, \{a, b, c, d\}.$$

In other words, a belongs to exactly half of the subsets of A. The same is true of each element. For example, the element c belongs to the following eight subsets

$$\{c\}, \{a, c\}, \{b, c\}, \{c, d\}, \{a, b, c\}, \{a, c, d\}, \{b, c, d\}, \{a, b, c, d\}.$$

Why is this? If we think about an element of A and a subset of A, there are only two options: either the element belongs to the subset or it does not. Thus, to form a subset of A, we may consider each element in turn and make a choice: either include it in the subset or don't. Each set of choices gives a different subset. For example, the set of choices 'include, exclude, include, include' gives rise to the subset $\{a, c, d\}$. Counting the number of different sets of choices therefore counts the number of subsets and this, finally, gives us an elementary way of proving our conjecture. By 'elementary', we mean a way of proving the conjecture relying on minimal background knowledge.

Theorem 5.5
Let A be a finite set. If A has cardinality n, then its power set has cardinality 2^n.

Proof. Let A be a set with cardinality n. Then we may write $A = \{a_1, a_2, \ldots, a_n\}$.

To form a subset of A, consider each element a_r in turn and either include it in, or exclude it from, the subset. For each element there are two choices, include or exclude, and the choice for each element is independent of the choices for the other elements. Therefore there are $2 \times 2 \times \cdots \times 2$ (n times) $= 2^n$ choices in total.

Each set of choices gives a unique subset of A and every subset arises in this way. Therefore there are 2^n subsets of A so $|\mathbb{P}(A)| = 2^n$. □

We have noted that there is an alternative proof that involves counting the number of r-element sets for each r and adding these using the binomial theorem. For those readers who possess the appropriate background knowledge, we leave this as an exercise. It is interesting to note that our two different ways of analysing the example where $A = \{a, b, c, d\}$ lead to these two different methods of proving the result. In fact, there is another approach to proving the theorem, based on the Principle of Mathematical Induction, which we will consider in chapter 8.

We now turn attention to how the relation of subset interacts with the concept of power set. As is common, to get a feel for the situation, we consider examples. In fact, the examples given in example 3.8.1 and repeated above will be sufficient to make a conjecture. Considering the list of sets and their power sets on page 189 above, it is clear that if A is a subset of B, then $\mathbb{P}(A)$ is also a subset of $\mathbb{P}(B)$. For example, taking $A = \{a, b\}$ and $B = \{a, b, c, d\}$, so that $A \subseteq B$, we have

$$\mathbb{P}(A) = \{\varnothing, \{a\}, \{b\}, \{a, b\}\}$$

and
$$\mathbb{P}(B) = \{\varnothing, \{a\}, \{b\}, \{c\}, \{d\}, \{a, b\}, \{a, c\}, \{a, d\},$$
$$\{b, c\}, \{b, d\}, \{c, d\}, \{a, b, c\}, \{a, b, d\},$$
$$\{a, c, d\}, \{b, c, d\}, \{a, b, c, d\}\}$$

and clearly $\mathbb{P}(A) \subseteq \mathbb{P}(B)$. Thus we may make the following conjecture where A and B are sets:

$$\text{if } A \subseteq B \text{ then } \mathbb{P}(A) \subseteq \mathbb{P}(B).$$

In order to see how we might prove this, we need to understand *why* the conjecture is true. Recall that, for any sets X and Y, X is a subset of Y precisely when every element of X is also an element of Y — see page 87. Symbolically,

$$X \subseteq Y \text{ if and only if } x \in X \Rightarrow x \in Y \text{ for all } x.$$

In the case of power sets, $X \in \mathbb{P}(A)$ means $X \subseteq A$. (This is just the definition of power set of A: it is the set that contains all the subsets of A.) Thus, to prove that $\mathbb{P}(A) \subseteq \mathbb{P}(B)$, we need to show that $X \subseteq A$ implies $X \subseteq B$. For this, we will use a direct proof.

Theorem 5.6
For all sets A and B, if $A \subseteq B$, then $\mathbb{P}(A) \subseteq \mathbb{P}(B)$.

Proof. Suppose that $A \subseteq B$. To prove that $\mathbb{P}(A) \subseteq \mathbb{P}(B)$, we need to show that $X \in \mathbb{P}(A) \Rightarrow X \in \mathbb{P}(B)$ for all sets X.

Let $X \in \mathbb{P}(A)$. This means that $X \subseteq A$. We need to show that $X \subseteq B$, so that $X \in \mathbb{P}(B)$.

Let $x \in X$. Since $X \subseteq A$ it follows that $x \in A$. However, we also have that $A \subseteq B$, so it now follows that $x \in B$. We have shown that $x \in X \Rightarrow x \in B$ for all elements x, which means that $X \subseteq B$. Hence $X \in \mathbb{P}(B)$.

Overall we have now shown that, for all sets X, $X \in \mathbb{P}(A) \Rightarrow X \in \mathbb{P}(B)$. Hence $\mathbb{P}(A) \subseteq \mathbb{P}(B)$ as required. $\qquad\square$

This proof is not long and none of the individual reasoning steps are hard. Nevertheless many readers will find this proof quite difficult both to understand and, more especially, to construct. The reason for this is that the proof is 'multi-layered'. We may illustrate the layers in the proof as follows.

Assume $A \subseteq B$

Need to deduce $\mathbb{P}(A) \subseteq \mathbb{P}(B))$

$\qquad \mathbb{P}(A) \subseteq \mathbb{P}(B)$ means $X \in \mathbb{P}(A) \Rightarrow X \in \mathbb{P}(B)$

$\qquad$ Assume $X \in \mathbb{P}(A)$

$\qquad$ Need to deduce $X \in \mathbb{P}(B)$

$\qquad X \in \mathbb{P}(A)$ means $X \subseteq A$

$\qquad X \in \mathbb{P}(B)$ means $X \subseteq B$ so need to deduce $X \subseteq B$

$\qquad\qquad X \subseteq B$ means $x \in X \Rightarrow x \in B$

$\qquad\qquad$ Assume $x \in X$

$$\vdots$$

$\qquad\qquad$ Deduce $x \in B$

$\qquad$ Hence $X \subseteq B$

$\qquad$ Hence $X \in \mathbb{P}(B)$

Hence $\mathbb{P}(A) \subseteq \mathbb{P}(B)$

The structure of the proposition to be proved is a conditional, $A \subseteq B \Rightarrow$ $\mathbb{P}(A) \subseteq \mathbb{P}(B)$. At the highest level, the structure of the proof is that of a direct proof of a conditional:

$\qquad$ assume $A \subseteq B$ and deduce $\mathbb{P}(A) \subseteq \mathbb{P}(B)$.

However, the required deduction $\mathbb{P}(A) \subseteq \mathbb{P}(B)$ is really another conditional statement $X \in \mathbb{P}(A) \Rightarrow X \in \mathbb{P}(B)$, so we use another direct proof

$\qquad$ assume $X \in \mathbb{P}(A)$ and deduce $X \in \mathbb{P}(B)$

nested inside the top-level direct proof. Again the required deduction, $X \in$ $\mathbb{P}(B)$ is equivalent to another conditional statement, $x \in X \Rightarrow x \in B$, which requires a further nested direct proof,

$\qquad$ assume $x \in X$ and deduce $x \in B$.

When we have multi-layered proofs such as this, is is often sensible to see whether the proof may be 'modularised' by pulling out part of the proof into a separate result that may be proved independently of the main theorem. If we can do this, we will often refer to the separate 'interim' results as lemmas.

In the example of theorem 5.6, there is an interim 'stand-alone' result, called the transitive property of subset, that we can prove separately. We describe this approach below. Although there are now two separate proofs to be constructed, 'modularising' the proof in this way almost certainly provides a more understandable route to the proof of theorem 5.6.

Lemma 1 (Transitive property of subset)
For all sets X, Y, and Z, if $X \subseteq Y$ and $Y \subseteq Z$, then $X \subseteq Z$.

The statement to be proved is a conditional $(X \subseteq Y \wedge Y \subseteq Z) \Rightarrow X \subseteq Z$ and we will use the method of direct proof: assume $X \subseteq Y \wedge Y \subseteq Z$ and deduce $X \subseteq Z$.

Proof. Let X, Y, and Z be sets such that $X \subseteq Y$ and $Y \subseteq Z$.

We need to show that $X \subseteq Z$. In other words, we must show that, for all elements x, if $x \in X$, then $x \in Z$.

So suppose that $x \in X$. Since $X \subseteq Y$ it follows that $x \in Y$. But $Y \subseteq Z$ also, so it follows that $x \in Z$.

We have shown that $x \in X \Rightarrow x \in Z$, which means that $X \subseteq Z$ as required. $\square$

Alternative proof of theorem 5.6 *using lemma* 1. Suppose that $A \subseteq B$.

To prove that $\mathbb{P}(A) \subseteq \mathbb{P}(B)$, we need to show that $X \in \mathbb{P}(A) \Rightarrow X \in \mathbb{P}(B)$ for all sets X.

Suppose that $X \in \mathbb{P}(A)$. This means that $X \subseteq A$. Then we have both $X \subseteq A$ and $A \subseteq B$ so we may deduce $X \subseteq B$, by lemma 1. Hence $X \in \mathbb{P}(B)$.

Overall we have now shown that, for all sets X, $X \in \mathbb{P}(A) \Rightarrow X \in \mathbb{P}(B)$. Hence $\mathbb{P}(A) \subseteq \mathbb{P}(B)$ as required. $\square$

Injective and surjective functions and cardinality

For our next collection of example proofs, we consider injective and surjective functions and what the injectivity and surjectivity properties tell us about the cardinalities of the domain and codomain. The injective and surjective properties are given in definitions 3.3 and 3.4, respectively. Informally, recall that an injective function is one where different elements of the domain have different images, and a surjective function is one where every element of the codomain is the image of something in the domain.

Example 5.3
We begin by considering the function with rule

$$f(x) = \frac{3x + 5}{x + 4}$$

where x is a real number. Note that $f(x)$ is not defined when $x = -4$, so we will take the domain for f to be the set $\mathbb{R} - \{-4\} = \{x \in \mathbb{R} : x \neq -4\}$. A little thought should reveal that $f(x)$ cannot equal 3. One way of seeing this is to re-write the expression for $f(x)$ as

$$f(x) = \frac{3(x + 5/3)}{x + 4};$$

since $x + 5/3 \neq x + 4$, it is clear that $f(x) \neq 3$. As a result, we will take the codomain for f to be the set $\mathbb{R} - \{3\} = \{x \in \mathbb{R} : x \neq 3\}$. Our aim is to prove that the resulting function f is both injective and surjective. We capture this as the following claim.

Claim

The function

$$f : \mathbb{R} - \{-4\} \to \mathbb{R} - \{3\} \text{ defined by } f(x) = \frac{3x + 5}{x + 4}$$

is both injective and surjective.

There are two things to prove here, so we give separate proofs that f is injective and f is surjective. Recalling the definitions, for injective we need to prove

$$f(x) = f(y) \Rightarrow x = y \quad \text{for all } x, y \in \mathbb{R} - \{-4\}$$

and for surjective we need to prove

for all $y \in \mathbb{R} - \{3\}$ there exists $x \in \mathbb{R} - \{-4\}$ such that $f(x) = y$.

For injectivity, we will use a direct proof of a conditional, so we will assume $f(x) = f(y)$ and deduce $x = y$. For surjectivity, given $y \in \mathbb{R} - \{3\}$, we need to *find* an appropriate $x \in \mathbb{R} - \{-4\}$ and then show that $f(x) = y$. The discovery of an appropriate x is obviously vital to the proof, but is not part of the proof itself, so we will consider this first. Given $y \in \mathbb{R} - \{3\}$, we are seeking x such that

$$f(x) = \frac{3x + 5}{x + 4} = y$$

so to find x we just need to solve this equation for x, as follows:

$$\frac{3x + 5}{x + 4} = y \Rightarrow 3x + 5 = y(x + 4) \Rightarrow 3x - xy = 4y - 5 \Rightarrow x = \frac{4y - 5}{3 - y}.$$

It is this value of x, $x = \dfrac{4y - 5}{3 - y}$, that we will use in the proof itself.

Proof. Firstly we show that f is injective.

Let $x, y \in \mathbb{R} - \{-4\}$. Then

$$f(x) = f(y) \quad \Rightarrow \quad \frac{3x + 5}{x + 4} = \frac{3y + 5}{y + 4}$$

$$\Rightarrow \quad (3x + 5)(y + 4) = (3y + 5)(x + 4)$$

$$\Rightarrow \quad 3xy + 12x + 5y + 20 = 3xy + 5x + 12y + 20$$

$$\Rightarrow \quad 12x + 5y = 5x + 12y$$

$$\Rightarrow \quad 7x = 7y$$

$$\Rightarrow \quad x = y.$$

We have shown that $f(x) = f(y) \Rightarrow x = y$ for all x and y in the domain of f. Therefore f is injective.

Secondly, we show that f is surjective.

Let $y \in \mathbb{R} - \{3\}$. Then $y \neq 3$, so let $x = \dfrac{4y - 5}{3 - y}$.

To show $x \in \mathbb{R} - \{-4\}$, we need to show that $x \neq -4$. Writing x as $x = \dfrac{4(y - 5/4)}{-(y - 3)}$ and noting that $y - 5/4 \neq y - 3$, it follows that $x \neq -4$.

We now have

$$f(x) = f\left(\frac{4y - 5}{3 - y}\right)$$

$$= \frac{3\left(\dfrac{4y - 5}{3 - y}\right) + 5}{\left(\dfrac{4y - 5}{3 - y}\right) + 4}$$

$$= \frac{3(4y - 5) + 5(3 - y)}{(4y - 5) + 4(3 - y)} \qquad \text{(multiplying through by } 3 - y)$$

$$= \frac{12y - 15 + 15 - 5y}{4y - 5 + 12 - 4y}$$

$$= \frac{7y}{7} = y.$$

We have shown that, for all y in the codomain of f there exists x in the domain such that $f(x) = y$. Therefore f is surjective. $\qquad \qquad \square$

Two-phase proof discovery

It is worth reflecting further on the nature and proof of the surjectivity condition. There are two phases to finding a proof: the discovery phase and the proof phase. In the discovery phase, given an element y in the codomain, we need to discover an appropriate x in the domain such that $f(x) = y$. This discovery phase is not part of the proof itself. During the proof phase, we only need to verify that the value of x found in the discovery phase does indeed satisfy the appropriate property, namely $f(x) = y$.

This two-phase proof process is common when proving statements with two quantifiers which have the following structure

$$\forall a\, \exists b \bullet P(a, b).$$

The statement that a function $f : A \to B$ is surjective has this structure:

for all $y \in B$ there exists $x \in A$ such that $f(x) = y$.

For the discovery phase, we assumed that y was an arbitrary element of the codomain. This is because y is universally quantified so that our y needs to be arbitrary. Then we found an appropriate x in the domain, that *depended on* y and satisfied the required property.

Generalising this process to the generic statement $\forall a\, \exists b \bullet P(a, b)$, the two phases in finding the proof are described as follows.

> **Discovery phase.** For an arbitrary a, find some b, which is likely to depend on a, such that $P(a, b)$ is true.

> **Proof phase.** In the proof, let a be an arbitrary element of its domain. Let b have the value found during the discovery phase. Verify that, for the chosen value of b, the statement $P(a, b)$ is true.

We will see this two-phase process in operation in section 5.4 where we prove properties of limits which also involve doubly quantified propositional functions of the form $\forall a\, \exists b \bullet P(a, b)$.

We conclude this section with a consideration of the relationship between the properties of injectivity and surjectivity and the cardinalities of the domain and codomain (where these are finite sets). Figure 5.4 shows the arrow mappings for a typical injective and a typical surjective function, each with a small finite domain and codomain.

For the injective function, there needs to be at least as many elements in the codomain as the domain in order that different elements of the domain can have different images in the codomain. For the surjective function, there needs to be at least as many elements in the domain as the codomain in order that each element in the codomain can be 'hit' by an arrow; more formally, so that each element in the codomain can be the image of some element in the domain. The following theorem expresses these observations more formally. Each of the two parts of the proof is a conditional proposition and we use the method direct proof in each case.

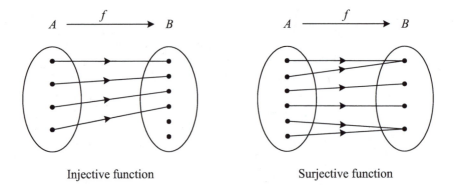

Injective function   Surjective function

FIGURE 5.4: Injective and surjective functions with finite domain and codomain.

Theorem 5.7

Let $f : A \rightarrow B$ be a function, where A and B are finite sets.

(i) If f is injective, then $|A| \leq |B|$.

(ii) If f is surjective, then $|A| \geq |B|$.

Proof. Let A and B be finite sets, say $|A| = n$ and $|B| = m$. Then we may write $A = \{a_1, a_2, \ldots, a_n\}$ and $B = \{b_1, b_2, \ldots, b_m\}$ where there are no repeated elements listed in each case.

(i) Suppose that f is injective. Then different elements of A have different images so that the elements $f(a_1), f(a_2), \ldots, f(a_n)$ of B are all distinct. Hence B contains at least n elements, so $|B| \geq n = |A|$, as required.

(ii) Suppose that f is surjective. Then the images of the elements of A, $f(a_1), f(a_2), \ldots, f(a_n)$, must contain all the elements of B. However, this list may contain repeats, so B contains at most n elements. Hence $|B| \leq n = |A|$, as required.

$\square$

There are some results that follow 'almost immediately' from a particular theorem. It is common to refer to such a result as a corollary of the particular theorem. In the case of theorem 5.7 there is an obvious corollary concerning bijective functions obtained by putting the two parts of the theorem together. The proof is a simple direct proof.

Corollary

Let $f : A \rightarrow B$ be a bijective function, where A and B are finite sets. Then A and B have the same cardinality, $|A| = |B|$.

Proof. Suppose that $f : A \to B$ is a bijective function.

Then f is injective, so $|A| \leq |B|$ by theorem 5.4 (i). Also f is surjective, so $|A| \geq |B|$ by theorem 5.4 (ii).

Hence $|A| = |B|$. $\qquad\qquad\qquad\qquad\qquad\qquad\qquad\qquad\qquad\qquad\qquad\qquad$ □

Exercises 5.2

1. Prove that, for all finite sets A and B, if $A \subseteq B$ then $|A| \leq |B|$.

2. Give an alternative proof of theorem 5.5 by counting the number of subsets of A with no elements, one element, two elements, and so on and then using the Binomial Theorem.

3. Describe the 'layers' in the proof of lemma 1 and the alternative proof of theorem 5.6 in a similar way to the description of the original proof of theorem 5.6 given on page 192.

4. Prove that for all sets A and B, $\mathbb{P}(A \cap B) = \mathbb{P}(A) \cap \mathbb{P}(B)$.

5. Prove that, for all sets A and B, $\mathbb{P}(A \cup B) \subseteq \mathbb{P}(A) \cup \mathbb{P}(B)$.

   Are there any circumstances where $\mathbb{P}(A \cup B) = \mathbb{P}(A) \cup \mathbb{P}(B)$? If so, can you prove it?

6. This question explores how the Cartesian product of sets interacts with the subset relation and with intersections and unions of sets.

   (i) Prove that, for all sets A, B, X, and Y, if $A \subseteq X$ and $B \subseteq Y$, then $A \times B \subseteq X \times Y$.

   (ii) Prove that, for all sets A, B, X, and Y, $(A \cap B) \times (X \cap Y) = (A \times X) \cap (B \times Y)$.

   (iii) Prove that, for all sets A, B, X, and Y, $(A \cup B) \times (X \cup Y) = (A \times X) \cup (A \times Y) \cup (B \times X) \cup (B \times Y)$.

7. The **symmetric difference** $A * B$ of two sets A and B is defined by

$$A * B = (A - B) \cup (B - A).$$

   Using the Set Theory Laws, page 99, prove each of the following identities.

   (i) For all sets A, $A * \varnothing = A$ and $A * A = \varnothing$.

   (ii) For all sets A and B, $A * B = (A \cup B) - (A \cap B)$.

   (iii) For all sets A, B, and C, $A \cap (B * C) = (A \cap B) * (A \cap C)$.

8. Show that each of the following functions is a bijection.

   (i) $f : \mathbb{R} \to \mathbb{R}$, $f(x) = \dfrac{5x + 3}{8}$

   (ii) $f : \mathbb{R} - \{-1\} \to \mathbb{R} - \{3\}$, $f(x) = \dfrac{3x}{x + 1}$

   (iii) $f : [1, 3] \to [-2, 2]$, $f(x) = 2x - 4$

   (iv) $f : \mathbb{R}^+ \to (0, 2)$, $f(x) = \dfrac{4}{x + 2}$.

   (v) $f : \mathbb{R}^2 \to \mathbb{R}^2$, $f(x, y) = (2x - 1, 5y + 3)$.

   (vi) $f : \mathbb{Z}^+ \to \mathbb{Z}$, $f(n) = \begin{cases} \dfrac{n}{2} & \text{if } n \text{ is even} \\ \dfrac{1 - n}{2} & \text{if } n \text{ is odd.} \end{cases}$

9. Let $f : A \to B$ be a function. Prove each of the following.

   (i) For all subsets C_1 and C_2 of A, if $C_1 \subseteq C_2$, then $f(C_1) \subseteq f(C_2)$.
   (ii) For all subsets D_1 and D_2 of A, if $D_1 \subseteq D_2$, then $f^{-1}(D_1) \subseteq f^{-1}(D_2)$.
   (iii) For all subsets C of A, $C \subseteq f^{-1}(f(C))$.
   (iv) If f is injective, then for all subsets C of A, $C = f^{-1}(f(C))$.
   (v) If f is surjective, then for all subsets D of B, $f(f^{-1}(D)) = D$.

10. A function $f : \mathbb{R} \to \mathbb{R}$ is said to be **even** if $f(-x) = f(x)$ for all $x \in \mathbb{R}$. Similarly f is **odd** if $f(-x) = -f(x)$ for all $x \in \mathbb{R}$.

   (i) Let $g : \mathbb{R} \to \mathbb{R}$ be any function. Prove that the function $f : \mathbb{R} \to \mathbb{R}$ defined by $f(x) = g(x) + g(-x)$ is even.
   (ii) Prove that any function $f : \mathbb{R} \to \mathbb{R}$ may be written as $f = g + h$ where g is an even function and h is an odd function.

5.3 Examples from algebra

5.3.1 Group theory

An important algebraic construction in mathematics is that of a 'group'.[3] Groups are important because the concept is simple and has examples and

---

[3] The notion of a group evolved gradually in nineteenth-century mathematics. The French mathematician Évariste Galois (1811–1832) first coined the term 'groupe' and he used the permutation group of the roots of polynomial equations to show that polynomials of degree at least 5 are not solvable with radicals.

applications in a wide variety of contexts. A group is simply a collection of objects together with a rule for combining those objects satisfying three simple conditions. The rule for combining objects takes two objects and 'combines' them to produce another object of the same kind. For example, addition is a rule for combining integers: given two integers m and n, their sum $m + n$ is another integer. A rule for combining is called a *binary operation* and the formal definition is the following.

Definition 5.3

Let S be a set. A **binary operation** on S is a mapping, or function, $\mu :$ $S \times S \to S$. For $x, y \in S$, we will write $\mu(x, y)$ as $x * y$.

The domain of the binary operation is the Cartesian product $S \times S$ which contains all ordered pairs of elements of S. The term *binary* operation reflects the fact that the domain comprises *pairs* of elements. The fact that the codomain of the operation is also the set S means that 'combining' x and y in S produces another element $x * y$ that is *also an element of S*. Some authors refer to this as the **closure property** of the binary operation.

Familiar examples of binary operations are addition and multiplication of integers as $\mu(m, n) = m + n$ and $\mu(m, n) = mn$, each of which defines a mapping $\mathbb{Z} \times \mathbb{Z} \to \mathbb{Z}$. Notice that division $\mu(m, n) = m/n$ does *not* define a binary operation on $\mathbb{Z}$ for two reasons. Most obviously, the codomain is not $\mathbb{Z}$ because, in general, $m/n \notin \mathbb{Z}$; for example $3/2$ is not an integer. Also the domain is not $\mathbb{Z} \times \mathbb{Z}$ either because division by zero is not defined, so that $m/0$ is meaningless. We could define integer division as a mapping $\mathbb{Z} \times \mathbb{Z}^* \to \mathbb{Q}$, where $\mathbb{Z}^*$ denotes the set of non-zero integers, but even this does not define a binary operation according to definition 5.3.

Definition 5.4

A **group** is a pair $(G, *)$, where G is a set and $*$ is a binary operation on G that satisfies the following three properties.

(G1) **Associative property.**

For all $x, y, z \in G$, $\quad x * (y * z) = (x * y) * z$.

(G2) **Existence of an identity element.**

There is an element $e \in G$, called an **identity element** for $*$, such that

$$e * x = x = x * e \quad \text{for all } x \in G.$$

(G3) **Existence of inverses.**

For each $x \in G$ there is an element $x^{-1} \in G$ called an **inverse** for x, such that

$$x * x^{-1} = e = x^{-1} * x.$$

The properties (G1) – (G3) are often referred to as the 'group axioms' as they are the basic properties that define what we mean by a group. Whenever we refer to 'a group $(G, *)$' we may assume that these properties are satisfied. Interested readers may refer to section 4.5 for a more extensive discussion of the nature of axioms. Some authors include a fourth property or axiom called the 'closure property' of the binary operation: this says that, if x and y belong to G, then so, too, does $x * y$. However, we have included this in the *definition* of a binary operation on G as a mapping whose codomain is the set G.

Note that we do *not* assume the **commutative property**:

$$x * y = y * x \quad \text{for all } x, y \in G.$$

If a group $(G, *)$ also satisfies the commutative property, it is called an **Abelian group**[4] (or **commutative group**).

Examples 5.4

1. The set of integers $\mathbb{Z}$ with binary operation of addition forms a group $(\mathbb{Z}, +)$. The identity element is 0, since $0 + n = n = n + 0$ for all n, and the inverse of n is $-n$, since $n + (-n) = 0 = (-n) + n$ for all n. The associative property of addition,

$$m + (n + p) = (m + n) + p \quad \text{for all } m, n, p \in \mathbb{Z},$$

   is a familiar property (whose proof depends on giving a precise definition of the set of integers and the operation of addition).

   Note that the set of integers $\mathbb{Z}$ with the binary operation of multiplication does not form a group. Multiplication is an associative binary operation on $\mathbb{Z}$ and 1 is the identity element. However, most elements do not have an inverse. For example, 2 does not have an inverse since there is no integer n such that $2n = 1$. In fact, 1 and -1 are self-inverse (since $1 \times 1 = 1$ and $(-1) \times (-1) = 1$) but no other integers have an inverse under multiplication.

2. If we wish to create a group under multiplication, the previous example indicates that we will need to include fractions in the set. So we might conjecture that the set of rational numbers $\mathbb{Q}$ forms a group under multiplication. As in the previous example, multiplication is an associative binary operation and 1 is the identity element. The inverse of $p/q \in \mathbb{Q}$ is q/p since

$$\frac{p}{q} \times \frac{q}{p} = 1.$$

---

[4] Named after the Danish mathematician Niels Abel (1802–1829), one of pioneers of group theory.

However q/p is only defined when $p \neq 0$. In other words, every *non-zero* rational number has an inverse under multiplication. Let

$$\mathbb{Q}^* = \{q \in \mathbb{Q} : q \neq 0\} = \left\{ \frac{p}{q} : p, q \in \mathbb{Z} \text{ and } p, q \neq 0 \right\}$$

denote the set of non-zero rational numbers. Since the product of non-zero rational numbers is again a *non-zero* rational number, multiplication is also a binary operation on the set $\mathbb{Q}^*$. Since every element of $\mathbb{Q}^*$ has an inverse,

$$\left(\frac{p}{q} \right)^{-1} = \frac{1}{p/q} = \frac{q}{p},$$

it follows that $(\mathbb{Q}^*, \times)$ is a group.

In a similar way, the set of non-zero real numbers

$$\mathbb{R}^* = \{x \in \mathbb{R} : x \neq 0\}$$

is also a group under multiplication, again with inverses given by with $x^{-1} = 1/x$.

3. A **symmetry** of a plane figure F is a mapping $F \to F$ such that F looks the same before and after the mapping. The symmetries of a (non-square) rectangle are:

- the identity mapping that 'does nothing', this is denoted e;
- $r =$ a rotation anticlockwise by π radians (or 180°) about the centre;
- $v =$ a reflection in a vertical line though the centre;
- $h =$ a reflection in a horizontal line though the centre.

These symmetries are illustrated in figure 5.5.

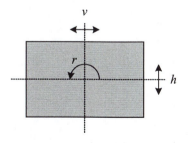

FIGURE 5.5: Illustrating the symmetries of a rectangle.

We can define composition of symmetries: if a and b are symmetries, we define ab to be the symmetry 'a followed by b'. This is just composition of functions but with a slight change of notation. If f and g are mappings, their composite 'f followed by g' is denoted $g \circ f$ (see page 113). In the context of groups, we regard symmetries more as algebraic entities (rather than functions), which is why the composite 'a followed by b' is denoted ab. However, since composition of functions in general is known to be associative, $(f \circ g) \circ h = f \circ (g \circ h)$ when either side is defined, we will assume that composition of symmetries is associative.

Figure 5.6 illustrates the composite rv, r followed by v. In order to keep track of what is happening to the rectangle, we have labelled its corners. From the figure we can see that rv has the same effect on the rectangle as the horizontal reflection, so we write $rv = h$.

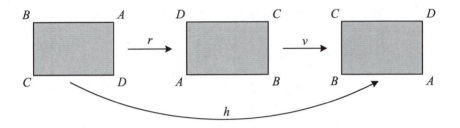

FIGURE 5.6: Composing symmetries of a rectangle.

More generally, figure 5.6 also illustrates that composing two symmetries of the rectangle produces another symmetry of the rectangle. Thus if we let $S(\square) = \{e, r, v, h\}$ denote the set of symmetries of the rectangle, then composition defines a binary operation on $S(\square)$. In order to determine whether or not $S(\square)$ forms a group under composition, we first show all of the possible composites of symmetries in the following **Cayley table**.[5] The rows represent the first element and the columns represent the second element of the composite. Hence the h in the row labelled r and column labelled v represents $rv = h$ as illustrated in figure 5.6.

	e	r	v	h
e	e	r	v	h
r	r	e	h	v
v	v	h	e	r
h	h	v	r	e

We can now see that $S(\square)$ is a group under the operation of composition of symmetries. As mentioned above, composition is associative.

---

[5] Named after the English mathematician Arthur Cayley (1821–1879) who was the first person to introduce the concept of an abstract group.

Clearly the identity symmetry e is the identity element for the binary operation. Also each element is self-inverse because $ss = e$ for every symmetry $s \in S(\square)$. Note also that the operation is commutative — the Cayley table is symmetric about the top-left to bottom-right diagonal — so this is an Abelian group. This group is called the **Klein 4-group** after the German mathematician Felix Klein (1849–1925).

4. Let $S(\triangle)$ denote the symmetries of an equilateral triangle. The symmetries are the following, illustrated in figure 5.7.

e	the identity symmetry
r_1	rotation by $\frac{2\pi}{3} = 120°$ anti-clockwise
r_2	rotation by $\frac{4\pi}{3} = 240°$ anti-clockwise
s	reflection in the line ℓ_1
t	reflection in the line ℓ_2
u	reflection in the line ℓ_3

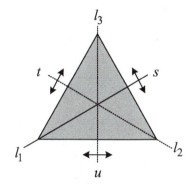

FIGURE 5.7: Symmetries of an equilateral triangle.

With the binary operation of composition of symmetries, as in example 2, the Cayley table for $S(\triangle) = \{e, r_1, r_2, s, s, u\}$ is the following.

	e	r_1	r_2	s	t	u
e	e	r_1	r_2	s	t	u
r_1	r_1	r_2	e	t	u	s
r_2	r_2	e	r_1	u	s	t
s	s	u	t	e	r_2	r_1
t	t	s	u	r_1	e	r_2
u	u	t	s	r_2	r_1	e

As in the previous example, composition is an associative binary operation with identity element e. Since $r_1 r_2 = e = r_2 r_1$ (from the Cayley table), it follows that $r_1^{-1} = r_2$ and $r_2^{-1} = r_1$. Each reflection x satisfies $xx = e$ and hence is self-inverse. Thus every element has an inverse. (It is always the case that the identity is self-inverse $e^{-1} = e$.) Therefore $S(\triangle)$ is a group under composition of symmetries.

Note that the group is non-Abelian. For example, from the Cayley table we have

$$r_1 s = t \quad \text{but} \quad s r_1 = u.$$

In fact, this is the smallest example of a non-Abelian group.

Having defined a group and given some examples, it is now time to turn our attention to proving some elementary results about groups.

When solving equations in 'ordinary' algebra, by which we mean the algebra of the real numbers, the following *Cancellation Property* is used frequently:

$$ax = ay \implies x = y.$$

For example, if $2x = 2y$, then we may deduce that $x = y$. The cancellation property is only valid for $a \neq 0$. For example, $0x = 0y$ is true for all real numbers x and y. The Cancellation Property should be stated more precisely as:

$$(a \neq 0 \text{ and } ax = ay) \implies x = y.$$

In fact, this property is a group theory property. It holds for the non-zero real numbers $\mathbb{R}^*$ because $(\mathbb{R}^*, \times)$ is a group; see example 5.4.2. In a group, we do not assume the binary operation is commutative, so there are two cancellation properties, which are called left cancellation and right cancellation.

Theorem 5.8 (Cancellation properties for groups)
Let $(G, *)$ be a group and let $g, x, y \in G$.

(i) Left cancellation: $g * x = g * y \implies x = y$.

(ii) Right cancellation: $x * g = y * g \implies x = y$.

Thinking about the proof, to proceed from $g * x = g * y$ to $x = y$ we need to find a way of 'removing' g from both sides, noting that we may only assume the properties of a group as given in definition 5.4. Since $g^{-1} * g = e$ (G3) and $e * x = x$ (G2), we may 'remove' g from both sides of $g * x = g * y$ by combining both sides on the left by g^{-1}.

Proof.

(i) Let G be a group and let $g, x, y \in G$. Then

$$\begin{aligned}
g * x = g * y \;\Rightarrow\; & g^{-1} * (g * x) = g^{-1} * (g * y) \\
\Rightarrow\; & (g^{-1} * g) * x = (g^{-1} * g) * y \quad \text{(by (G1))} \\
\Rightarrow\; & e * x = e * y \quad\quad\quad\quad\quad\; \text{(by (G3))} \\
\Rightarrow\; & x = y \quad\quad\quad\quad\quad\quad\;\; \text{(by (G2)).}
\end{aligned}$$

(ii) The proof of right cancellation is similar and is left as an exercise.

$\square$

Our second elementary property of groups concerns the interaction between the binary operation and taking inverses, and gives an expression for the inverse of $x * y$. Since the cancellation properties of the previous example were motivated by the cancellation properties of the real numbers, we might again look to the real numbers where

$$(xy)^{-1} = \frac{1}{xy} = \frac{1}{x} \times \frac{1}{y} = x^{-1} y^{-1}.$$

From this we might be tempted to conjecture that, in a group, $(x * y)^{-1} = x^{-1} * y^{-1}$. However this is incorrect. We can see this by considering the group $S(\triangle)$ in example 5.4.4. In this group

$$(r_1 s)^{-1} = t^{-1} = t \quad \text{but} \quad r_1^{-1} s^{-1} = r_2 s = u.$$

The problem with our conjectured generalisation is that multiplication of real numbers is commutative but, in general, the binary operation in a group is not. The correct result for groups is the following.

Theorem 5.9 ('Shoes and Socks' theorem for groups)
Let $(G, *)$ be a group and let $x, y \in G$. Then

$$(x * y)^{-1} = y^{-1} * x^{-1}.$$

Before considering the proof of the theorem, we first reflect on its name. When putting on a pair of shoes and socks, one first puts on the socks and then the shoes. When reversing this process, the order is reversed: the shoes are removed before the socks. So it is when taking the inverse of $x * y$: the inverses of x and y are combined in the opposite order.

Turning to the proof, we first need to understand what we *mean* by the inverse $(x * y)^{-1}$. The inverse of an element, as defined in (G3) in definition 5.4, is the element that combines with it (each way around) to produce the identity e. Thus the inverse of $x * y$ is whatever element $g \in G$ satisfies the property $(x * y) * g = e$ and $g * (x * y) = e$. So the proof of the theorem just involves checking that $g = y^{-1} * x^{-1}$ satisfies this property.

Proof. Let $(G, *)$ be a group and let $x, y \in G$. Then

$$
\begin{aligned}
(x * y) * (y^{-1} * x^{-1}) &= x * (y * (y^{-1} * x^{-1})) &\text{(by (G1))}\\
&= x * ((y * y^{-1}) * x^{-1}) &\text{(by (G1))}\\
&= x * (e * x^{-1}) &\text{(by (G3))}\\
&= x * x^{-1} &\text{(by (G2))}\\
&= e &\text{(by (G3)).}
\end{aligned}
$$

The proof that $(y^{-1} * x^{-1}) * (x * y) = e$ is similar and is left as an exercise.

Therefore, by the definition of inverse (G3),

$$
(x * y)^{-1} = y^{-1} * x^{-1}.
$$

□

Our last result about groups in this chapter concerns groups where each element x satisfies $x * x = e$. An example of such a group is the Klein 4-group given in example 5.4.3.

Before we consider the theorem, we wish to simplify our notation for groups. In the previous two examples, it has been a little cumbersome to write the binary operation as $x * y$. It is very common to omit the binary operation symbol and write $x * y = xy$. This is sometimes referred to as writing the group operation 'multiplicatively'. We may extend this idea and use a power notation as follows:

$$
x^2 = x * x, \quad x^3 = x * x^2 = x * (x * x), \quad \ldots \ .
$$

We will also simplify the way we refer to a group and use the phrase 'a group G' as shorthand for 'a group $(G, *)$'.

Theorem 5.10
Let G be a group such that $x^2 = e$ for every element $x \in G$. Then G is Abelian.

Recall that an Abelian group is one that satisfies the commutative property: $xy = yx$ for all $x, y \in G$. To obtain a direct proof of the theorem, we need to assume

$$
x^2 = e \text{ for every element } x \in G
$$

and from this deduce
$$xy = yx \text{ for all } x, y \in G.$$

There are two key observations that allow us to construct an appropriate chain of deductions. The first is the observation that, if $x^2 = e$, then x is self-inverse $x^{-1} = x$. This is because the inverse of x is precisely the element y satisfying $xy = e$ and $yx = e$. The second key observation is the 'Shoes and socks' theorem 5.9 which, in our simplified notation, gives $(xy)^{-1} = y^{-1}x^{-1}$. If every element satisfies $x^2 = e$, then every element is self-inverse so the equation $(xy)^{-1} = y^{-1}x^{-1}$ simplifies to $xy = yx$, which is what we needed to deduce. We can now organise these considerations into a coherent argument as follows.

Proof. Let G be a group such that $x^2 = e$ for every element $x \in G$.

Let $x, y \in G$. Then
$$x^2 = e, \quad y^2 = e \quad \text{and} \quad (xy)^2 = e$$

so each of these elements is self-inverse
$$x^{-1} = x, \quad y^{-1} = y \quad \text{and} \quad (xy)^{-1} = xy.$$

Hence $\quad xy = (xy)^{-1} \quad$ (since xy is self-inverse)

$\qquad\quad = y^{-1}x^{-1} \quad$ (by theorem 5.9)

$\qquad\quad = yx \qquad$ (since x and y are self-inverse).

Therefore G is Abelian. $\qquad\qquad\qquad\qquad\qquad\qquad\qquad\qquad\qquad$ □

5.3.2 Linear algebra

The area of mathematics known as linear algebra studies objects called **vector spaces** which generalise the notion of vectors in two dimensions. Linear algebra is closely related to the theory of matrices. A vector in two dimensions is a quantity that has magnitude and direction; it can be represented as an ordered pair of real numbers or as a **column vector**,

$$\mathbf{a} = (a_1, a_2) = \begin{pmatrix} a_1 \\ a_2 \end{pmatrix}$$

as shown in figure 5.8. The **magnitude** of $\mathbf{a}$, which is denoted $\|\mathbf{a}\|$, is represented by the length of the line and, by Pythagoras' theorem, is given by

$$\|\mathbf{a}\| = \sqrt{a_1^2 + a_2^2}.$$

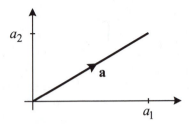

FIGURE 5.8: A vector in $\mathbb{R}^2$.

There are two natural operations that can be performed on vectors in $\mathbb{R}^2$. Two vectors **a** and **b** may be added according to the rule

$$\mathbf{a} + \mathbf{b} = \begin{pmatrix} a_1 \\ a_2 \end{pmatrix} + \begin{pmatrix} b_1 \\ b_2 \end{pmatrix} = \begin{pmatrix} a_1 + b_1 \\ a_2 + b_2 \end{pmatrix}.$$

Also, a vector **a** may be multiplied by a scalar λ, which is just a real number, according to the rule

$$\lambda \mathbf{a} = \lambda \begin{pmatrix} a_1 \\ a_2 \end{pmatrix} = \begin{pmatrix} \lambda a_1 \\ \lambda a_2 \end{pmatrix}.$$

Geometrically, these operations of addition and scalar multiplication are illustrated in figure 5.9.

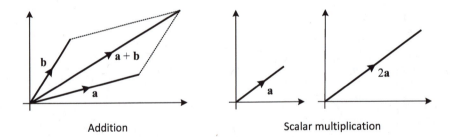

Addition   Scalar multiplication

FIGURE 5.9: Addition and scalar multiplication of vectors in $\mathbb{R}^2$.

The vectors in $\mathbb{R}^2$ satisfy various properties. For example, addition is commutative: $\mathbf{a} + \mathbf{b} = \mathbf{b} + \mathbf{a}$ for all vectors $\mathbf{a}, \mathbf{b}$. If we collect a 'reasonable' collection of properties that the vectors in $\mathbb{R}^2$ satisfy, we obtain the definition of a vector space. In essence, a vector space is any set of objects, which we call 'vectors', that may be added and multiplied by scalars (real numbers) and which satisfy properties similar to those of vectors in $\mathbb{R}^2$. The formal definition is the following.

Definition 5.5

A **vector space** comprises:

- a set V whose elements are called **vectors**;
- a rule of addition of vectors: if $\mathbf{u}, \mathbf{v} \in V$, then $\mathbf{u} + \mathbf{v} \in V$;
- a rule for multiplying vectors by real numbers (called **scalars**): if $\mathbf{v} \in S$ and $\lambda \in \mathbb{R}$, then $\lambda \mathbf{v} \in V$

such that the following properties are satisfied.

Addition properties

(A1)   For all $\mathbf{u}, \mathbf{v} \in V$, $\mathbf{u} + \mathbf{v} = \mathbf{v} + \mathbf{u}$.

(A2)   For all $\mathbf{u}, \mathbf{v}, \mathbf{w} \in V$, $(\mathbf{u} + \mathbf{v}) + \mathbf{w} = \mathbf{u} + (\mathbf{v} + \mathbf{w})$.

(A3)   There is a **zero vector** $\mathbf{0} \in V$ such that, for all $\mathbf{v} \in V$,
$\mathbf{0} + \mathbf{v} = \mathbf{v} = \mathbf{v} + \mathbf{0}$.

(A4)   For every $\mathbf{v} \in V$ there is a vector $-\mathbf{v} \in V$, called the **negative** of $\mathbf{x}$,
such that $\mathbf{v} + (-\mathbf{v}) = \mathbf{0}$.

Scalar multiplication properties

(M1)   For all $\mathbf{v} \in V$, $\lambda, \mu \in \mathbb{R}$, $\lambda(\mu \mathbf{v}) = (\lambda \mu) \mathbf{v}$.

(M2)   For all $\mathbf{u}, \mathbf{v} \in V$, $\lambda \in \mathbb{R}$, $\lambda(\mathbf{u} + \mathbf{v}) = \lambda \mathbf{u} + \lambda \mathbf{v}$.

(M3)   For all $\mathbf{v} \in V$, $\lambda, \mu \in \mathbb{R}$, $(\lambda + \mu)\mathbf{v} = (\lambda \mathbf{v}) + (\mu \mathbf{v})$.

(M4)   For all $\mathbf{v} \in V$, $1\mathbf{v} = \mathbf{v}$.

The properties (A1) – (A4) and (M1) – (M4) are often referred to as the vector space axioms. Readers who are familiar with group theory, or who have read the previous section, may have noticed that addition of vectors is a binary operation on V and that the conditions (A1) to (A4) define an Abelian group — see definition 5.4. In other words, forgetting about scalar multiplication, the pair $(V, +)$ is an Abelian group. This means that anything we can prove about groups will apply to V under the operation of addition. Scalar multiplication is not a binary operation as it combines two different *kinds* of things: real numbers (scalars) and elements of V (vectors). In fact, scalar multiplication defines a mapping,

$$\mathbb{R} \times V \to V, \ (\lambda, \mathbf{x}) \mapsto \lambda \mathbf{x}.$$

We now consider some examples of vector spaces and then prove some fairly straightforward properties of vector spaces.

Examples 5.5

1. The set
$$\mathbb{R}^2 = \left\{ \begin{pmatrix} x_1 \\ x_2 \end{pmatrix} : x_1, x_2 \in \mathbb{R} \right\}$$
is a vector space with addition and scalar multiplication defined above.

Similarly, vectors in $\mathbb{R}^3$, which have three components, also form a vector space with similar definitions for addition and scalar multiplication. More generally, define

$$\mathbb{R}^n = \left\{ \begin{pmatrix} x_1 \\ x_2 \\ \vdots \\ x_n \end{pmatrix} : x_1, x_2, \ldots, x_n \in \mathbb{R} \right\}$$

to be the set of n-dimensional vectors. Algebraically, there is nothing very special about having vectors with n components rather than two or three. Geometrically, it is harder to visualise n-dimensional space, of course. Addition and scalar multiplication are defined in $\mathbb{R}^n$ in the obvious way. If

$$\mathbf{x} = \begin{pmatrix} x_1 \\ x_2 \\ \vdots \\ x_n \end{pmatrix}, \mathbf{y} = \begin{pmatrix} y_1 \\ y_2 \\ \vdots \\ y_n \end{pmatrix} \in \mathbb{R}^n \quad \text{and} \quad \lambda \in \mathbb{R},$$

then

$$\mathbf{x} + \mathbf{y} = \begin{pmatrix} x_1 + y_1 \\ x_2 + y_2 \\ \vdots \\ x_n + y_n \end{pmatrix} \quad \text{and} \quad \lambda \mathbf{x} = \begin{pmatrix} \lambda x_1 \\ \lambda x_2 \\ \vdots \\ \lambda x_n \end{pmatrix}.$$

With these definitions, $\mathbb{R}^n$ is a vector space. The zero vector is

$$\mathbf{0} = \begin{pmatrix} 0 \\ 0 \\ \vdots \\ 0 \end{pmatrix}$$

and the negative of $\mathbf{x}$ is

$$-\mathbf{x} = \begin{pmatrix} -x_1 \\ -x_2 \\ \vdots \\ -x_n \end{pmatrix}.$$

2. Let

$$M_{2 \times 2}(\mathbb{R}) = \left\{ \begin{pmatrix} a & b \\ c & d \end{pmatrix} : a, b, c, d \in \mathbb{R} \right\}$$

be the set of all 2×2 real matrices (matrices with real number entries). Matrices in $M_{2 \times 2}(\mathbb{R})$ may be added and multiplied by scalars:

$$\begin{pmatrix} a_{11} & a_{12} \\ a_{21} & a_{22} \end{pmatrix} + \begin{pmatrix} b_{11} & b_{12} \\ b_{21} & b_{22} \end{pmatrix} = \begin{pmatrix} a_{11} + b_{11} & a_{12} + b_{12} \\ a_{21} + b_{21} & a_{22} + b_{22} \end{pmatrix}$$

and

$$\lambda \begin{pmatrix} a_{11} & a_{12} \\ a_{21} & a_{22} \end{pmatrix} = \begin{pmatrix} \lambda a_{11} & \lambda a_{12} \\ \lambda a_{21} & \lambda a_{22} \end{pmatrix}.$$

With these definitions, $M_{2\times 2}(\mathbb{R})$ forms a vector space. (We may also multiply matrices in $M_{2\times 2}(\mathbb{R})$, but this is not relevant so far as forming a vector space is concerned.)

3. Let $\mathscr{F} = \{\text{functions } \mathbb{R} \to \mathbb{R}\}$ be the set of all functions $\mathbb{R} \to \mathbb{R}$. We may add two functions and multiply a function by a scalar. Given functions f and g, define $f + g$ by

$$f + g : \mathbb{R} \to \mathbb{R} \quad \text{by} \quad (f + g)(x) = f(x) + g(x).$$

Similarly, given $f \in \mathscr{F}$ and $\lambda \in \mathbb{R}$, we define the function λf by

$$\lambda f : \mathbb{R} \to \mathbb{R}, \ (\lambda f)(x) = \lambda f(x).$$

For example, if f and g are defined by $f(x) = x^2$ and $g(x) = \sin x$, then $(f + g)(x) = f(x) + g(x) = x^2 + \sin x$, $(3f)(x) = 3f(x) = 3x^2$, and so on.

With these definitions $\mathscr{F}$ is a vector space. The 'vectors' are the functions in $\mathscr{F}$. The zero function, satisfying axiom (A3) in definition 5.5, is zero $\in \mathscr{F}$ defined by

$$\text{zero} : \mathbb{R} \to \mathbb{R}, \ \text{zero}(x) = 0.$$

Similarly, the negative of $f \in \mathscr{F}$ is the function $-f \in \mathscr{F}$ defined by

$$-f : \mathbb{R} \to \mathbb{R}, \ (-f)(x) = -f(x).$$

We begin by considering some elementary properties of vector spaces. The following properties of vector spaces give simple interactions between addition and scalar multiplications of vectors.

Theorem 5.11 (Elementary properties of a vector space)
Let V be a vector space. For all $\mathbf{v} \in V$ and $\lambda \in \mathbb{R}$, we have:

(i) $0\mathbf{v} = \mathbf{0}$

(ii) $\lambda \mathbf{0} = \mathbf{0}$

(iii) $(-1)\mathbf{v} = -\mathbf{v}$

(iv) $\lambda \mathbf{v} = \mathbf{0} \ \Rightarrow \ \lambda = 0$ or $\mathbf{v} = \mathbf{0}$ or both.

Consider property (i). This says that taking the scalar multiple of $0 \in \mathbb{R}$ with any vector $\mathbf{v} \in V$ gives the zero vector $\mathbf{0} \in V$. This might seem very obvious, especially if we are thinking of $\mathbb{R}^2$ where

$$0 \begin{pmatrix} x_1 \\ x_2 \end{pmatrix} = \begin{pmatrix} 0 \\ 0 \end{pmatrix} = \mathbf{0}.$$

This is not a proof, of course, because it refers to a specific vector space. So, how do we prove property (i)?

By axiom (A3) of the definition of a vector space, the zero vector satisfies

$$\mathbf{0} + \mathbf{v} = \mathbf{v}.$$

We first show that $0\mathbf{v}$ also satisfies this property:

$$
\begin{aligned}
0\mathbf{v} + \mathbf{v} &= 0\mathbf{v} + 1\mathbf{v} & \text{(by axiom (M4))} \\
&= (0 + 1)\mathbf{v} & \text{(by axiom (M2))} \\
&= 1\mathbf{v} & \text{(property of } \mathbb{R}) \\
&= \mathbf{v} & \text{(by axiom (M4))}.
\end{aligned}
$$

Now, *if we knew* that the zero vector was unique, then the proof is just about complete. In other words, if we knew that the zero vector was the *only* vector satisfying $\mathbf{0} + \mathbf{v} = \mathbf{v}$, then we could deduce $0\mathbf{v} = \mathbf{0}$. In fact, the zero vector is unique, but unfortunately we have not proved it. We will consider uniqueness proofs in section 7.5. This means we need to work a little harder to complete the proof. The idea is to add $-\mathbf{v}$ to both sides of the equation $0\mathbf{v} + \mathbf{v} = \mathbf{v}$ established above, as this will give the zero vector on the right-hand side. We have:

$$
\begin{aligned}
(0\mathbf{v} + \mathbf{v}) - \mathbf{v} &= \mathbf{v} - \mathbf{v} \\
\Rightarrow \quad 0\mathbf{v} + (\mathbf{v} - \mathbf{v}) &= \mathbf{v} - \mathbf{v} & \text{(by axiom (A2))} \\
\Rightarrow \quad 0\mathbf{v} + \mathbf{0} &= \mathbf{0} & \text{(by axiom (A4))} \\
\Rightarrow \quad 0\mathbf{v} &= \mathbf{0} & \text{(by axiom (A3))}.
\end{aligned}
$$

We have now found a chain of reasoning that establishes the result in part (i) above. We can now present it as a coherent proof as follows.

Proof of (i). Let V be a vector space and let $\mathbf{v} \in V$. Then

$$
\begin{aligned}
0\mathbf{v} + \mathbf{v} &= 0\mathbf{v} + 1\mathbf{v} & \text{(by axiom (M4))} \\
&= (0 + 1)\mathbf{v} & \text{(by axiom (M2))} \\
&= 1\mathbf{v} & \text{(property of } \mathbb{R}) \\
&= \mathbf{v} & \text{(by axiom (M4))}.
\end{aligned}
$$

Now, adding $-\mathbf{v}$ to both sides gives

$$(0\mathbf{v} + \mathbf{v}) - \mathbf{v} = \mathbf{v} - \mathbf{v}$$

$\Rightarrow \qquad 0\mathbf{v} + (\mathbf{v} - \mathbf{v}) = \mathbf{v} - \mathbf{v}$ \qquad (by axiom (A2))

$\Rightarrow \qquad \qquad 0\mathbf{v} + \mathbf{0} = \mathbf{0}$ \qquad (by axiom (A4))

$\Rightarrow \qquad \qquad \qquad 0\mathbf{v} = \mathbf{0}$ \qquad (by axiom (A3)).

Therefore $0\mathbf{v} = \mathbf{0}$, as required. □

We will leave the proof of parts (ii) and (iii) as exercises and consider, instead, part (iv).

To prove $\lambda\mathbf{v} = \mathbf{0} \Rightarrow \lambda = 0$ or $\mathbf{v} = \mathbf{0}$ by the method of direct proof, we assume $\lambda\mathbf{v} = \mathbf{0}$ and, from this, deduce $\lambda = 0$ or $\mathbf{v} = \mathbf{0}$. To establish the conclusion we show that if $\lambda \neq 0$, then we must have $\mathbf{v} = \mathbf{0}$. In other words, what we actually prove is

$$(\lambda\mathbf{v} = \mathbf{0} \text{ and } \lambda \neq 0) \Rightarrow \mathbf{v} = \mathbf{0}.$$

Essentially we are using the logical equivalence

$$P \Rightarrow (Q \vee R) \equiv (P \wedge Q) \Rightarrow R,$$

which is easily established using truth tables.

So suppose $\lambda\mathbf{v} = \mathbf{0}$ and $\lambda \neq 0$. Since λ is non-zero, it has a reciprocal $1/\lambda$. The idea is to multiply both sides of $\lambda\mathbf{v} = \mathbf{0}$ by $1/\lambda$ to obtain $\mathbf{v} = \mathbf{0}$. Having found the overall structure of the proof, we can fill in the precise details as follows.

Proof of (iv). Let V be a vector space and let $\mathbf{v} \in V$ and $\lambda \in \mathbb{R}$ be such that $\lambda\mathbf{v} = \mathbf{0}$.

Suppose $\lambda \neq 0$. Then $\dfrac{1}{\lambda} \in \mathbb{R}$, so:

$$\lambda\mathbf{v} = \mathbf{0} \quad \Rightarrow \quad \frac{1}{\lambda}(\lambda\mathbf{v}) = \frac{1}{\lambda}\mathbf{0}$$

$$\Rightarrow \quad \left(\frac{1}{\lambda}\lambda\right)\mathbf{v} = \frac{1}{\lambda}\mathbf{0} \qquad \text{(by axiom (M1))}$$

$$\Rightarrow \quad 1\mathbf{v} = \frac{1}{\lambda}\mathbf{0} \qquad \text{(property of } \mathbb{R}\text{)}$$

$$\Rightarrow \quad \mathbf{v} = \frac{1}{\lambda}\mathbf{0} \qquad \text{(by axiom (M4))}$$

$$\Rightarrow \quad \mathbf{v} = \mathbf{0} \qquad \text{(by property (ii)).}$$

□

In order to look at some more advanced — and more interesting — aspects of vector spaces, we will first need to define some additional terminology. We have chosen to explore the properties of subspaces. A subspace of a vector space V is simply a subset of V that is also itself a vector space.

Definition 5.6

Let V be a vector space and let S be a subset of V, $S \subseteq V$. Then S is a **subspace** of V if S is itself a vector space with the same operations of addition and scalar multiplication that are defined in V.

If we regard $\mathbb{R}^2$ as being the subset of $\mathbb{R}^3$ comprising those vectors with final coordinate equal to 0,

$$\mathbb{R}^2 = \left\{ \begin{pmatrix} x_1 \\ x_2 \\ 0 \end{pmatrix} : x_1, x_2 \in \mathbb{R} \right\},$$

then $\mathbb{R}^2$ is a subspace of $\mathbb{R}^3$ since we have noted, in example 5.5.1, that $\mathbb{R}^2$ is a vector space.

Example 5.6

Show that

$$S = \left\{ \begin{pmatrix} s - t \\ -s - t \\ 2t \end{pmatrix} : s, t \in \mathbb{R} \right\}$$

is a subspace of $\mathbb{R}^3$. The set S comprises the points lying on a plane in $\mathbb{R}^3$ passing through the origin, illustrated in figure 5.10.

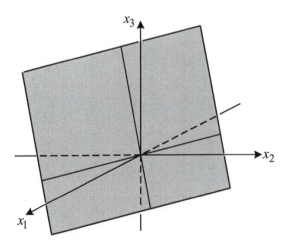

FIGURE 5.10: A plane in $\mathbb{R}^3$ passing through the origin.

Proof. To show that S is a subspace of $\mathbb{R}^3$, we need to verify that S is a vector space with the usual operations of addition and scalar multiplication of vectors in $\mathbb{R}^3$. Before we start checking the conditions (A1)–(A4) and (M1)–M4) in definition 5.5, we must verify that addition and scalar multiplication are valid operations *in* S. For addition, this means that adding two vectors in S gives a vector that also lies in S.

Let $\mathbf{v}, \mathbf{w} \in S$. Then, for some real numbers p, q, s, t we have

$$\mathbf{v} = \begin{pmatrix} p - q \\ -p - q \\ 2q \end{pmatrix} \quad \text{and} \quad \mathbf{w} = \begin{pmatrix} s - t \\ -s - t \\ 2t \end{pmatrix}.$$

Then

$$\mathbf{v} + \mathbf{w} = \begin{pmatrix} p - q \\ -p - q \\ 2q \end{pmatrix} + \begin{pmatrix} s - t \\ -s - t \\ 2t \end{pmatrix}$$

$$= \begin{pmatrix} p - q + s - t \\ -p - q - s - t \\ 2p + 2t \end{pmatrix} = \begin{pmatrix} (p + s) - (q + t) \\ -(p + s) - (q + t) \\ 2(q + t) \end{pmatrix}.$$

Hence $\mathbf{v} + \mathbf{w}$ belongs to S because is can be expressed as

$$\mathbf{v} + \mathbf{w} = \begin{pmatrix} a - b \\ -a - b \\ 2b \end{pmatrix},$$

where $a = p + s \in \mathbb{R}$ and $b = q + t \in \mathbb{R}$. Hence addition is defined as an operation on the set S.

Next we verify that scalar multiplication is a well-defined operation on S; in other words, if $\mathbf{v} \in S$ and $\lambda \in \mathbb{R}$, then $\lambda \mathbf{v} \in S$. With $\mathbf{v}$ defined as above, we have

$$\lambda \mathbf{v} = \lambda \begin{pmatrix} p - q \\ -p - q \\ 2q \end{pmatrix} = \begin{pmatrix} \lambda p - \lambda q \\ -\lambda p - \lambda q \\ 2\lambda q \end{pmatrix},$$

which is an element of S since $\lambda p, \lambda q \in \mathbb{R}$.

Having shown that addition and scalar multiplication are *bona fide* operations on S, we now need to verify that the axioms (A1)–(A4) and (M1)–M4) in definition 5.5 are satisfied.

(A1) The equation $\mathbf{u} + \mathbf{v} = \mathbf{v} + \mathbf{u}$ is satisfied for all vectors in S because is is satisfied for all vectors in $\mathbb{R}^3$ and $S \subseteq \mathbb{R}^3$.

(A2) Similarly, $(\mathbf{u} + \mathbf{v}) + \mathbf{w} = \mathbf{u} + (\mathbf{v} + \mathbf{w})$ is satisfied for all vectors in S because is is satisfied in $\mathbb{R}^3$.

(A3) The zero vector belongs to S since

$$\mathbf{0} = \begin{pmatrix} 0 - 0 \\ -0 - 0 \\ 2 \times 0 \end{pmatrix}.$$

Clearly, $\mathbf{0} + \mathbf{v} = \mathbf{v} = \mathbf{v} + \mathbf{0}$ is satisfied for vectors in S because it is satisfied for vectors in $\mathbb{R}^3$.

(A4) If $\mathbf{v} \in S$ is defined as above, then its negative

$$-\mathbf{v} = -\begin{pmatrix} p - q \\ -p - q \\ 2q \end{pmatrix} = \begin{pmatrix} -p - (-q) \\ -(-p) - (-q) \\ 2(-q) \end{pmatrix} \in S$$

since $-p, -q \in \mathbb{R}$. Then it follows that $\mathbf{v} + (-\mathbf{v}) = \mathbf{0}$ because this equation holds in $\mathbb{R}^3$.

(M1) In a similar way to condition (A1), the equation $\lambda(\mu\mathbf{v}) = (\lambda\mu)\mathbf{v}$ holds for vectors in S (and scalars in $\mathbb{R}$) because it holds in $\mathbb{R}^3$.

(M2) For all $\mathbf{u}, \mathbf{v} \in S$, $\lambda \in \mathbb{R}$, $\lambda(\mathbf{u} + \mathbf{v}) = \lambda\mathbf{u} + \lambda\mathbf{v}$, since this holds in $\mathbb{R}^3$.

(M3) Similarly, for all $\mathbf{v} \in S$, $\lambda, \mu \in \mathbb{R}$, $(\lambda + \mu)\mathbf{v} = (\lambda\mathbf{v}) + (\mu\mathbf{v})$, since this holds in $\mathbb{R}^3$.

(M4) Finally, for all $\mathbf{v} \in S$, $1\mathbf{v} = \mathbf{v}$, again because this holds in $\mathbb{R}^3$.

Since addition and scalar multiplication are defined in S and all the axioms in definition 5.5 are satisfied, S is a vector space. Therefore S is a subspace of $\mathbb{R}^3$. □

Much of the proof that S is a subspace of $\mathbb{R}^3$ in the previous example can be categorised as 'this condition holds in the subset S because it holds in $\mathbb{R}^3$'. Essentially, once we had established that addition and scalar multiplication were well defined in S, the vector space axioms follow for S because they hold in the larger space $\mathbb{R}^3$. This is formalised in the following theorem, which we will not prove. Any vector space must be non-empty since it has to contain the zero vector, and this is the reason for the first condition in the theorem.

Theorem 5.12 (Subspace Test)
Let $S \subseteq V$ where V is a vector space. If S is

   (i) non-empty: $S \neq \varnothing$,

  (ii) closed under addition: $\mathbf{v}, \mathbf{w} \in S \Rightarrow \mathbf{v} + \mathbf{w} \in S$,

 (iii) closed under scalar multiplication: $\mathbf{v} \in S$ and $\lambda \in \mathbb{R} \Rightarrow \lambda\mathbf{v} \in S$,

then S is a subspace of V.

The second and third properties are usually summarised by stating that S is **closed** under addition and scalar multiplication. We will now use theorem 5.12 to explore the properties of subspaces.

Theorem 5.13 (Intersection of subspaces)
Let S and T be subspaces of a vector space V. Then their intersection $S \cap T$ is a subspace of V.

Theorem 5.12 provides a generic framework for proving that a subset of a vector space is a subspace. If we can verify that the conditions (i), (ii), and (iii) of the theorem are satisfied for a particular set, then it follows that the set is a subspace.

Proof. Let S and T be subspaces of a vector space V. Recall that $S \cap T = \{\mathbf{v} : \mathbf{v} \in S \text{ and } \mathbf{v} \in T\}$.

(i) Each subspace contains the zero vector, $\mathbf{0} \in S$ and $\mathbf{0} \in T$. Hence $\mathbf{0} \in S \cap T$, from which it follows that $S \cap T$ is non-empty, $S \cap T \neq \emptyset$.

(ii) Let $\mathbf{v}, \mathbf{w} \in S \cap T$.

Then $\mathbf{v}, \mathbf{w} \in S$ and hence, as S is a vector space, $\mathbf{v} + \mathbf{w} \in S$. Similarly, $\mathbf{v}, \mathbf{w} \in T$ and so, as T is a vector space, $\mathbf{v} + \mathbf{w} \in T$.

Therefore $\mathbf{v} + \mathbf{w} \in S \cap T$.

(iii) Let $\mathbf{v} \in S \cap T$ and $\lambda \in \mathbb{R}$.

Then $\mathbf{v} \in S$ and hence, as S is a vector space, $\lambda\mathbf{v} \in S$. Similarly, $\mathbf{v} \in T$ and so, as T is a vector space, $\lambda\mathbf{v} \in T$.

Therefore $\lambda\mathbf{v} \in S \cap T$.

It now follows, from theorem 5.17, that $S \cap T$ is a subspace of V. □

Example 5.7
Let $\mathbf{A}$ be an $n \times n$ square matrix and let $\lambda \in \mathbb{R}$. Consider the equation

$$\mathbf{A}\mathbf{x} = \lambda\mathbf{x},$$

where $\mathbf{x} \in \mathbb{R}^n$. For most values of λ, the equation will only have the zero solution, $\mathbf{x} = \mathbf{0}$.

For example, let $\mathbf{A} = \begin{pmatrix} 3 & -1 \\ 4 & -2 \end{pmatrix}$ and let $\lambda = 5$. Then

$$\mathbf{A}\mathbf{x} = \begin{pmatrix} 3 & -1 \\ 4 & -2 \end{pmatrix}\begin{pmatrix} x_1 \\ x_2 \end{pmatrix} = 5\begin{pmatrix} x_1 \\ x_2 \end{pmatrix} \Rightarrow \begin{cases} 3x_1 - x_2 = 5x_1 \\ 4x_1 - 2x_2 = 5x_2 \end{cases}$$

$$\Rightarrow \begin{cases} 2x_1 - x_2 = 0 & \text{(i)} \\ 4x_1 - 7x_2 = 0 & \text{(ii)} \end{cases}$$

$$\Rightarrow \quad 5x_2 = 0 \quad (2 \times \text{(i)} - \text{(ii)})$$

$$\Rightarrow \quad x_2 = 0$$

$$\Rightarrow \quad x_1 = 0 \quad \text{(from (i))}.$$

Hence, when $\lambda = 5$, the only solution to $\mathbf{A}\mathbf{x} = \lambda\mathbf{x}$ is $\mathbf{x} = \mathbf{0}$.

However there may be particular values for λ for which the equation $\mathbf{A}\mathbf{x} = \lambda\mathbf{x}$ has non-zero solutions $\mathbf{x} \neq \mathbf{0}$. For example,

$$\begin{pmatrix} 3 & -1 \\ 4 & -2 \end{pmatrix} \begin{pmatrix} 1 \\ 1 \end{pmatrix} = \begin{pmatrix} 2 \\ 2 \end{pmatrix} = 2 \begin{pmatrix} 1 \\ 1 \end{pmatrix}$$

so, for $\lambda = 2$ there is a non-zero solution to the equation $\mathbf{A}\mathbf{x} = \lambda\mathbf{x}$, namely

$$\mathbf{x} = \begin{pmatrix} 1 \\ 1 \end{pmatrix}.$$

Here the value $\lambda = 2$ is called an **eigenvalue** of the matrix $\mathbf{A}$ and any non-zero vector satisfying $\mathbf{A}\mathbf{x} = 2\mathbf{x}$ is called an **eigenvector** corresponding to eigenvalue 2.

In fact, any vector of the form

$$\mathbf{x} = \begin{pmatrix} t \\ t \end{pmatrix}$$

where $t \in \mathbb{R}$ and $t \neq 0$ is an eigenvector corresponding to $\lambda = 2$ since it also satisfies $\mathbf{A}\mathbf{x} = 2\mathbf{x}$:

$$\begin{pmatrix} 3 & -1 \\ 4 & -2 \end{pmatrix} \begin{pmatrix} t \\ t \end{pmatrix} = \begin{pmatrix} 2t \\ 2t \end{pmatrix} = 2 \begin{pmatrix} t \\ t \end{pmatrix}.$$

The matrix $\mathbf{A}$ has another eigenvalue. Since

$$\begin{pmatrix} 3 & -1 \\ 4 & -2 \end{pmatrix} \begin{pmatrix} 1 \\ 4 \end{pmatrix} = \begin{pmatrix} -1 \\ -4 \end{pmatrix} = - \begin{pmatrix} 1 \\ 4 \end{pmatrix},$$

it follows that $\lambda = -1$ is an eigenvalue and

$$\mathbf{x} = \begin{pmatrix} 1 \\ 4 \end{pmatrix}$$

is a corresponding eigenvector. Again, it is easy to verify that any vector of the form

$$\mathbf{x} = \begin{pmatrix} t \\ 4t \end{pmatrix}$$

where $t \in \mathbb{R}$ and $t \neq 0$ is an eigenvector corresponding to $\lambda = -1$ since it also satisfies $\mathbf{A}\mathbf{x} = -\mathbf{x}$. Although we shall not prove this, the values 2 and -1 are the only eigenvalues for the matrix $\mathbf{A}$; for any other value of λ, the equation $\mathbf{A}\mathbf{x} = \lambda\mathbf{x}$ only has the zero solution $\mathbf{x} = \mathbf{0}$.

Note that, for each eigenvector λ, the set $E(\lambda)$, of all eigenvectors (together with $\mathbf{0}$) is a line in $\mathbb{R}^2$ through the origin:

$$E(2) = \left\{ \begin{pmatrix} t \\ t \end{pmatrix} : t \in \mathbb{R} \right\} \quad \text{and} \quad E(-1) = \left\{ \begin{pmatrix} t \\ 4t \end{pmatrix} : t \in \mathbb{R} \right\}.$$

These sets are shown in figure 5.11.

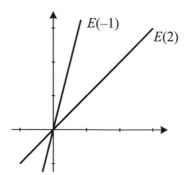

FIGURE 5.11: The sets of eigenvectors $E(2)$ and $E(-1)$.

Recall that the plane passing through the origin given in example 5.6 was a subspace of $\mathbb{R}^3$. It is straightforward to use theorem 5.12 to show that each of the sets $E(2)$ and $E(-1)$ is a subspace of $\mathbb{R}^2$; this is left as an exercise. Instead, we consider the corresponding result for any eigenvalue of any square matrix.

Theorem 5.14 (Space of eigenvectors)
Let $\mathbf{A}$ be an $n \times n$ matrix and let λ be an eigenvalue of $\mathbf{A}$. In other words, the equation $\mathbf{Ax} = \lambda\mathbf{x}$ has non-zero solutions for $\mathbf{x} \in \mathbb{R}^n$. Then the set of all eigenvectors corresponding to λ (together with $\mathbf{0}$),

$$E(\lambda) = \{\mathbf{x} \in \mathbb{R}^n : \mathbf{Ax} = \lambda\mathbf{x}\},$$

is a subspace of $\mathbb{R}^n$.

The set $E(\lambda)$ is called the **eigenspace** corresponding to λ.

The Subspace Test, theorem 5.12, again provides the framework for the proof. This says that, to show $E(\lambda)$ is a subspace of $\mathbb{R}^n$, we need to show that it is non-empty, closed under addition, and closed under scalar multiplication. Consider the condition of closure under addition: if $\mathbf{x}, \mathbf{y} \in E(\lambda)$, then $\mathbf{x} + \mathbf{y} \in E(\lambda)$. We will give a direct proof of this, so begin by assuming $\mathbf{x}, \mathbf{y} \in E(\lambda)$. We need to find a chain of deductions ending with $\mathbf{x} + \mathbf{y} \in E(\lambda)$:

$$\mathbf{x}, \mathbf{y} \in E(\lambda) \implies \cdots$$
$$\implies \mathbf{x} + \mathbf{y} \in E(\lambda).$$

The first step in the deduction is to understand what it means to say $\mathbf{x}$ and $\mathbf{y}$ belong to $E(\lambda)$. It is often the case that the first step in the deduction amounts to an 'unpacking' of the initial assumption. By the definition of $E(\lambda)$, this means that $\mathbf{Ax} = \lambda\mathbf{x}$ and $\mathbf{Ay} = \lambda\mathbf{y}$. To prove that the sum $\mathbf{x} + \mathbf{y} \in E(\lambda)$, we need to show that $\mathbf{x} + \mathbf{y}$ also satisfies the same equation defining $E(\lambda)$:

$\mathbf{A}(\mathbf{x} + \mathbf{y}) = \lambda(\mathbf{x} + \mathbf{y})$. This follows by adding the two equations for $\mathbf{x}$ and $\mathbf{y}$ and using the following properties of matrices and vectors:

$$\mathbf{Ax} + \mathbf{Ay} = \mathbf{A}(\mathbf{x} + \mathbf{y}) \quad \text{and} \quad \lambda\mathbf{x} + \lambda\mathbf{y} = \lambda(\mathbf{x} + \mathbf{y}).$$

These considerations allow us to construct a simple chain of deductions from $\mathbf{x}, \mathbf{y} \in E(\lambda)$ to $\mathbf{x} + \mathbf{y} \in E(\lambda)$, which we include in the proof below.

The proof that $E(\lambda)$ is closed under scalar multiplication is similar to that for addition, so we include it directly in the proof without the preamble given here.

Proof. Let $\mathbf{A}$ be an $n \times n$ matrix and let λ be an eigenvalue of $\mathbf{A}$.

First note that $\mathbf{A0} = \mathbf{0} = \lambda\mathbf{0}$, so $\mathbf{0} \in E(\lambda)$. Hence $E(\lambda) \neq \varnothing$.

Let $\mathbf{x}, \mathbf{y} \in E(\lambda)$. Then $\mathbf{Ax} = \lambda\mathbf{x}$ and $\mathbf{Ay} = \lambda\mathbf{y}$. Hence

$$\begin{aligned}
\mathbf{A}(\mathbf{x} + \mathbf{y}) &= \mathbf{Ax} + \mathbf{Ay} && \text{(property of matrix multiplication)} \\
&= \lambda\mathbf{x} + \lambda\mathbf{y} && \text{(since } \mathbf{Ax} = \lambda\mathbf{x} \text{ and } \mathbf{Ay} = \lambda\mathbf{y}) \\
&= \lambda(\mathbf{x} + \mathbf{y}) && \text{(property of scalar multiplication).}
\end{aligned}$$

It now follows that $\mathbf{x} + \mathbf{y} \in E(\lambda)$, so $E(\lambda)$ is closed under addition.

Now let $\mathbf{x} \in E(\lambda)$, as above, and let $\alpha \in \mathbb{R}$. We wish to show that $\alpha\mathbf{x} \in E(\lambda)$:

$$\begin{aligned}
\mathbf{A}(\alpha\mathbf{x}) &= \alpha(\mathbf{Ax}) && \text{(property of matrix multiplication)} \\
&= \alpha(\lambda\mathbf{x}) && \text{(since } \mathbf{Ax} = \lambda\mathbf{x}) \\
&= \lambda(\alpha\mathbf{x}) && \text{(property of scalar multiplication).}
\end{aligned}$$

Hence $\alpha\mathbf{x} \in E(\lambda)$ so $E(\lambda)$ is closed under scalar multiplication.

We have shown that $E(\lambda)$ is non-empty, closed under addition, and closed under scalar multiplication. Therefore, by the Subspace Test, theorem 5.12, $E(\lambda)$ is a subspace of $\mathbb{R}^n$. $\qquad\square$

Note that, in showing $E(\lambda)$ is closed under addition and scalar multiplication, we are using the properties of a vector space for $\mathbb{R}^n$. For addition, the equation $\lambda\mathbf{x} + \lambda\mathbf{y} = \lambda(\mathbf{x} + \mathbf{y})$ is just condition (M2) of definition 5.5. For scalar multiplication, our proof left out the details of establishing the equation $\alpha(\lambda\mathbf{x}) = \lambda(\alpha\mathbf{x})$. It would be quite common to leave out the details but, for completeness, here is the full deduction:

$$\begin{aligned}
\alpha(\lambda\mathbf{x}) &= (\alpha\lambda)\mathbf{x} && \text{(property (M1) of a vector space)} \\
&= (\lambda\alpha)\mathbf{x} && \text{(commutativity of multiplication in } \mathbb{R}) \\
&= \lambda(\alpha\mathbf{x}) && \text{(property (M1) of a vector space).}
\end{aligned}$$

We conclude this section by proving one further result about eigenvalues and eigenvectors.

Theorem 5.15 (Eigenspace of the square of a matrix)

Let $\mathbf{A}$ be an $n \times n$ matrix and let λ be an eigenvalue of $\mathbf{A}$. Then λ^2 is an eigenvalue of the matrix $\mathbf{A}^2$.

Furthermore, the eigenspace of λ for the matrix $\mathbf{A}$, $E_{\mathbf{A}}(\lambda)$, is a subspace of the eigenspace of λ^2 for the matrix $\mathbf{A}^2$, $E_{\mathbf{A}^2}(\lambda^2)$.

Before we consider the proof, it is worth remarking that we have had to extend our previous notation for eigenspaces to take account of the fact that there are two different matrices involved here. Since we need to consider the eigenspaces of different matrices, we need to extend the previous notation for the eigenspace to refer also to the matrix.

For the first part of the theorem, we need to find a chain of deductions from 'λ is an eigenvalue of $\mathbf{A}$' to 'λ^2 is an eigenvalue of the matrix $\mathbf{A}^2$'. Now 'λ is an eigenvalue of $\mathbf{A}$' means that $\mathbf{A}\mathbf{x} = \lambda\mathbf{x}$ for some non-zero vector $\mathbf{x}$. Similarly, 'λ^2 is an eigenvalue of $\mathbf{A}^2$' means that $\mathbf{A}^2\mathbf{y} = \lambda^2\mathbf{y}$ for some non-zero vector $\mathbf{y}$. We have used different letters for the vectors $\mathbf{x}$ and $\mathbf{y}$ as we cannot assume that the same vector satisfies both equations. Thus a partial proof has the following structure:

$$\lambda \text{ is an eigenvalue of } \mathbf{A}$$
$$\Rightarrow \quad \mathbf{A}\mathbf{x} = \lambda\mathbf{x} \text{ for some non-zero vector } \mathbf{x}$$
$$\Rightarrow \quad \cdots$$
$$\Rightarrow \quad \mathbf{A}^2\mathbf{y} = \lambda^2\mathbf{y} \text{ for some non-zero vector } \mathbf{y}$$
$$\Rightarrow \quad \lambda^2 \text{ is an eigenvalue of } \mathbf{A}^2.$$

To complete the missing steps, consider $\mathbf{A}^2\mathbf{x}$. Without, at this stage, worrying too much about the detailed justifications, we have

$$\mathbf{A}^2\mathbf{x} = \mathbf{A}(\mathbf{A}\mathbf{x}) = \mathbf{A}(\lambda\mathbf{x}) = \lambda(\mathbf{A}\mathbf{x}) = \lambda(\lambda\mathbf{x}) = \lambda^2\mathbf{x}. \qquad (*)$$

Provided we can justify each of these equations, this completes the deduction chain.

We now consider the second part of the theorem. The eigenspaces are

$$E_{\mathbf{A}}(\lambda) = \{\mathbf{x} \in \mathbb{R}^n : \mathbf{A}\mathbf{x} = \lambda\mathbf{x}\},$$
$$E_{\mathbf{A}^2}(\lambda^2) = \{\mathbf{x} \in \mathbb{R}^n : \mathbf{A}^2\mathbf{x} = \lambda^2\mathbf{x}\}.$$

From the previous theorem, theorem 5.14, we know that each of these is a subspace of $\mathbb{R}^n$. Hence, to show that $E_{\mathbf{A}}(\lambda)$ is a sub*space* of $E_{\mathbf{A}^2}(\lambda^2)$, we only need to show that $E_{\mathbf{A}}(\lambda)$ is a sub*set* of $E_{\mathbf{A}^2}(\lambda^2)$, $E_{\mathbf{A}}(\lambda) \subseteq E_{\mathbf{A}^2}(\lambda^2)$.

Recall that, for two sets A and B, $A \subseteq B$ means that, for all x, $x \in A \Rightarrow x \in B$.

Taking into account the definitions of the eigenspaces, this gives the following structure to the proof:

$$\mathbf{x} \in E_{\mathbf{A}}(\lambda) \ \Rightarrow \ \mathbf{A}\mathbf{x} = \lambda\mathbf{x}$$
$$\Rightarrow \ \cdots$$
$$\Rightarrow \ \mathbf{A}^2\mathbf{x} = \lambda^2\mathbf{x}$$
$$\Rightarrow \ \mathbf{x} \in E_{\mathbf{A}^2}(\lambda^2).$$

However, the missing part of the reasoning is captured by the equations (*) above. We can now piece together the different parts of the reasoning and give a proof of the theorem.

Proof. Let $\mathbf{A}$ be an $n \times n$ matrix and let λ be an eigenvalue of $\mathbf{A}$. Then, by definition, there exists an eigenvector $\mathbf{x} \in \mathbb{R}^n$ which is a non-zero vector such that $\mathbf{A}\mathbf{x} = \lambda\mathbf{x}$. Therefore

$$
\begin{aligned}
\mathbf{A}^2\mathbf{x} &= \mathbf{A}(\mathbf{A}\mathbf{x}) \\
&= \mathbf{A}(\lambda\mathbf{x}) && \text{(since } \mathbf{A}\mathbf{x} = \lambda\mathbf{x}) \\
&= \lambda(\mathbf{A}\mathbf{x}) && \text{(properties of matrix multiplication)} \\
&= \lambda(\lambda\mathbf{x}) && \text{(since } \mathbf{A}\mathbf{x} = \lambda\mathbf{x}) \\
&= \lambda^2\mathbf{x}.
\end{aligned}
$$

This shows that $\mathbf{x}$ is also an eigenvector corresponding to λ^2 for the matrix $\mathbf{A}^2$. In particular, λ^2 is an eigenvalue of $\mathbf{A}^2$, which is the first part of the theorem.

The reasoning above shows that

$$\mathbf{A}\mathbf{x} = \lambda\mathbf{x} \Rightarrow \mathbf{A}^2\mathbf{x} = \lambda^2\mathbf{x},$$

so $E_{\mathbf{A}}(\lambda) \subseteq E_{\mathbf{A}^2}(\lambda^2)$. Since each of these sets is a subspace of $\mathbb{R}^n$ (by theorem 5.14), it follows that $E_{\mathbf{A}}(\lambda)$ is a subspace of $E_{\mathbf{A}^2}(\lambda^2)$, which is the second part of the theorem. $\square$

Exercises 5.3

1. Prove theorem 5.8 (ii), the Right Cancellation Property for groups.

2. Prove the identity missing from the proof of theorem 5.9: for elements x and y of a group $(G, *)$, $(y^{-1} * x^{-1}) * (x * y) = e$.

3. Similar to the definition of a subspace of a vector space, definition 5.6, we may define a **subgroup** of a group $(G, *)$ to be a subset $H \subset G$ that is also a group under the same binary operation; $(H, *)$ is a group.

   Under this definition, both $(\{e\}, *)$ and $(G, *)$ are subgroups of $(G, *)$. Any other subgroup different from these is called a **proper subgroup** of $(G, *)$.

   Prove the following subgroup test.

   Theorem 5.16 (Subgroup Test)
   Let $(G, *)$ and let H be a subset of G that satisfies:

   (i) H is non-empty: $H \neq \varnothing$;
   (ii) H is closed under $*$: for all $x, y \in H$, $x * y \in H$;
   (iii) H is closed under taking inverses: for all $x \in H$, $x^{-1} \in H$.

   Then $(H, *)$ is a subgroup of $(G, *)$.

4. Let H and K be two subgroups of a group G. Prove that their intersection $H \cap K$ is a subgroup of G.

   Hint: you may use the subgroup test, theorem 5.16, from the previous example.

5. A set S has a binary operation $*$ that satisfies the following two axioms.
   (A1)   There is an identity element $e \in S$.
   (A2)   For all $x, y, z \in S$, $(x * y) * z = x * (z * y)$.

   Show that $*$ is both commutative and associative on S.

6. Prove properties (ii) and (iii) in theorem 5.11.

7. Let S and T be two subspaces of a vector space U. Define the **sum** of S and T to be

   $$S + T = \{\mathbf{x} + \mathbf{y} : \mathbf{x} \in S \text{ and } \mathbf{y} \in T\}.$$

   Prove that $S + T$ is a subspace of U.

8. Using theorem 5.12, show that each of the sets $E(2)$ and $E(-1)$, given in example 5.7, is a subspace of $\mathbb{R}^2$.

9. Let λ be an eigenvalue of an $n \times n$ matrix $\mathbf{A}$.

   (i) Prove that $\lambda + 1$ is an eigenvalue of the matrix $\mathbf{A} + \mathbf{I}_n$.

   The matrix $\mathbf{I}_n$ is the $n \times n$ matrix with 1s along the leading top-left to bottom-right diagonal and 0s elsewhere. It is called the $n \times n$ **identity matrix** as it satisfies the property that, for any $n \times n$ matric $\mathbf{M}$, $\mathbf{M}\mathbf{I}_n = \mathbf{I}_n\mathbf{M} = \mathbf{M}$.

(ii) Let $E_{\mathbf{A}}(\lambda)$ denote the eigenspace of λ for the matrix $\mathbf{A}$ and let $E_{\mathbf{A}+\mathbf{I}_n}(\lambda+1)$ denote the eigenspace of $\lambda+1$ for the matrix $\mathbf{A}+\mathbf{I}_n$. Show that $E_{\mathbf{A}}(\lambda) = E_{\mathbf{A}+\mathbf{I}_n}(\lambda+1)$.

(iii) Let k be a positive integer. Generalise the result of part (i) for the eigenvalue $\lambda + k$. Prove your generalisation.

10. Let $\mathbf{A}$ be an $m \times n$ matrix. The **null space** of $\mathbf{A}$ is $N(\mathbf{A}) = \{\mathbf{x} \in \mathbb{R}^n : \mathbf{Ax} = \mathbf{0}\}$.

Show that the $N(A)$ is a subspace of $\mathbb{R}^n$.

11. Let $\mathbf{A}$ be an $n \times n$ matrix. Writing $\mathbf{x} \in \mathbb{R}^n$ as a column vector, we may define a mapping

$$T : \mathbb{R}^n \to \mathbb{R}^n \quad \text{by} \quad T(\mathbf{x}) = \mathbf{Ax}.$$

The mapping T is called a **linear transformation**;[6] it satisfies the following two properties:

(LT1)   for all $\mathbf{x}, \mathbf{y} \in \mathbb{R}^n$, $T(\mathbf{x} + \mathbf{y}) = T(\mathbf{x}) + T(\mathbf{y})$;
(LT2)   for all $\mathbf{x} \in \mathbb{R}^n$ and $\alpha \in \mathbb{R}$, $T(\alpha \mathbf{x}) = \alpha T(\mathbf{x})$.

(i) Prove that the set $\ker T = \{\mathbf{x} \in \mathbb{R}^n : T(\mathbf{x}) = \mathbf{0}\}$ is a subspace of $\mathbb{R}^n$. The space $\ker T$ is called the **kernel** of T.

(ii) Prove that the **image** of T, $\operatorname{im} T = \{T(\mathbf{x}) : x \in \mathbb{R}^n\}$, is a subspace of $\mathbb{R}^n$.

5.4   Examples from analysis

In this section we will consider two important topics in real analysis: convergence of sequences and limits of functions. The key notion of a limit is similar in each of the two contexts. The definition of a limit involves a statement with two quantifiers which is of the form

$$\forall x \, \exists y \bullet P(x, y).$$

It is probably the interaction between the two quantifiers that makes proofs involving limits somewhat tricky. As we discussed in section 5.2, when considering the proof of surjectivity of a function, proofs of statements of this form with a universally and existentially quantified variable, often require some

---

[6] Any mapping between arbitrary vector spaces that satisfies the properties (LT1) and (LT2) is called a linear transformation.

preliminary 'discovery work' before the proof itself can commence. As this discovery work is not part of the proof itself, it is sometimes omitted in an explanation of the proof and, when this occurs, some of the details of the proof itself look somewhat arbitrary or even mysterious.

5.4.1   Sequences

Intuitively a sequence of real numbers is an infinite list of real numbers $a_1, a_2, a_3, a_4, \dots$. It is often helpful to capture the idea of an infinite list as a single mathematical concept. With this in mind, consider how we might define a function $f : \mathbb{Z}^+ \to \mathbb{R}$. We might have a rule for $f(n)$ but, if not, we would need to specify each of the images $f(1), f(2), f(3), \dots$. In other words, defining a function $\mathbb{Z}^+ \to \mathbb{R}$ is akin to specifying an infinite list of real numbers. Hence it is usual to define a sequence more formally as a function, as follows.

Definition 5.7
A **sequence** (of real numbers) is a function $a : \mathbb{Z}^+ \to \mathbb{R}$. The image of $n \in \mathbb{Z}^+$, $a(n)$, is frequently denoted a_n. The whole sequence is often abbreviated as (a_n).

Often, we are interested in the long-term behaviour of sequences. The graph of a sequence $a : \mathbb{Z}^+ \to \mathbb{R}$ just plots the points a_n; it is often referred to as a **sequence diagram**. Figure 5.12 shows some possible behaviours of sequences.

- Sequence 1 is **periodic**: it endlessly cycles between a collection of values, in this case $\frac{1}{2}, \frac{1}{2}, 1, \frac{1}{2}, \frac{1}{2}, 1, \frac{1}{2}, \frac{1}{2}, 1, \dots$.

- Sequences 2 and 4 **converge**: the values get closer and closer to some limiting value, 1 in the case of Sequence 2 and $\frac{1}{2}$ in the case of Sequence 4.

- Sequence 3 is **chaotic**: there is no discernible pattern to its values.

We aim to prove some properties of convergent sequences. In order to do so, we first need a precise definition of *convergent*. Intuitively, a sequence (a_n) converges to a limit ℓ if the terms eventually get 'as close as possible' to ℓ. To make this notion precise we need to quantify 'eventually' and 'as close as possible'. Saying that the terms come 'as close as possible' to ℓ means that given *any* positive quantity ε (no matter how small) we can make the distance between a_n and ℓ less than ε. Since $|a_n - \ell|$ represents the distance between a_n and ℓ, this means that 'eventually' we have $|a_n - \ell| < \varepsilon$. By 'eventually' we just mean that this inequality is satisfied provided we go beyond some point N in the sequence. These considerations lead to the following definition.

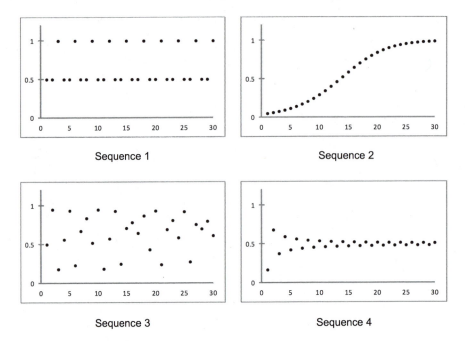

FIGURE 5.12: Possible behaviour of sequences.

Definition 5.8

The sequence (a_n) **converges** to limit ℓ if, given any $\varepsilon > 0$, there exists $N \in \mathbb{Z}^+$ such that

$$n > N \implies |a_n - \ell| < \varepsilon.$$

In this case we write either $a_n \to \ell$ as $n \to \infty$ or $\displaystyle\lim_{n \to \infty} a_n = \ell$.

Definition 5.8 is illustrated in figure 5.13. The inequality $|a_n - \ell| < \varepsilon$ is satisfied provided $\ell - \varepsilon < a_n < \ell + \varepsilon$; in other words, when a_n lies in a 'band' of width ε either side of ℓ. The figure is intended to represent the fact that we can ensure that the terms of the sequence lie in the given the 'ε-band' about the limit ℓ provided we take n greater than the specified value N.

Example 5.8

A sequence (a_n) is defined by $a_n = \dfrac{3n^2 - 4n}{(n+1)(n+2)}$ for $n \in \mathbb{Z}^+$. Show that $\displaystyle\lim_{n \to \infty} a_n = 3$.

Solution

Here the limit is $\ell = 3$. Let $\varepsilon > 0$ be an arbitrary positive real number. We

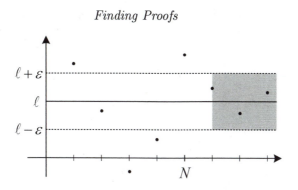

FIGURE 5.13: Illustrating the definition of a convergent sequence.

need to find a positive integer N such that, for all $n > N$,

$$|a_n - \ell| = \left| \frac{3n^2 - 4n}{(n+1)(n+2)} - 3 \right| < \varepsilon.$$

The question is: how can we identify a suitable integer N?

To answer this, we manipulate the expression for $|a_n - \ell|$, making various choices for n until we can be assured that the expression is less than any positive ε. We proceed as follows:

$$\left| \frac{3n^2 - 4n}{(n+1)(n+2)} - 3 \right| = \left| \frac{3n^2 - 4n - 3(n+1)(n+2)}{(n+1)(n+2)} \right|$$

$$= \left| \frac{3n^2 - 4n - (3n^2 + 9n + 6)}{(n+1)(n+2)} \right|$$

$$= \left| \frac{-13n - 6}{(n+1)(n+2)} \right|.$$

$$= \frac{13n + 6}{(n+1)(n+2)}.$$

Now we wish to show that the expression

$$\frac{13n + 6}{(n+1)(n+2)}$$

is less than or equal to some simpler fraction which can still be made less than ε. For example, we could replace the numerator $13n + 6$ with $13n + 6n = 19n$ since $n \geq 1$. Just as replacing the numerator with a larger value increases the size of the expression, so does replacing the denominator with a smaller value. Hence we have:

$$\frac{13n + 6}{(n+1)(n+2)} \leq \frac{19n}{(n+1)(n+2)} \leq \frac{19n}{n^2} = \frac{19}{n}.$$

We can ensure that the $19/n$ is as small as we please by choosing n sufficiently large. In particular, to ensure $19/n < \varepsilon$ we may take $n > 19/\varepsilon$.

It is tempting to believe that we have now found the required value of N, namely $19/\varepsilon$. However there is one further complication. Definition 5.8 requires the value of N to be a positive integer and there is no guarantee that $19/\varepsilon$ will be an integer.

However, this is easy to fix: we take N to be the **integer part** or **floor** of $19/\varepsilon$. This is defined to be the largest integer less than or equal to $19/\varepsilon$ and is denoted

$$\left\lfloor \frac{19}{\varepsilon} \right\rfloor .$$

Now that we have found a suitable value for N, we can at last proceed to the proof itself. The proof amounts to checking that definition 5.8 is satisfied for the given sequence.

Proof. Let $\varepsilon > 0$. Now let $N = \left\lfloor \dfrac{19}{\varepsilon} \right\rfloor \in \mathbb{Z}^+$. Then, for all integers $n > N$, we have

$$\left| \frac{3n^2 - 4n}{(n+1)(n+2)} - 3 \right| = \left| \frac{3n^2 - 4n - 3(n+1)(n+2)}{(n+1)(n+2)} \right|$$

$$= \left| \frac{3n^2 - 4n - (3n^2 + 9n + 6)}{(n+1)(n+2)} \right|$$

$$= \left| \frac{-13n - 6}{(n+1)(n+2)} \right|$$

$$= \frac{13n + 6}{(n+1)(n+2)}$$

$$\leq \frac{13n + 6n}{(n+1)(n+2)}$$

$$\leq \frac{19n}{n^2}$$

$$= \frac{19}{n}$$

$$< \varepsilon \quad \text{since } n > \frac{19}{\varepsilon}.$$

Therefore $\displaystyle\lim_{n \to \infty} \frac{3n^2 - 4n}{(n+1)(n+2)} = 3$. □

It is worth remarking that, taken in isolation, the beginning of the proof

$$\text{Let } \varepsilon > 0. \text{ Now let } N = \left\lfloor \frac{19}{\varepsilon} \right\rfloor \in \mathbb{Z}^+$$

does look strange. Without the algebraic manipulation before the proof, the choice of $N = \lfloor 19/\varepsilon \rfloor$ looks curious in the extreme. This is often the case in proofs of this type. If we do not show the 'discovery phase', which is not part of the proof itself, then the choices made in the proof can seem somewhat arbitrary.

Notice that the overall structure of the proof is simply to verify that the conditions of definition 5.8 for a convergent sequence are satisfied by the given sequence. As we have noted, the definition of convergence involved two quantifiers and is of the form

$$\forall \varepsilon \, \exists N \bullet P(\varepsilon, N)$$

where $P(\varepsilon, N)$ is a propositional function and the universes for ε and N are $\mathbb{R}^+$ and $\mathbb{Z}^+$, respectively. The structure of the proof starts with an arbitrary ε (in its universe) and then selects a particular N (in its universe) which may depend on ε. To complete the proof we show that the propositional function $P(\varepsilon, N)$, which is itself a conditional, is satisfied for the arbitrary ε and particular N. The principal difficulty in the proof is in selecting an appropriate N, depending on ε, but this occurs outside the proof itself in the preliminary discovery phase of the work.

We now prove a general theorem about limits of sequences. This theorem says, roughly, that limits 'behave well' with respect to addition of sequences. Given two sequences (a_n) and (b_n), we may form their sum $(a_n + b_n)$, which is also a sequence. More precisely, the theorem says that if (a_n) and (b_n) are both convergent, then their sum is also convergent and, furthermore, the limit of $(a_n + b_n)$ is the sum of the limits of (a_n) and (b_n).

Theorem 5.17 (Sum of convergent sequences)
Let (a_n) and (b_n) be two convergent sequences with $\lim\limits_{n \to \infty} a_n = \ell$ and $\lim\limits_{n \to \infty} b_n = m$.

Then the sum $(a_n + b_n)$ is convergent and $\lim\limits_{n \to \infty} a_n + b_n = \ell + m$.

As in the previous example, we will need to carry out some preliminary work to understand how the proof will proceed. According to definition 5.8, we need to ensure that, given any $\varepsilon > 0$,

$$|(a_n + b_n) - (\ell + m)| < \varepsilon$$

provided we take n greater than some positive integer N. The triangle inequality for the modulus function (theorem 4.6, page 146) gives

$$|(a_n + b_n) - (\ell + m)| = |(a_n - \ell) + (b_n - m)| \le |a_n - \ell| + |b_n - m|.$$

Since $\lim_{n \to \infty} a_n = \ell$ and $\lim_{n \to \infty} b_n = m$, we are able to control the sizes of both $|a_n - \ell|$ and $|b_n - m|$ by taking n sufficiently large. In particular, we can ensure each of these terms is less than $\varepsilon/2$ by 'going far enough' along the sequences (a_n) and (b_n):

$$n > N_1 \Rightarrow |a_n - \ell| < \frac{\varepsilon}{2} \text{ and } n > N_2 \Rightarrow |b_n - m| < \frac{\varepsilon}{2}$$

for some $N_1, N_2 \in \mathbb{Z}^+$. However, we need to ensure both of these inequalities hold, so we may take our integer N to be the larger of N_1 and N_2. We are now ready to embark on the proof.

Proof. Suppose that (a_n) and (b_n) are two convergent sequences with $\lim\limits_{n \to \infty} a_n = \ell$ and $\lim\limits_{n \to \infty} b_n = m$.

Let $\varepsilon > 0$.

Then $\varepsilon/2 > 0$, so there exist $N_1, N_2 \in \mathbb{Z}^+$ such that

$$n > N_1 \Rightarrow |a_n - \ell| < \frac{\varepsilon}{2} \text{ and } n > N_2 \Rightarrow |b_n - m| < \frac{\varepsilon}{2}.$$

Let $N = \max\{N_1, N_2\}$. Then

$$
\begin{aligned}
n > N \Rightarrow\ & n > N_1 \text{ and } n > N_2 \\
\Rightarrow\ & |a_n - \ell| < \frac{\varepsilon}{2} \text{ and } |b_n - m| < \frac{\varepsilon}{2} \\
\Rightarrow\ & |(a_n + b_n) - (\ell + m)| = |(a_n - \ell) + (b_n - m)| \\
& \qquad\qquad\qquad\ \ \leq |a_n - \ell| + |b_n - m| \\
& \qquad\qquad\qquad\ \ < \frac{\varepsilon}{2} + \frac{\varepsilon}{2} = \varepsilon.
\end{aligned}
$$

Therefore $(a_n + b_n)$ is convergent and $\lim\limits_{n \to \infty} a_n + b_n = \ell + m$. $\qquad\square$

5.4.2   Limits of functions

The notion of a limit of a function has similarities with the convergence of a sequence. Suppose that $f : A \to B$ is a function where the domain and codomain are subsets of the real numbers, $A \subseteq \mathbb{R}$ and $B \subseteq \mathbb{R}$. Let $a \in A$ be an element of the domain.[7] Intuitively, we say that f converges to a limit ℓ as x tends to a provided we may make $f(x)$ come 'as close as we choose' to ℓ by taking x 'sufficiently close' to a. Just as for sequences, saying that the value of $f(x)$ is 'as close as we choose' to ℓ means ensuring that $|f(x) - \ell|$ is less than some specified positive quantity ε. Similarly, taking x 'sufficiently

---

[7] In fact, it is not strictly necessary that $a \in A$. What is required is that the domain contains open intervals on either side of a: $(p, a) \cup (a, q) \subseteq A$ for some $p < a$ and $q > a$.

close' to a means requiring that $|x - a|$ is less than some (other) positive value, traditionally labelled δ. There is one small modification, however. In the limiting process as x tends towards a, we will not allow x actually to be equal to a; indeed, as we indicated in the footnote, it may be that f is not actually defined at a. These considerations lead to the following definition.

Definition 5.9

Let $f : A \to B$ be a function where $A \subseteq \mathbb{R}$ and $B \subseteq \mathbb{R}$. Let $a \in A$. Then $f(x)$ tends towards a **limit** ℓ as x tends towards a if, given any $\varepsilon > 0$, there exists $\delta > 0$ such that

$$0 < |x - a| < \delta \implies |f(x) - \ell| < \varepsilon.$$

In this case, we either write $f(x) \to \ell$ as $x \to a$ or we write $\lim_{x \to a} f(x) = \ell$.

The values of x satisfying $0 < |x - a| < \delta$ lie in the union of open intervals either side of a:

$$(a - \delta, a) \cup (a, a + \delta).$$

This set is often referred to as a **punctured neighbourhood** of a. Definition 5.9 is illustrated in figure 5.14. In order to ensure that the values of $f(x)$ lie in the horizontal band of width ε either side of the limit ℓ, we need to take x to lie in the punctured neighbourhood of width δ either side of a.

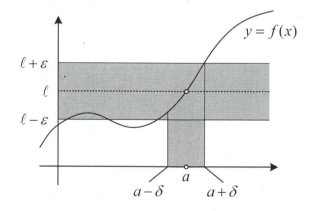

FIGURE 5.14: Illustrating the definition of a limit of a function.

Example 5.9

Show that $\lim_{x \to 2} 2x^2 - 5x = -2$.

Solution

In this example, we need to show that definition 5.9 is satisfied for $f(x) =$

$2x^2 - 5x$, $\ell = -2$ and $a = 2$. As with the proofs for sequences, we need to do some preliminary 'discovery work'. We need to ensure that

$$|f(x) - \ell| = |2x^2 - 5x + 2|$$

is less than any specified positive ε by taking x such that $0 < |x - 2| < \delta$. Hence we begin by considering $|2x^2 - 5x + 2|$:

$$|2x^2 - 5x + 2| = |(x - 2)(2x - 1)| = |x - 2||2x - 1|.$$

We will be able to control the size of the term $|x - 2|$ since we can choose δ such that $|x - 2| < \delta$. The term $|2x - 1|$ causes a little more difficulty, but we can also control its size if we re-write it in terms of $|x - 2|$ using the triangle inequality (theorem 4.6, page 146) as follows.

$$
\begin{aligned}
|2x - 1| &= |2(x - 2) + 3| \\
&\leq |2(x - 2)| + |3| \quad \text{(by the triangle inequality)} \\
&= 2|x - 2| + 3.
\end{aligned}
$$

Therefore

$$|2x^2 - 5x + 2| = |x - 2||2x - 1| \leq |x - 2|(2|x - 2| + 3)$$

and we consider each of the terms $|x - 2|$ and $2|x - 2| + 3$ separately. Provided $|x - 2| < 1$, say, the second term $2|x - 2| + 3 < 5$. (Taking $|x - 2|$ to be less than 1 here is somewhat arbitrary; we really just need it to be no greater than *some* value. Later we will consider other choices at this step.) If, in addition, $|x - 2| < \varepsilon/5$, then the product $|x - 2|(2|x - 2| + 3)$ will be less than ε, which is what we are trying to ensure. We can now embark on the proof itself.

Proof. Let $\varepsilon > 0$. Now let $\delta = \min\left\{1, \dfrac{\varepsilon}{5}\right\}$.

Then

$$
\begin{aligned}
&0 < |x - 2| < \delta \\
\Rightarrow\quad &|x - 2| < 1 \text{ and } |x - 2| < \frac{\varepsilon}{5} \\
\Rightarrow\quad &|2x^2 - 5x + 2| = |x - 2||2x - 1| \\
& \leq |x - 2|(2|x - 2| + 3) \quad \text{(triangle inequality)} \\
& < 5|x - 2| \quad \text{(since } |x - 2| < 1) \\
& < \varepsilon \quad \left(\text{since } |x - 2| < \frac{\varepsilon}{5}\right).
\end{aligned}
$$

Therefore $\lim\limits_{x \to 2} 2x^2 - 5x = -2$. $\qquad\qquad\square$

Just as with example 5.8, the first line of the proof would look mysterious without having seen the pre-proof 'discovery' process to find a suitable choice for δ. In fact, because there are two requirements for δ, there are other choices that will work. For example, suppose that, at the stage where we had the inequality $|2x^2 - 5x + 2| = |x - 2||2x - 1| \leq |x - 2|(2|x - 2| + 3)$, we had taken $|x-2| < \frac{1}{2}$ (instead of $|x-2| < 1$). Then the term $2|x-2|+3 < 2 \times \frac{1}{2}+3 = 4$, so we have $|2x^2 - 5x + 2| < 4|x - 2|$, which means we also require $|x - 2| < \varepsilon/4$. This means that an alternative choice for δ is $\delta = \min\{\frac{1}{2}, \varepsilon/4\}$. With this choice, the last part of the proof becomes

$$
\begin{aligned}
|2x^2 - 5x + 2| &= |x - 2||2x - 1| \\
&\leq |x - 2|(2|x - 2| + 3) \quad \text{(triangle inequality)} \\
&< 4|x - 2| \quad \left(\text{since } |x - 2| < \tfrac{1}{2}\right) \\
&< \varepsilon \quad \left(\text{since } |x - 2| < \frac{\varepsilon}{4}\right).
\end{aligned}
$$

In Theorem 5.17 we proved that the limit of sequences behaves well with respect to addition; more specifically, the limit of the sum of sequences is the sum of their limits. There is a similar result for limits of functions and the proof follows a similar pattern. We leave this as an exercise. Instead we prove that the limit of a function behaves well with respect to taking multiples.

Theorem 5.18 (Limit of a multiple of a function)
Let f be a function such that $\lim_{x \to a} f(x) = \ell$ and let $\alpha \in \mathbb{R}$. Then the function αf satisfies $\lim_{x \to a} \alpha f(x) = \alpha\ell$.

Again we will carry out some preliminary work. In the proof we will need to show that $|\alpha f(x) - \alpha\ell| < \varepsilon$ and, because $\lim_{x \to a} f(x) = \ell$, we can control the size of $|f(x) - \ell|$. Now $|\alpha f(x) - \alpha\ell| = |\alpha||f(x) - \ell|$, which will be less than ε provided $|f(x) - \ell| < \varepsilon/|\alpha|$. Since $\varepsilon/|\alpha|$ is only defined when $\alpha \neq 0$, we will consider $\alpha = 0$ as a special case.

We are now ready to embark on the proof.

Proof. Suppose $\lim_{x \to a} f(x) = \ell$ and let $\alpha \in \mathbb{R}$.

If $\alpha = 0$, then $\alpha f(x) = 0$ for all $x \in \mathbb{R}$. Hence, for any $\varepsilon > 0$,

$$
|\alpha f(x) - \alpha\ell| = 0 < \varepsilon
$$

for all x in the domain of f. Therefore $\lim_{x \to a} \alpha f(x) = 0 = \alpha\ell$ in this case.

Now suppose $\alpha \neq 0$.

Let $\varepsilon > 0$. Then $\dfrac{\varepsilon}{|\alpha|} > 0$.

Since $\lim\limits_{x \to a} f(x) = \ell$ it follows that there exists $\delta > 0$ such that

$$0 < |x - a| < \delta \;\Rightarrow\; |f(x) - \ell| < \frac{\varepsilon}{|\alpha|}.$$

Hence $\quad 0 < |x - a| < \delta \;\Rightarrow\; |f(x) - \ell| < \dfrac{\varepsilon}{|\alpha|}$

$$\Rightarrow \;\; |\alpha f(x) - \alpha \ell| = |\alpha||f(x) - \ell| < |\alpha| \times \frac{\varepsilon}{|\alpha|} = \varepsilon.$$

Therefore $\lim\limits_{x \to a} \alpha f(x) = \alpha \ell$ in this case too.   □

In fact, there is a slightly different proof that does not require separate consideration of the case where $\alpha = 0$. Note that $|\alpha| + 1 > 0$ for all α. Thus, if we replace $\varepsilon/|\alpha|$ by $\varepsilon/(|\alpha| + 1)$ in the second part of the proof, we can avoid the special case. This leads to the following alternative proof.

Alternative Proof. Suppose $\lim\limits_{x \to a} f(x) = \ell$ and let $\alpha \in \mathbb{R}$.

Let $\varepsilon > 0$. Then $\dfrac{\varepsilon}{|\alpha| + 1} > 0$.

Since $\lim\limits_{x \to a} f(x) = \ell$ it follows that there exists $\delta > 0$ such that

$$0 < |x - a| < \delta \;\Rightarrow\; |f(x) - \ell| < \frac{\varepsilon}{|\alpha| + 1}.$$

Hence $\quad 0 < |x - a| < \delta$

$$\Rightarrow \; |f(x) - \ell| < \frac{\varepsilon}{|\alpha| + 1}$$

$$\Rightarrow \; |\alpha f(x) - \alpha \ell| = |\alpha||f(x) - \ell| < |\alpha| \times \frac{\varepsilon}{|\alpha| + 1} < \varepsilon.$$

Therefore $\lim\limits_{x \to a} \alpha f(x) = \alpha \ell$.   □

Exercises 5.4

1. For each of the following sequences, give an 'ε-N' argument to prove $\lim\limits_{n \to \infty} a_n = \ell$.

(i) $(a_n) = \left(\dfrac{3n + 4}{5n + 1}\right); \quad \ell = 3$

(ii) $(a_n) = \left(\dfrac{n^3 + 3n}{4n^3 + 7n^2} \right); \quad \ell = \frac{1}{4}$

(iii) $(a_n) = \left(\dfrac{2n^2 + 4}{(n+1)(3n+2)} \right); \quad \ell = \frac{2}{3}$

(iv) $(a_n) = \left(\dfrac{2n^4 - 6}{n^4 - 2n} \right); \quad \ell = 2$

2. Let (a_n) be a convergent sequence with $\lim\limits_{n\to\infty} a_n = \ell$. Also let $\lambda \in \mathbb{R}$.

   Prove that (λa_n) converges and $\lim\limits_{n\to\infty} \lambda a_n = \lambda \ell$.

3. Use the 'ε-δ' definition to prove each of the following limits.

   (i) $\lim\limits_{x\to 1} 3x - 7 = -4$ (v) $\lim\limits_{x\to 2} \dfrac{1}{x} = \dfrac{1}{2}$

   (ii) $\lim\limits_{x\to 3} x^2 - 4x = -3$ (vi) $\lim\limits_{x\to 2} \sqrt{x} = \sqrt{2}$

   (iii) $\lim\limits_{x\to -1} 2x^2 + x = 1$ (vii) $\lim\limits_{x\to a} x^3 = a^3$

   (iv) $\lim\limits_{x\to 1} 3x^2 + 5x = 8$ (viii) $\lim\limits_{x\to 0} x \sin x = 0$

4. Prove that the limit of functions behaves well with respect to addition. More precisely, suppose that f and g are two functions such that

$$\lim_{x\to a} f(x) = \ell \text{ and } \lim_{x\to a} g(x) = m.$$

   Prove that $\lim\limits_{x\to a} f(x) + g(x) = \ell + m$.

5. In this question we will prove that, if f and g are two functions such that

$$\lim_{x\to a} f(x) = \ell \text{ and } \lim_{x a r a a} g(x) = m,$$

   then $\lim\limits_{x\to a} f(x)g(x) = \ell m$.

   (i) Show that $|f(x)g(x) - \ell m| \le |g(x)||f(x) - \ell| + |\ell||g(x) - m|$.

   (ii) Show that there exists $\delta_1 > 0$ such that

$$0 < |x - a| < \delta_1 \Rightarrow |g(x)| < 1 + |m|.$$

   (iii) Show that, given $\varepsilon > 0$, there exists $\delta_2 > 0$ such that

$$0 < |x - a| < \delta_2 \Rightarrow |f(x) - \ell| < \frac{\varepsilon}{2(1 + |m|)}.$$

   (iv) Suppose $\ell \ne 0$. Show that, given $\varepsilon > 0$, there exists $\delta_3 > 0$ such that

$$0 < |x - a| < \delta_3 \Rightarrow |g(x) - m| < \frac{\varepsilon}{2|\ell|}.$$

(v) Use parts (i) to (iv) to prove $\lim_{x \to a} f(x)g(x) = \ell m$ in the case where $\ell \neq 0$.

(vi) Modify your proof in part (v) for the case where $\ell = 0$.

Chapter 6

Direct Proof: Variations

6.1 Introduction

In some situations it is not possible, or not easy, to find a direct proof of a theorem. The nature of the statement of the theorem may make it difficult to find a chain of deductions ending with the appropriate conclusion. A simple example will illustrate this. In example 1.3 (page 11), we considered the statement *'for all integers n, if n is odd, then n^2 is odd'* and we gave what we now recognise as a direct proof. Consider now the converse statement (where, for simplicity, we have restricted the universe to be the positive integers):

for all positive integers n, if n^2 is odd, then n is odd.

How might we prove this? A direct proof would start by assuming 'n^2 is odd' for an arbitrary positive integer n and, from this, attempt to deduce 'n is odd'. Thus a direct proof would have the following structure.

> *Proof.* Suppose that n is a positive integer such that n^2 is odd. Then $n^2 = 2k + 1$ for some integer k (where $k \geq 0$).
>
> $\vdots$
>
> some chain of deductions
>
> $\vdots$
>
> Therefore n is odd. □

The problem here is that it is not at all clear what might be a sensible deduction from the statement '$n^2 = 2k + 1$'. Knowing that we need to make a conclusion about n itself (namely n is odd), we might be tempted to take square roots and write

$$n = \sqrt{2k + 1} \text{ for some integer } k \geq 0.$$

We have now hit a dead end: there is no obvious way of showing that $\sqrt{2k + 1}$ is an odd positive integer. The problem here is not so much in our ability to make appropriate deductions as the *overall structure* of the attempted proof.

This is because it is much more difficult to deduce information about an integer n (such as n is odd) from information about its square than it is the other way around. In other words, starting from 'n^2 is odd' it is very difficult to find a chain of deductions that allows us to deduce 'n is odd'.

In this chapter, we will introduce methods of proof that may be useful when the method of direct proof appears problematic. In each case, the method will follow the pattern:

> . instead of proving a statement P, we may instead prove a statement P' that is logically equivalent to P.

Recall that two propositions are logically equivalent if they are true under the same conditions. Thus is it perfectly reasonable to prove a logically equivalent statement if this is easier to achieve.

6.2 Proof using the contrapositive

Recall the Contrapositive Law from section 2.3, page 40, that states that a conditional proposition $P \Rightarrow Q$ is logically equivalent to its contrapositive $\neg Q \Rightarrow \neg P$:

$$P \Rightarrow Q \equiv \neg Q \Rightarrow \neg P.$$

This means that, to prove a conditional proposition $P \Rightarrow Q$, we may instead prove its contrapositive $\neg Q \Rightarrow \neg P$ if this is easier. The proof of the contrapositive itself will be a direct proof:

> assume $\neg Q$ and, from this, deduce $\neg P$.

The following examples illustrate the method. The first of these proves the result that we discussed in the previous section.

Examples 6.1

1. In the introduction, we indicated that a direct proof of the following theorem is problematic because it is not clear what sensible deductions we can make from the statement 'n^2 is odd'. This makes it a prime candidate for a proof using the contrapositive. We have also removed the restriction that n should be positive.

   Theorem 6.1
   For all integers n, if n^2 is odd, then n is odd.

Proof. We will prove the contrapositive: if n is even, then n^2 is even.

Let n be an even integer. Then $n = 2m$ for some integer m.

Hence $n^2 = (2m)^2 = 4m^2 = 2(2m^2)$ where $2m^2$ is an integer. Therefore n^2 is even.

We have proved that, for all integers n, if n is even, then n^2 is even. Hence the contrapositive follows: for all integers n, if n^2 is odd then n is odd. $\square$

Commentary

There are a few things worth noting about this proof. Firstly, it is good practice to signal at the beginning that the proof is using the contrapositive. Without this first sentence, the reader may not see why the main body of the proof starts by assuming that n is even. Secondly, we have not explicitly mentioned that the negation of 'n is odd' is 'n is even' as this was felt to be sufficiently obvious. Sometimes, when the negation of one or both of the component propositions of the conditional statements is less obvious, then it would be sensible to explain the negation in the proof itself.

Finally, note that the proposition being proved is a universally quantified propositional function

$$\forall n \bullet O(n^2) \Rightarrow O(n)$$

where O is the predicate 'is odd' and where the universe is $\mathbb{Z}^+$. We have used the Principle of Universal Generalisation given in section 4.3 to reduce the proof requirement to $O(n^2) \Rightarrow O(n)$ for an arbitrary positive integer and then applied the Contrapositive Law, to change the proof requirement to $\neg O(n) \Rightarrow \neg O(n^2)$ for the arbitrary integer n. In other words, we have used both the Principle of Universal Generalisation and the Contrapositive Law but it is only the Contrapositive Law that is explicitly referred to in the proof itself.

2. In this example, we prove a similar result to the previous one. Since the negation of one of the statements in the conditional is a little less obvious, the proof explicitly 'unravels' the negation.

Theorem 6.2
For all integers n and m, if mn is even, then m is even or n is even.

Proof. We will prove the contrapositive.

By De Morgan's Law, the negation of 'm is even or n is even' is 'm is odd *and* n is odd'.

So suppose that m and n are both odd integers. Then $m = 2k + 1$ and $n = 2\ell + 1$ for some integers k and ℓ.

Hence $mn = (2k + 1)(2\ell + 1) = 4k\ell + 2k + 2\ell + 1 = 2(2k\ell + k + \ell) + 1$ where $2k\ell + k + \ell$ is an integer. Therefore mn is odd.

We have proved: if m and n are both odd, then mn is odd. Therefore the contrapositive follows: if mn is even, then m is even or n is even. $\square$

3. The next result we consider is somewhat trickier to prove than those in the two previous examples. It will also require a little more background knowledge. In particular, we will assume the following formula for the sum of a geometric progression. If a is a real number and $a \neq 1$, then for all positive integers n,

$$1 + a + a^2 + \cdots + a^{n-1} = \frac{a^n - 1}{a - 1}.$$

We will give a proof of this result in chapter 8; see example 8.1.4.

Theorem 6.3
For all positive integers n, if $2^n - 1$ is prime, then n is prime.

Before embarking on the proof, consider, for a moment, how a direct proof of this result might progress. We would firstly need to assume that $2^n - 1$ is prime and, from this, deduce that n itself is prime. The difficulty here is that it is not clear what deductions can usefully be made from the assumption $2^n - 1$ is prime. Certainly $2^n - 1$ would satisfy any known properties of prime numbers, but it is not clear how this might provide useful information about n itself.

For example, assuming that $2^n - 1$ is prime we might deduce that $2^n - 1$ is either equal to 2 or is an odd integer (not necessarily prime). The first case gives $2^n = 3$, which is not possible where n is an integer. Hence $2^n - 1$ is an odd integer, which implies that 2^n is even — but we knew this anyway! At this stage, it appears that a direct proof is not leading anywhere.

We therefore turn our attention to the contrapositive: if n is not prime, then $2^n - 1$ is not prime. The statement 'n is not prime', or is 'n is composite', means that $n = rs$ for some positive integers r and s such that $1 < r < n$ and $1 < s < n$. (Note that these two statements are frequently combined into a single statement $1 < r, s < n$ which is more compact but, strictly speaking, is an abuse of notation.) Thus a proof using the contrapositive will assume n is composite and, from this, deduce that $2^n - 1$ is composite.

The geometric progression above with $a = 2$ just gives

$$2^n - 1 = 1 + 2 + 2^2 + \cdots + 2^{n-1},$$

which is not very helpful. But this is for an arbitrary integer n and we are assuming that n is composite. In this case $n = rs$ so we have

$$2^n - 1 = 2^{rs} - 1 = (2^r)^s - 1$$

and we can apply the geometric progression formula with $a = 2^r$ and $n = s$. This gives

$$1 + 2^r + (2^r)^2 + \cdots + (2^r)^{s-1} = \frac{(2^r)^s - 1}{2^r - 1} = \frac{2^n - 1}{2^r - 1}.$$

Rearranging this equation, we have

$$2^n - 1 = (2^r - 1)(1 + 2^r + (2^r)^2 + \cdots + (2^r)^{s-1}),$$

which expresses $2^n - 1$ as a product and hence shows that $2^n - 1$ is composite, which is what we were seeking to achieve. Having now found a chain of reasoning from 'n is composite' to '$2^n - 1$ is composite', we can now organise this into a coherent proof.

Proof. We will prove the contrapositive.

Suppose that n is not prime. Then $n = rs$ where $1 < r < n$ and $1 < s < n$. Then $2^n = 2^{rs} = (2^r)^s$.

We will use the result

$$1 + a + a^2 + \cdots + a^{n-1} = \frac{a^n - 1}{a - 1}$$

with $a = 2^r$ and $n = s$. This gives

$$1 + (2^r) + (2^r)^2 + \cdots + (2^r)^{s-1} = \frac{(2^r)^s - 1}{2^r - 1} = \frac{2^n - 1}{2^r - 1}$$

so

$$(2^r - 1)(1 + (2^r) + (2^r)^2 + \cdots + (2^r)^{s-1}) = 2^n - 1.$$

This last equation expresses $2^n - 1$ as the product of two integers, namely $2^r - 1$ and $1 + (2^r) + (2^r)^2 + \cdots + (2^r)^{s-1}$. Hence $2^n - 1$ is not prime.

This completes the proof of the contrapositive: n is not prime $\Rightarrow 2^n - 1$ is not prime. Hence the result follows. $\qquad \square$

As is always the case with mathematical proof, we are unable to give precise and foolproof rules concerning when a particular method of proof will be applicable. However, considering the structure of the propositions $P \Rightarrow Q$ that we proved in the three previous examples, we might informally characterise P as containing 'compound information' but Q as containing 'simple information' of a similar type. In example 6.1.1, P contained information about the parity (evenness/oddness) of a *square* of an integer, whereas Q contained information about the parity of the integer itself. example 6.1.2 was similar: P was a statement about the parity of a *product* whereas Q was a statement about the parity of the component integers. Finally, in example 6.1.3, P was a statement about the primality of $2^n - 1$ whereas Q was simply a statement about the primality of n itself. In each case, the statement P has given 'compound' information (about a square, a product of an expression involving a power) whilst Q has provided the corresponding information about an integer or integers.

It is often simpler to 'build up' to more complex, or 'compound', information by combining simpler pieces of information than it is to go in the opposite direction by 'splitting down' compound information into constituent parts. It is in situations such as these that a proof using the contrapositive might be considered because the 'direction' of the implication is reversed.

To illustrate this, consider the following theorem. Each part is a conditional statement $P \Rightarrow Q$ where P refers to a composite function, and so represents 'compound' information, and Q refers to similar information but about just one of the functions. Given our previous discussion, this suggests that a proof using the contrapositive may be fruitful. For example, in part (i), to prove the contrapositive we assume that f is not injective and then deduce that the composite $g \circ f$ is not injective.

Theorem 6.4
Let $f : A \to B$ and $g : B \to C$ be functions.

(i) If the composite $g \circ f$ is injective, then so, too, is the function f.

(ii) If the composite $g \circ f$ is surjective, then so, too, is the function g.

We will prove part (i) and leave part (ii) as an exercise (see exercise 6.1.5).

We first need to define what it means for a function to be injective. Recall from section 3.6 that a function is injective if different elements (of the domain) have different images. This was formalised in definition 3.3 which says that a function f is injective if, for all elements of its domain,

$$f(a_1) = f(a_2) \implies a_1 = a_2.$$

Since we are aiming to prove the contrapositive (f is not injective $\Rightarrow g \circ f$ is not injective) we will need the negation of this statement. In section 2.4, when considering the negation of quantified propositional functions, we established the following general logical equivalence (see page 64)

$$\neg \forall x \bullet P(x) \Rightarrow Q(x) \equiv \exists x \bullet P(x) \wedge \neg Q(x).$$

Thus, saying that f is *not* injective means that there exist elements a_1 and a_2 in its domain such that $f(a_1) = f(a_2)$ but $a_1 \neq a_2$. Thus, if we assume that f is not injective, it follows that there are distinct (that is, not equal) elements $a_1, a_2 \in A$ such that $f(a_1) = f(a_2)$. It then follows immediately that $g(f(a_1)) = g(f(a_2))$ for these distinct elements, so the composite function $g \circ f$ is also not injective. We can now organise these thoughts into a quite a short and straightforward proof.

Proof of (i). We will prove the contrapositive: f is not injective $\Rightarrow g \circ f$ is not injective. So suppose that $f : A \to B$ and $g : B \to C$ are functions and that f is not injective.

Then there exist elements $a_1, a_2 \in A$ such that $a_1 \neq a_2$ but $f(a_1) = f(a_2)$. It follows immediately that $g(f(a_1)) = g(f(a_2))$, which means $(g \circ f)(a_1) = (g \circ f)(a_2)$.

Hence $g \circ f$ is not injective, which completes the proof of the contrapositive. $\square$

Exercises 6.1

1. Prove each of the following using the contrapositive. Is it possible to prove any of these results (easily) without using the contrapositive?

   (i) For all integers n, if $5n + 6$ is even, then n is even.

   (ii) For all integers n and all odd integers k, if $kn + (k + 1)$ is even, then n is even.

   (iii) For all integers n, if n^2 is divisible by 3, then n is divisible by 3.

   (iv) For all integers n, if n^2 is not divisible by 7, then n is not divisible by 7.

   (v) For all integers m and n, if mn is divisible by 3, then m is divisible by 3 or n is divisible by 3.

2. Prove each of the following using the contrapositive.

   (i) If m and n are positive integers such that $mn = 100$, then $m \leq 10$ or $n \leq 10$.

   (ii) If m, n and k are positive integers such that $mn = k^2$, then $m \leq k$ or $n \leq k$.

3. Prove that, if a is an odd integer, then the quadratic equation $x^2 - x - a = 0$ has no roots which are integers.

4. In section 6.5, we will prove that $\sqrt{2}$ is irrational.

   Use the Prime Factorisation Theorem 4.3 to prove that, if n is a positive integer that is not a perfect square, then $\sqrt{n}$ is irrational.

5. Prove part (ii) of theorem 6.4.

6. Prove that, if x is a real number such that $0 < x < 1$, then $x > x^2$.

7. Prove each of the following using the contrapositive. Are any of the following statements easily proved directly?

(i) For all sets A, B, X, and Y, if $A \times B \subseteq X \times Y$ then $A \subseteq X$ or $B \subseteq Y$.

(ii) For all non-empty sets A, B, X, and Y, if $A \times B \subseteq X \times Y$, then $A \subseteq X$ and $B \subseteq Y$.

(iii) For all sets A and B, if $\mathbb{P}(A) \subseteq \mathbb{P}(B)$, then $A \subseteq B$.

8. Let $\mathbf{A} = \begin{pmatrix} a_{11} & a_{12} & a_{13} \\ a_{21} & a_{22} & a_{23} \\ a_{31} & a_{32} & a_{33} \end{pmatrix}$ be a 3×3 real matrix.

Prove that, if $\mathbf{A}^2 = \mathbf{0}$, the 3×3 zero matrix, then at least one of the 'upper diagonal' elements, $a_{11}, a_{12}, a_{13}, a_{22}, a_{23}, a_{33}$ is non-zero.

6.3 Proof of biconditional statements

Many theorems in mathematics are expressed in the form of biconditional statements 'P if and only if Q'. Here are a few examples.

1. For all integers n, n^2 is odd if and only if n is odd.

2. The line $y = mx - 2$ intersects the parabola $y = 3x^2 + 1$ if and only if $|m| \geq 6$.

3. For all sets A and B, $\mathbb{P}(A) \subseteq \mathbb{P}(B)$ if and only if $A \subseteq B$.

4. A group G is Abelian if and only if, for all $g, h \in G$, $(gh)^{-1} = g^{-1}h^{-1}$.

In section 2.3, we established the following logical equivalence, which we called the Biconditional Law.

$$P \Leftrightarrow Q \equiv (P \Rightarrow Q) \wedge (Q \Rightarrow P).$$

This provides a method for proving a biconditional statement of the form $P \Leftrightarrow Q$: prove each of the conditional statements $P \Rightarrow Q$ and $Q \Rightarrow P$ separately. For each of the 'sub-proofs', $P \Rightarrow Q$ and $Q \Rightarrow P$, we will typically use either a direct proof or a proof using the contrapositive.

The examples above are clearly signalled as biconditional statements through the phrase 'if and only if'. Of course, the biconditional connective is 'symmetric', in the sense that $P \Leftrightarrow Q \equiv Q \Leftrightarrow P$, so that it does not matter whether we write 'P if and only if Q' or 'Q if and only if P'. Sometimes 'if and only if' is abbreviated as 'iff' so that the first example above may be expressed as 'for all integers n, n^2 is odd iff n is odd'. Another way of signalling $P \Leftrightarrow Q$ is to use the expression 'P is a necessary and sufficient condition for Q'. Thus the third example above could have been expressed as 'for all sets A and B, $A \subseteq B$ is a necessary and sufficient condition for $\mathbb{P}(A) \subseteq \mathbb{P}(B)$'. Regardless of how we express a biconditional $P \Leftrightarrow Q$, the standard proof technique is to prove separately $P \Rightarrow Q$ and its converse $Q \Rightarrow P$.

Occasionally it will be possible to prove $P \Leftrightarrow Q$ by providing a sequence of intermediate propositions which can be linked using the biconditional

$$P \Leftrightarrow P_1 \Leftrightarrow P_2 \Leftrightarrow \cdots \Leftrightarrow Q.$$

However, this is unusual and separate proofs of $P \Rightarrow Q$ and $Q \Rightarrow P$ are far more common.

Examples 6.2

1. For our first proof, we will combine two proofs that we have given previously. In example 1.3, we gave what we can now recognise as a direct proof of the proposition 'for all integers n, if n is odd, then n^2 is odd'. In example 6.1.1 we used the contrapositive to prove 'for all integers n, if n^2 is odd, then n is odd'. We now combine those proofs to give a proof of the following theorem.

Theorem 6.5
For all integers n, n is odd if and only if n^2 is odd.

Proof.

($\Rightarrow$) Let n be an odd integer. Then $n = 2m + 1$ for some integer m. Now

$$
\begin{aligned}
n^2 &= (2m+1)^2 \\
&= 4m^2 + 4m + 1 \\
&= 2(2m^2 + 2m) + 1 \\
&= 2M + 1 \quad \text{where } M = 2m^2 + 2m \text{ is an integer.}
\end{aligned}
$$

Therefore n^2 is odd.

($\Leftarrow$) We will prove the contrapositive of the converse. Since the converse is 'if n^2 is odd, then n is odd', its contrapositive is 'if n is even, then n^2 is even'.

Let n be an even integer. Then $n = 2m$ for some integer m. Now

$$n^2 = (2m)^2$$
$$= 4m^2$$
$$= 2(2m^2)$$
$$= 2M \quad \text{where } M = 2m^2 \text{ is an integer.}$$

Therefore n^2 is even.

This completes the proof of the contrapositive. Hence, for all integers n, if n^2 is odd, then n is odd.

$\square$

Commentary

Compared with the previous proofs, we have modified slightly how we have written the two parts of the proof to be consistent with one another. However, we have not changed the structure of the previous proofs. Note that the proof is laid out clearly in two parts with each part signalled either by ($\Rightarrow$) or by ($\Leftarrow$). Although it is not necessary to lay out the proof in this way, it is good policy to identify clearly each of the two halves of the proof.

2. Our second example comes from set theory. For the proof we will need to recall from section 3.2 the meaning of the subset relation: $X \subseteq Y$ if and only if, for all x, $x \in X \Rightarrow x \in Y$. The following theorem is easily visualised by drawing a Venn-Euler diagram.

Theorem 6.6
For all sets A, B, and C, $A \subseteq (B - C)$ if and only if both $A \subseteq B$ and $A \cap C = \varnothing$.

Proof. Firstly, suppose that A, B, and C are sets such that $A \subseteq (B-C)$. We need to establish both $A \subseteq B$ and $A \cap C = \varnothing$.

Let $x \in A$. Then $x \in B - C$ since $A \subseteq (B - C)$. This implies both that $x \in B$ and that $x \notin C$. We have shown $x \in A \Rightarrow x \in B$; therefore $A \subseteq B$. But we have also shown $x \in A \Rightarrow x \notin C$ so that A and C have no elements in common, $A \cap C = \varnothing$.

Conversely, suppose that A, B, and C are sets such that $A \subseteq B$ and $A \cap C = \varnothing$.

We need to show that $A \subseteq (B - C)$, so let $x \in A$. Since $A \subseteq B$ it follows that $x \in B$. We also know that A and C have no elements in common, so $x \notin C$. Therefore $x \in B - C$ (since we have shown $x \in B$ and $x \notin C$). We have proved that $x \in A \Rightarrow x \in B - C$; it follows that $A \subseteq (B - C)$.

$\square$

Commentary

In this proof, we have not used the feature in the previous example of labelling each half of the proof with ($\Rightarrow$) or ($\Leftarrow$). Instead we have signalled the two halves of the proof using 'Firstly, suppose ...' and 'Conversely, suppose ...'.

As with many proofs about sets, we need to 'drill down' to considering elements. The overall structure of the statement to be proved is $P \Leftrightarrow (Q \wedge R)$. However the propositions P and Q identify subset relations and hence each is equivalent to a conditional statement involving elements. For example, the statement P is $A \subseteq (B - C)$, which is equivalent to $x \in A \Rightarrow x \in B - C$ (for an arbitrary element x).

3. In this example, we illustrate that it is sometimes more natural (or more efficient) not to separate the proof of a biconditional into two parts. Consider the second of our initial examples.

Theorem 6.7

The line $y = mx - 2$ intersects the parabola $y = 3x^2 + 1$ if and only if $|m| \geq 6$.

Before we embark on the proof, there are two pieces of background knowledge that we will need. Although we anticipate that each of these will be familiar to our readers, it is probably helpful to identify them explicitly. Firstly, to say that two curves (and a line is an example of a curve) intersect means that there is at least one point that lies on each. This is true if and only if the pair of simultaneous equations defining the curves has a solution. The second piece of background knowledge concerns quadratic equations: the quadratic equation $ax^2 + bx + c = 0$, where $a \neq 0$, has a (real) solution if and only if $b^2 - 4ac \geq 0$. This of course follows from the quadratic formula

$$x = \frac{-b \pm \sqrt{b^2 - 4ac}}{2a}.$$

Proof. The line intersects the parabola if and only the pair of simultaneous equations $y = mx - 2$, $y = 3x^2 + 1$ has a solution. Now

$$\left. \begin{array}{l} y = mx - 2 \\ y = 3x^2 + 1 \end{array} \right\} \text{ has a solution} \quad \Leftrightarrow \quad 3x^2 + 1 = mx - 2$$

$$\Leftrightarrow \quad 3x^2 - mx + 3 = 0.$$

The quadratic equation $3x^2 - mx + 3 = 0$ has a solution if and only if its discriminant '$b^2 - 4ac$' is non-negative. Hence

$$3x^2 - mx + 3 = 0 \text{ has a solution} \quad \Leftrightarrow \quad m^2 - 4 \times 3 \times 3 \geq 0$$

$$\Leftrightarrow \quad m^2 \geq 36$$

$$\Leftrightarrow \quad |m| \geq 6.$$

Hence the line $y = mx - 2$ intersects the parabola $y = 3x^2 + 1$ if and only $|m| \geq 6$.   □

Commentary

Rather than proving two separate conditional propositions, in this case we have linked the two statements 'the line $y = mx - 2$ intersects the parabola $y = 3x^2 + 1$' and '$|m| \geq 6$' directly via a sequence of intermediate statements. We have been careful to ensure that each statement in the chain is connected to those before and after it by the biconditional 'if and only if' connective. Thus the structure of the proof can be described as follows.

$$\text{The line } y = mx - 2 \text{ intersects the parabola } y = 3x^2 + 1$$

$\Leftrightarrow \quad \ldots$

$\Leftrightarrow \quad \ldots$ [a sequence of intermediate statements]

$\Leftrightarrow \quad \ldots$

$\Leftrightarrow \quad |m| \geq 6.$

It is possible in this example to write the proof as two separate sub-proofs:

$$y = mx - 2 \text{ intersects } y = 3x^2 + 1 \implies |m| \geq 6$$

and

$$|m| \geq 6 \implies y = mx - 2 \text{ intersects } y = 3x^2 + 1.$$

However, in this case, the two sub-proofs would essentially be the same argument in reverse. It is therefore both more efficient to write and easier to understand if the two sub-proofs are combined as we have done.

Exercises 6.2

1. Let $x, y \in \mathbb{R}$. Prove that $|x + y| = |x| + |y|$ if and only if $xy \geq 0$.

2. Prove each of the following properties of the divisibility relation on $\mathbb{Z}^+$. Recall from exercise 5.1.4 that $m|n$ denotes 'm divides n.

   (i) For all positive integers m and n, $m|n$ and $n|m$ if and only if $m = n$.

   (ii) For all positive integers a, m, and n, $m|n$ if and only if $am|an$.

3. Prove that, for all integers m and n and all primes p, p is a factor of the product mn if and only if p is a factor of m or p is a factor of n.

4. Prove each of the following about a function $f : A \to B$.

    (i) f is injective if and only if, for all subsets C_1 and C_2 of A, $f(C_1 \cap C_2) = f(C_1) \cap f(C_2)$.

    (ii) f is injective if and only if, for all subsets C of A, $C = f^{-1}(f(C))$.

    (iii) f is surjective if and only if, for all subsets D of B, $f(f^{-1}(D)) = D$.

5. Let A and B be subsets of $\mathbb{R}$. A function $f : A \to B$ is said to be **increasing** if, for all $x, y \in A$, $x < y \Rightarrow f(x) \leq f(y)$. The function f is **strictly increasing** if, for all $x, y \in A$, $x < y \Rightarrow f(x) < f(y)$. Similarly, we say that f is **decreasing** if, for all $x, y \in A$, $x < y \Rightarrow f(x) \geq f(y)$ and f is **strictly increasing** if, for all $x, y \in A$, $x < y \Rightarrow f(x) > f(y)$.

Prove each of the following for functions $A \to B$ where A and B are subsets of $\mathbb{R}$. For which of the statements is it possible to replace 'increasing' with 'strictly increasing' and 'decreasing' with 'strictly decreasing' (as appropriate)?

    (i) The function f is increasing if and only if the function $-f$ is decreasing. The function $-f : A \to B$ is defined by $(-f)(x) = -f(x)$.

    (ii) For all $\alpha \in \mathbb{R}^+$, f is increasing if and only if αf is increasing. The function $\alpha f : A \to B$ is defined by $(\alpha f)(x) = \alpha f(x)$.

    (iii) Suppose that $f(x) > 0$ for all $x \in A$. Then f is increasing if and only if the function $1/f$ is decreasing. The function $1/f : A \to B$ is defined by $(1/f)(x) = 1/f(x)$.

6. Prove that, for all positive integers n, n has exactly 5 factors (including 1 and n itself) if and only if $n = p^4$ for some prime p.

7. Let S and T be subspaces of a vector space U. Recall from exercise 5.3.7 that $S + T = \{\mathbf{x} + \mathbf{y} : \mathbf{x} \in S \text{ and } \mathbf{y} \in T\}$ is also a subspace of U.

Suppose that V is a subspace of U with the property that every $\mathbf{v} \in V$ can be written *uniquely* as $\mathbf{v} = \mathbf{x} + \mathbf{y}$ where $\mathbf{x} \in S$ and $\mathbf{y} \in T$. Then we write $V = S \oplus T$ and say that V is the **direct sum** of S and T.

Suppose $V = S + T$. Prove that $V = S \oplus T$ if and only if $S \cap T = \{\mathbf{0}\}$.

6.4 Proof of conjunctions

We begin this section with the simple observation that, to prove a conjunction $P \wedge Q$, it is sufficient to prove P and Q separately. This is intuitively obvious but may be formalised using the conjunction deduction rule from propositional logic given in section 2.5. This states that if we know P and we know Q, then we may deduce $P \wedge Q$.

Suppose we now combine this observation with the logical equivalence

$$P \Rightarrow (Q \wedge R) \equiv (P \Rightarrow Q) \wedge (P \Rightarrow R)$$

that we established in example 2.4.2 (page 36). We may then observe that, to prove a proposition of the form $P \Rightarrow (Q \wedge R)$, it is sufficient to prove separately the two conditional propositions $P \Rightarrow Q$ and $P \Rightarrow R$.

Examples 6.3

1. Consider the following theorem and its (simple) proof.

   Theorem 6.8
   For all real numbers x, $|x| \geq x$ and $x + |x| \geq 0$.

   Proof. Let x be a real number. We will consider two cases: $x \geq 0$ and $x < 0$.

   Suppose that $x \geq 0$. Then $|x| = x$ so, the inequalities $|x| \geq x$ and $x + |x| \geq 0$ both follow.

   Now suppose that $x < 0$. Then $|x| = -x > 0$, so again $|x| > 0 > x$ and $x + |x| = x + (-x) = 0$.

   In each case we have established $|x| \geq x$ and $x + |x| \geq 0$, completing the proof. $\qquad\qquad$ □

   Commentary

   We could regard this theorem as two separate theorems: 'for all real numbers x, $|x| \geq x$' and 'for all real numbers x, $x + |x| \geq 0$'. In the interests of efficiency, it is often sensible to combine simple related theorems (and their proofs). In this example, each of the separate proofs would still require two cases, $x \geq 0$ and $x < 0$, so it is more efficient to combine the two proofs.

2. In this example, the result to be proved is of the form $P \Rightarrow (Q \wedge R)$, so the two separate subproofs will be $P \Rightarrow Q$ and $P \Rightarrow R$.

   Theorem 6.9
   For all sets A and B, if A and B are disjoint, then $A \subseteq \bar{B}$ and $B \subseteq \bar{A}$.

   Proof. Suppose that A and B are disjoint sets. By definition, this means their intersection is empty, $A \cap B = \varnothing$.

To prove that $A \subseteq \bar{B}$, we need to show that, for all x, $x \in A \Rightarrow x \in \bar{B}$. Let $x \in A$. Since $A \cap B = \varnothing$, this implies that $x \notin B$. Hence $x \in \bar{B}$. We have shown that, for all x, if $x \in A$, then $x \in \bar{B}$. Therefore $A \subseteq \bar{B}$.

The proof of $B \subseteq \bar{A}$ follows from the previous argument by interchanging A and B. □

Commentary

The two sub-proofs here are:

$$A \cap B = \varnothing \ \Rightarrow \ A \subseteq \bar{B}$$

and

$$A \cap B = \varnothing \ \Rightarrow \ B \subseteq \bar{A}.$$

Since $A \cap B = B \cap A$, the second statement can be obtained from the first by interchanging A and B. Hence the proof of the second statement can be obtained by repeating the proof of the first and also interchanging A and B. Of course, there is little to be gained by actually carrying out this re-writing, which is why we have simply indicated how the proof may be constructed.

Set equality

There is a particular situation where the proof of a conjunction $P \wedge Q$ is commonly achieved by proving each of P and Q separately, and that is in establishing the equality of two sets. Recall from theorem 3.1 that two sets A and B are equal if and only if $A \subseteq B$ and $B \subseteq A$. Thus, to prove $A = B$, it is common to prove separately $A \subseteq B$ and $B \subseteq A$. For sets X and Y, $X \subseteq Y$ means that, for all x, $x \in X \Rightarrow x \in Y$. Piecing these considerations together gives the following template for a proof that $A = B$ for two sets A and B.

To prove $A = B$:

1. $A \subseteq B$: to do this, assume $x \in A$ and deduce $x \in B$
2. $B \subseteq A$: to do this, assume $x \in B$ and deduce $x \in A$.

Examples 6.4

1. For our first example, we prove one of De Morgan's Laws for sets. This was one of the Set Theory Laws introduced in section 3.3; see page 99.

Theorem 6.10
For all sets A and B, $\overline{A \cup B} = \bar{A} \cap \bar{B}$.

We will give three proofs of this theorem to illustrate the different possible styles of proof. As always, anyone writing a proof has to decide the style they wish to adopt, how much detail to give, how much background knowledge to assume, and so forth.

Our first proof is quite descriptive and 'wordy'. The second proof is very symbolic but follows the same overall structure as the first. We expect that some readers will prefer the style of the first proof and some the second. Our third proof combines the two parts of the symbolic second proof into a single short proof.

Proof 1. First, let $x \in \overline{A \cup B}$. This means that $x \notin A \cup B$. Now $A \cup B$ contains all elements in A or in B or in both, so $x \notin A \cup B$ means that $x \notin A$ and $x \notin B$. Hence $x \in \bar{A} \cap \bar{B}$. We have shown that $x \in \overline{A \cup B} \Rightarrow x \in \bar{A} \cap \bar{B}$. Therefore $\overline{A \cup B} \subseteq \bar{A} \cap \bar{B}$.

Next we need to establish the subset relation the other way around, so let $x \in \bar{A} \cap \bar{B}$. Then $x \in \bar{A}$ and $x \in \bar{B}$, so $x \notin A$ and $x \notin B$. Since x belongs neither to A nor to B, we have $x \notin A \cup B$, so $x \in \overline{A \cup B}$. We have shown that $x \in \bar{A} \cap \bar{B} \Rightarrow x \in \overline{A \cup B}$. Therefore $\bar{A} \cap \bar{B} \subseteq \overline{A \cup B}$.

Since each set is a subset of the other, it follows that $\bar{A} \cap \bar{B} = \overline{A \cup B}$.

$\square$

Our second proof follows the same structure as the first proof, but the reasoning is expressed symbolically. Perhaps the drawback of this proof is that it does not explain in detail why each of the statements '$x \notin A \cup B$' and '$x \notin A$ and $x \notin B$' imply each other.

Proof 2. Firstly, $\quad x \in \overline{A \cup B} \quad \Rightarrow \quad x \notin A \cup B$

$$\Rightarrow \quad x \notin A \text{ and } x \notin B$$
$$\Rightarrow \quad x \in \bar{A} \text{ and } x \in \bar{B}$$
$$\Rightarrow \quad x \in \bar{A} \cap \bar{B}.$$

Therefore $\overline{A \cup B} \subseteq \bar{A} \cap \bar{B}$.

Also, $\quad x \in \bar{A} \cap \bar{B} \quad \Rightarrow \quad x \in \bar{A} \text{ and } x \in \bar{B}$

$$\Rightarrow \quad x \notin A \text{ and } x \notin B$$
$$\Rightarrow \quad x \notin A \cup B$$
$$\Rightarrow \quad x \in \overline{A \cup B}.$$

Therefore $\bar{A} \cap \bar{B} \subseteq \overline{A \cup B}$.

Since each set is a subset of the other, it follows that $\bar{A} \cap \bar{B} = \overline{A \cup B}$.

$\square$

Having written the proof symbolically, it is clear that the steps in the reasoning in the second half of the proof are just the reverse of the steps in the first half. In other words, each of the implications in the first half of the proof reverses. We may therefore combine the two halves of the proof using the 'if and only if' connective. This gives our third proof.

Proof 3. Note that, $x \in \overline{A \cup B} \quad \Leftrightarrow \quad x \notin A \cup B$

$$\Leftrightarrow \quad x \notin A \text{ and } x \notin B$$
$$\Leftrightarrow \quad x \in \bar{A} \text{ and } x \in \bar{B}$$
$$\Leftrightarrow \quad x \in \bar{A} \cap \bar{B}.$$

Therefore $\overline{A \cup B} = \bar{A} \cap \bar{B}$. □

2. In exercise 3.2.5 we defined the symmetric difference $A * B$ of two sets A and B to be $A * B = (A - B) \cup (B - A)$. In this example we show that we could equally well have defined the symmetric difference to be $(A \cup B) - (A \cap B)$ and, indeed, some authors take this as the definition.

Theorem 6.11

For all sets A and B, $(A - B) \cup (B - A) = (A \cup B) - (A \cap B)$.

Proof. Firstly, let $x \in (A - B) \cup (B - A)$. Then $x \in A - B$ or $x \in B - A$.

If $x \in A - B$, then $x \in A$ and $x \notin B$. Hence $x \in A \cup B$ (since $x \in A$) and $x \notin A \cap B$ (since $x \notin B$). Therefore $x \in (A \cup B) - (A \cap B)$.

If $x \in B - A$, then $x \in B$ and $x \notin A$. Hence $x \in A \cup B$ (since $x \in B$) and $x \notin A \cap B$ (since $x \notin A$). Therefore $x \in (A \cup B) - (A \cap B)$.

In either case, $x \in (A \cup B) - (A \cap B)$. We have proved that $x \in (A - B) \cup (B - A) \Rightarrow x \in (A \cup B) - (A \cap B)$, so $(A - B) \cup (B - A) \subseteq (A \cup B) - (A \cap B)$.

Next we need to establish the subset relation the other way around, so let $x \in (A \cup B) - (A \cap B)$. Then $x \in A \cup B$ and $x \notin A \cap B$. Since $x \in A \cup B$, we know that $x \in A$ or $x \in B$. If $x \in A$ then, since $x \notin A \cap B$, we know that $x \notin B$. Hence $x \in A - B$. If $x \in B$ then, again since $x \notin A \cap B$, we know that $x \notin A$. Hence $x \in B - A$. We have shown that either $x \in A - B$ or $x \in B - A$. Therefore $x \in (A - B) \cup (B - A)$.

We have proved that $x \in (A \cup B) - (A \cap B) \Rightarrow x \in (A - B) \cup (B - A)$, so $(A \cup B) - (A \cap B) \subseteq (A - B) \cup (B - A)$.

Finally, since each set is a subset of the other, $(A - B) \cup (B - A) = (A \cup B) - (A \cap B)$. □

Exercises 6.3

1. Prove each of the following.

    (i) If x and y are positive real numbers such that $x + y = 1$, then
    $xy \leq \frac{1}{4}$ and $\left(x + \dfrac{1}{x}\right)^2 + \left(y + \dfrac{1}{y}\right)^2 \geq \frac{25}{2}$.

    (ii) If the roots of the quadratic equation $x^2 + ax + b = 0$ are odd integers, then a is an even integer and b is an odd integer.

    (iii) If $x \in \mathbb{R}$ is such that $-1 \leq x \leq 4$, then $-9 \leq x^2 - 6x \leq 7$ and $-5 \leq x^2 + 6x \leq 40$.

2. Prove each of the following set identities. In each case, the results are true for all sets.

    (i) $A \cup (B \cap C) = (A \cup B) \cap (A \cup C)$

    (ii) $\bar{A} \cup \bar{B} = \overline{A \cap B}$

    (iii) $(A \cup B) - C = (A - C) \cup (B - C)$

    (iv) $(A \cap B) \times (X \cap Y) = (A \times X) \cap (B \times Y)$

    (v) $(A \cup B) \times C = (A \times C) \cup (B \cup C)$

3. Prove each of the following identities involving the symmetric difference of two sets. From theorem 6.11, the symmetric difference $A * B$ may be taken to be either $(A - B) \cup (B - A)$ or $(A \cup B) - (A \cap B)$, whichever is most convenient. In each case the result is true for all sets.

    (i) $(A * B) \cap C = (A \cap C) * (B \cap C)$

    (ii) $(A \cup C) * (B \cup C) = (A * B) - C$

    (iii) $(A * B) \times C = (A \times C) * (A \times B)$

4. In each of the following cases, prove that the sets X and Y are equal.

    (i) $X = \{x \in \mathbb{R} : 2x^2 + 3x - 1 = x^2 + 8x - 5\}$, $Y = \{1, 4\}$

    (ii) $X = \left\{\dfrac{x}{x+1} : x \in \mathbb{R} \text{ and } x \neq -1\right\}$, $Y = \{x \in \mathbb{R} : x > -1\}$

    (iii) $X = \left\{\dfrac{x}{x^2+1} : x \in \mathbb{R} \text{ and } x \geq 1\right\}$, $Y = \{x \in \mathbb{R} : 0 < x \leq \frac{1}{2}\}$

5. Let n be a positive integer. The sets $A_1, A_2, \ldots, A_n$ are closed intervals defined by $A_r = [r, r + n]$ for $r = 1, 2, \ldots, n$.

    Prove that $A_1 \cap A_2 \cap \ldots \cap A_n = [n, n + 1]$.

6. Two elements x and y of a group G are said to commute if $xy = yx$.

    Prove that if x and y commute, then so do both x^{-1} and y^{-1} and $g^{-1}xg$ and $g^{-1}yg$ for all elements $g \in G$.

6.5 Proof by contradiction

Recall the Contradiction Law from section 2.3:

$$(\neg P \Rightarrow \mathsf{false}) \equiv P$$

where 'false' denotes any contradiction. This means that to prove a statement P, it is sufficient to prove the conditional $\neg P \Rightarrow$ false instead. The conditional is proved using direct proof: assume $\neg P$ and, from this, deduce false. In other words, to prove P we may start by assuming that P is false (that is, assume $\neg P$) and then deduce a contradiction.[1] This is the method of **proof by contradiction**, which is also known by the Latin *reductio ad absurdum* ('reduction to absurdity'). Initially it appears a slightly odd method of proof: to prove P we start by assuming that P is false! However, we only assume the falsity of P in order to obtain a contradiction. The contradiction obtained is usually of the form $Q \wedge \neg Q$ for some proposition Q; in other words, we deduce both Q and its negation $\neg Q$.

Before discussing the method further, we first give one of the classic proofs by contradiction: $\sqrt{2}$ is irrational. Recall from section 3.2 that a rational number is one that can be expressed as a fraction; that is, a rational number is a ratio of integers m/n, where $n \neq 0$. Therefore, to say that a number is irrational means that it cannot be expressed as the ratio of two integers.

Theorem 6.12
$\sqrt{2}$ is irrational.

Proof. The proof is by contradiction.

Suppose that $\sqrt{2}$ is rational. Then $\sqrt{2}$ can be expressed as a fraction in its lowest terms; that is, where numerator and denominator have no common factors. Hence $\sqrt{2}$ may be expressed as

$$\sqrt{2} = \frac{m}{n} \qquad (*)$$

where m and n have no common factors. Squaring both sides of the equation gives $2 = m^2/n^2$ so $m^2 = 2n^2$ which means that m^2 is even. It follows from

---

[1] In [9], the great mathematician G.H. Hardy described proof by contradiction as 'one of a mathematician's greatest weapons'. He went on to say: 'It is a far finer gambit than any chess gambit: a chess player may offer the sacrifice of a pawn or even a piece, but the mathematician offers *the game*.'

theorem 6.5 that m itself is even.[2] Thus $m = 2r$ for some integer r. Substituting $m = 2r$ into $m^2 = 2n^2$ gives $4r^2 = 2n^2$ so $n^2 = 2r^2$. This shows that r^2 is even, so again, by theorem 6.5, it follows that n itself is even.

We have established that both m and n are even, so they have 2 as a common factor. This contradicts the equation (*) in which m and n had no common factors. Hence, the assumption that $\sqrt{2}$ is rational is false.

Therefore $\sqrt{2}$ is irrational. □

Commentary

As always, there are different styles of expressing this proof. For example, we could have expressed some of the reasoning more symbolically such as '$\sqrt{2} = m/n \Rightarrow 2 = m^2/n^2 \Rightarrow m^2 = 2n^2 \Rightarrow m^2$ is even'. One aspect of this proof that is important, however, is to signal at the beginning that the proof uses the method of contradiction. Without this, the initial assumption 'suppose that $\sqrt{2}$ is rational' would look odd indeed.

We may think of theorem 6.12 as being a 'non-existence theorem': it asserts that there do *not* exist integers m and n such that $\sqrt{2} = m/n$. We will consider 'existence theorems' in the next chapter; here we reflect on the difficulty of proving a 'non-existence theorem', at least in the case where the universe is infinite. There are, of course, infinitely many rational numbers. To prove the irrationality of $\sqrt{2}$, we need to rule out the possibility that any of the infinitely many rational numbers could be equal to $\sqrt{2}$ and this seems a daunting task. Indeed, it is hard to imagine how we might go about constructing a direct proof of theorem 6.12 (although such direct proofs do exist).

The method of proof by contradiction will be a useful tool to have when proving non-existence where the universe is infinite. We illustrate this with our next theorem which asserts that there exist infinitely many prime numbers. This is not obviously a 'non-existence theorem' until we reflect that it is equivalent to asserting that there does not exist a largest prime number. The proof of theorem 6.13 is also regarded as a classic proof by contradiction. The proof is commonly attributed to Euclid, over 2000 years ago.

Theorem 6.13
There exist infinitely many prime numbers.

Before embarking on the proof itself, we begin by exploring the key idea using some small examples. Consider the list of the first few prime numbers: $2, 3, 5, 7, 11, 13, 17, 19, \ldots$. If we take the product of all of the primes up to some point and add 1, what happens?

---

[2] Theorem 6.5 says that 'n^2 is odd, if and only if n is odd' but this is clearly equivalent to 'n^2 is even if and only if n is even'.

$$2 + 1 = 3 \qquad\qquad\qquad \text{prime}$$
$$2 \times 3 + 1 = 7 \qquad\qquad \text{prime}$$
$$2 \times 3 \times 5 + 1 = 31 \qquad\qquad \text{prime}$$
$$2 \times 3 \times 5 \times 7 + 1 = 211 \qquad\qquad \text{prime}$$
$$2 \times 3 \times 5 \times 7 \times 11 + 1 = 2311 \qquad \text{prime}$$

In each case, the resulting number is a prime which is obviously greater than each of the primes used in its construction. This simple idea lies behind the proof which we can now give.

Proof. The proof is by contradiction.

Suppose that there are only finitely many primes. Then we may list all of them: $p_1, p_2, p_3, \ldots, p_n$. Now consider the integer N that is one larger than the product of all of the primes:

$$N = p_1 p_2 p_3 \ldots p_n + 1.$$

Since N is not equal to any of the primes in the list $p_1, p_2, \ldots, p_n$, and since we have assumed the list contains all primes, it follows that N is not prime.

Therefore, by the Prime Factorisation Theorem 4.3, N has a prime factor. In other words, for at least one of the primes p_k in the list, p_k is a factor of N.

However, $$p_1 p_2 p_3 \ldots p_n = p_k(p_1 \ldots p_{k-1} p_{k+1} \ldots p_n)$$
$$= p_k S \qquad \text{where } S = p_1 \ldots p_{k-1} p_{k+1} \ldots p_n$$
$$\Rightarrow \qquad N = p_k S + 1 \ \text{ where } S \in \mathbb{Z}.$$

Thus N gives a remainder of 1 when divided by p_k, so that p_k is not a factor of N.

This is a contradiction and completes the proof. $\qquad\qquad\qquad$ □

Exercises 6.4

1. Let x and y be real numbers such that x is rational and y is irrational, $x \in \mathbb{Q}$ and $y \notin \mathbb{Q}$. Prove that:

   (i) $x + y$ is irrational;

   (ii) xy is irrational.

2. Prove by contradiction that the set of positive integers $\mathbb{Z}^+$ is infinite.

3. Let $ax^2 + bx + c = 0$ be a quadratic equation where a, b, and c are odd integers. Prove by contradiction that the equation has no rational solutions.

4. Using the proof of theorem 6.12 as a guide, prove that $\sqrt{3}$ is irrational.

5. (i) Prove that, for all integers m and n where $n \neq 0$, $m + \sqrt{2}n$ is irrational.

    (ii) More generally, prove that if α is an irrational number then, for all integers m and n where $n \neq 0$, $m + \alpha n$ is irrational.

6. Prove that there is no rational number r such that $2^r = 3$.

7. Prove that there is no smallest real number that is strictly greater than $\frac{1}{2}$.

8. Prove that if the mean of four distinct integers is $n \in \mathbb{Z}$, then at least one of the integers is greater than $n + 1$.

9. Prove the following result, which can be used for 'primality testing'.

   If n is an integer greater than 1 which has no *prime* factor p satisfying $2 \leq p \leq \sqrt{n}$, then n is prime.

   Hint: use a proof by contradiction to prove the contrapositive.

10. Suppose there are $n \geq 3$ people at a party. Prove that at least two people know the same number of people at the party.

    Assume that if A knows B, then B knows A.

11. Prove that the set $A = \left\{ \dfrac{n-1}{n} : n \in \mathbb{Z}^+ \right\}$ does not have a largest element.

6.6   Further examples

In this section, we apply the methods introduced in this chapter to the contexts introduced in chapter 5, sections 5.3 and 5.4. In other words, we consider proofs in algebra and analysis that are variations on the method of direct proof.

6.6.1   Examples from algebra

Theorem 6.14 (Alternative Subspace Test)
Let S be a non-empty subset of a vector space V. Then S is a subspace of V if and only if it satisfies the following condition:

$$\text{for all } \mathbf{v}, \mathbf{w} \in S \text{ and } \lambda, \mu \in \mathbb{R}, \lambda\mathbf{v} + \mu\mathbf{w} \in S. \tag{*}$$

We use theorem 5.12, the Subspace Test proved in section 5.3.

Proof. Let S be a non-empty subset of a vector space V.

($\Rightarrow$) Suppose that S is a subspace of V. By definition 5.6, this means that S is itself a vector space with respect to the operations of addition of vectors and multiplication by scalars that are defined in V.

Let $\mathbf{v}, \mathbf{w} \in S$ and $\lambda, \mu \in \mathbb{R}$. Then, since scalar multiplication is a well-defined operation in S, we have $\lambda \mathbf{v} \in S$ and $\mu \mathbf{w} \in S$. Then, since addition of vectors is a well-defined operation in S, we have $\lambda \mathbf{v} + \mu \mathbf{w} \in S$.

Therefore S satisfies condition (*).

($\Leftarrow$) Suppose that S satisfies condition (*). We need to show that S is a subspace of V and to do this we use our previous Subspace Test, theorem 5.12.

First note that we are given that $S \neq \varnothing$, so S satisfies condition (i) of theorem 5.12.

Let $\mathbf{v}, \mathbf{w} \in S$. Then, using condition (*) with $\lambda = \mu = 1$ gives

$$1\mathbf{v} + 1\mathbf{w} = \mathbf{v} + \mathbf{w} \in S$$

since $1\mathbf{v} = \mathbf{v}$ and $1\mathbf{w} = \mathbf{w}$ by axiom (M4) of a vector space (see definition 5.5). Hence S satisfies condition (ii) of theorem 5.12.

Now let $\mathbf{v} \in S$ and $\lambda \in \mathbb{R}$. Using condition (*) with $\mu = 1$ and $\mathbf{w} = \mathbf{0}$ gives

$$\lambda \mathbf{v} + 1\mathbf{0} = \lambda \mathbf{v} + \mathbf{0} = \lambda \mathbf{v} \in S$$

using axioms (M4) and (A3) of a vector space. Hence S satisfies condition (iii) of theorem 5.12.

Since S satisfies all three conditions of theorem 5.12, it follows that S is a subspace of V, completing the proof.

$\square$

In exercise 5.3.7, we introduced the idea of the **sum** of two subspaces of a vector space. If S and T are subspaces of V, then their sum $S + T$ is the subspace of V defined by $S + T = \{\mathbf{x} + \mathbf{y} : \mathbf{x} \in S \text{ and } \mathbf{y} \in T\}$.

Theorem 6.15

Let S and T be subspaces of a vector space V. Then $S + T = T$ if and only if S is a subspace of T.

Proof. Let S and T be subspaces of a vector space V.

($\Rightarrow$) Suppose that $S + T = T$.

We need to show that S is a subspace of T. Since S and T are both vector spaces with the same operations — they are subspaces of V — to say that S is a subspace of T just means that S is a sub*set* of T.

Let $\mathbf{v} \in S$. Then we have $\mathbf{v} = \mathbf{v} + \mathbf{0} \in S + T$ since $\mathbf{v} \in S$ and $\mathbf{0} \in T$. This shows that $\mathbf{v} \in S \Rightarrow \mathbf{v} \in T$. Therefore S is a subset, and hence a subspace, of T.

($\Leftarrow$) Now suppose that S is a subspace of T. We need to show that $S + T = T$. As usual, to prove that these two sets are equal, we show that each is a subset of the other.

Let $\mathbf{v} + \mathbf{w} \in S + T$ where $\mathbf{v} \in S$ and $\mathbf{w} \in T$. Since S is a subspace of T, it follows that $\mathbf{v} \in T$. Hence we have $\mathbf{v}, \mathbf{w} \in T$ so $\mathbf{v} + \mathbf{w} \in T$. This shows that $\mathbf{v} + \mathbf{w} \in S + T \Rightarrow \mathbf{v} + \mathbf{w} \in T$, so $S + T \subseteq T$.

Now let $\mathbf{v} \in T$. Therefore $\mathbf{v} = \mathbf{0} + \mathbf{v} \in S + T$ since $\mathbf{0} \in S$ and $\mathbf{v} \in T$. This shows that $\mathbf{v} \in T \Rightarrow \mathbf{v} \in S + T$, so $T \subseteq S + T$.

Therefore $S + T = T$, as required.

□

Whenever we have a particular kind of algebraic structure, we may consider the mappings between them that 'preserve' or 'respect' the algebraic structure. For example, part of the structure of a vector space involves addition of vectors. To say that a mapping f between vector spaces 'preserves' addition of vectors means that $f(\mathbf{x} + \mathbf{y}) = f(\mathbf{x}) + f(\mathbf{y})$. The left-hand side of this equation adds vectors $\mathbf{x}$ and $\mathbf{y}$ and then applies the mapping f; the right-hand side first applies f to each vector and then adds the result. Thus, to say that f 'preserves addition' simply means that it does not matter whether we apply f before or after carrying out an addition of vectors. Of course, vector spaces have two operations — addition of vectors and multiplication by scalars. A mapping that preserves the algebraic structure of a vector space will need to preserve both operations. These are called linear transformations and their definition is given in exercise 5.3.11.

Our last example from the realm of algebra involves the corresponding idea from group theory. For groups, there is a single operation so the appropriate mapping just needs to preserve this. Such a mapping is called a 'morphism', or 'homomorphism', between groups. The formal definition is the following. Since there are two groups involved, we use different symbols for the binary operation in each group.

Definition 6.1
Let $(G, *)$ and $(H, \circ)$ be two groups. A mapping $\theta : G \to H$ is a **morphism**, or **homomorphism**, if, for all $x, y \in G$,

$$\theta(x * y) = \theta(x) \circ \theta((y).$$

The **kernel** of θ, ker θ, is the set of all elements of G that map to the identity element of H,

$$\ker \theta = \{g \in FG : \theta(g) = e\}.$$

For any morphism $\theta : G \to H$, its kernel is a subgroup of G. We leave it as an exercise to show this — see exercise 6.5.1. Our theorem gives a simple condition for a morphism to be injective.

Theorem 6.16

Let $(G, *)$ and $(H, \circ)$ be groups and let $\theta : G \to H$ be a morphism.

Then θ is injective if and only if ker $\theta = \{e\}$.

Proof. Suppose that θ is injective.

Let $x \in \ker \theta$. Then $\theta(x) = e$. But $\theta(e) = e$ (exercise 6.5.1 (i)), so $\theta(x) = \theta(e)$. Since θ is injective, we have $x = e$. In other words, e is the only element of ker θ, so ker $\theta = \{e\}$.

Conversely, suppose that ker $\theta = \{e\}$.

Let $x, y \in G$. Then

$$
\begin{aligned}
\theta(x) = \theta(y) \;\Rightarrow\;& \theta(x)\,(\theta(y))^{-1} = e \\
\Rightarrow\;& \theta(x)\theta(y^{-1}) = e \qquad \text{(exercise 6.5.1 (ii))} \\
\Rightarrow\;& \theta(xy^{-1}) = e \qquad\quad (\theta \text{ is a morphism}) \\
\Rightarrow\;& xy^{-1} \in \ker \theta \\
\Rightarrow\;& xy^{-1} = e \qquad\qquad (\text{since } \ker \theta = \{e\}) \\
\Rightarrow\;& x = y.
\end{aligned}
$$

We have shown that $\theta(x) = \theta(y) \Rightarrow x = y$, which is precisely the condition that θ is injective. $\qquad\qquad\square$

6.6.2 Examples from analysis

We will apply some of the techniques from this chapter to some proofs in analysis. First, however, we introduce two further concepts: the 'infimum' and 'supremum' of a non-empty set of real numbers. Let X be a non-empty subset of $\mathbb{R}$. An **upper bound** for X is a real number M such that $x \le M$ for all $x \in X$. If a set has an upper bound it is said to be **bounded above**.

For example, let $X = \{x \in \mathbb{R} : 0 < x < 1\}$ be the interval comprising all real numbers strictly between 0 and 1; this set is frequently denoted $X = (0, 1)$. An upper bound for X is 2 since $x \le 2$ for all $x \in X$. There are clearly many other upper bounds, including $\sqrt{2}$, π, 27.25, and so on. Note that saying that 2 is an upper bound for (any set) X tells us more about the set than saying that 27.25 is an upper bound. (If you had to guess a number to win a prize, say, would you prefer to be told that the number was less than 2 or less than 27.25? Clearly the former is preferable as it would save you 'wasting' any guesses between 2 and 27.25.) For our set $X = \{x \in \mathbb{R} : 0 < x < 1\}$, the upper bound that gives most information about X is 1. This is its smallest upper bound or 'least upper bound'; it is also called the 'supremum' of X, which we now define formally.

Definition 6.2

Let X be a non-empty subset of $\mathbb{R}$. A real number M satisfying:

(i) M is an upper bound for X: $x \leq M$ for all $x \in X$ and

(ii) if $a < M$, then a is not an upper bound for X:[3] if $a < M$, then there exists $x \in X$ such that $x > a$.

is called a **supremum** or **least upper bound** for X; it is denoted $M = \sup X$.

Suppose we reverse the inequalities in the previous discussion. This gives, first, the definition of a lower bound for a set: a **lower bound** for X is a real number m such that $m \leq x$ for all $x \in X$. If a set has a lower bound it is said to be **bounded below**; a set that is bounded above and below is said to be **bounded**. We can then model definition 6.2 to define an **infimum** or **greatest lower bound** for a set X as a lower bound with the additional property that any larger real number is not a lower bound. We leave it as an exercise to complete the definition.

Examples 6.5

1. For our first example, we will prove that 1 is, as expected, the supremum of the interval $X = (0,1) = \{x \in \mathbb{R} : 0 < x < 1\}$,

$$\sup\{x \in \mathbb{R} : 0 < x < 1\} = 1.$$

We need to show that $M = 1$ satisfies both the properties that define the supremum given in definition 6.2. The first property is clear: 1 is an upper bound for X since $x \leq 1$ for all $x \in X$. For the second property, let $a < 1$; we need to show that a is not an upper bound for X, which means finding an element of X that is larger than a. This is illustrated in figure 6.1. Probably the most obvious choice of $x \in X$ is the average

---

[3] Note that this is another instance where there is an implicit universal quantification. A more precise expression of this condition would be: for all $a \in \mathbb{R}$, if $a < M$, then a is not an upper bound for X.

of a and 1, $(a+1)/2$, which will lie strictly between a and 1 and therefore appears to belong to the set X. We need to be a little careful, however. The only restriction on a is that is must be a real number less that 1. Suppose $a = -3$, say, then the average of a and 1 is $(a+1)/2 = (-3+1)/2 = -1$ does not belong to X. (Recall that we are seeking $x \in X$ that is greater than a.) In this case, of course, every element of X is greater than $a = -3$. To cover all eventualities, we could take x to be the larger of 0.5, say, and $(a+1)/2$. However, it is probably simpler to consider the cases $a > 0$ and $a \leq 0$ separately, and this is the approach we will adopt. We are now in a position to commence the proof.

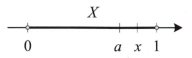

$$X$$

$$0 \qquad a \ \ x \ \ 1$$

FIGURE 6.1: Finding the supremum of $X = \{x \in \mathbb{R} : 0 < x < 1\}$.

Proof. Let $X = \{x \in \mathbb{R} : 0 < x < 1\}$.

First note that 1 is an upper bound for X since, by definition, $x < 1$ for all $x \in X$.

Now let $a < 1$. We consider two cases: $a > 0$ and $a \leq 0$.

Suppose $a > 0$ and let $x = (a+1)/2$. Now

$$0 < a < 1 \ \Rightarrow \ 1 < a+1 < 2 \ \Rightarrow \ \tfrac{1}{2} < \frac{a+1}{2} < 1.$$

In other words, $\tfrac{1}{2} \leq x < 1$, so $x \in X$. Also, since $a < 1$, we have $x = (a+1)/2 > (a+a)/2 = a$. We have shown that $x \in X$ and $x > a$. Therefore a is not an upper bound for X in this case.

Now suppose $a \leq 0$ and let $x = \tfrac{1}{2}$. Then $x \in X$ and $x > a$, so again, a is not an upper bound for X in this case.

Hence $\sup\{x \in \mathbb{R} : 0 < x < 1\} = 1$. $\qquad\square$

2. Let $X = \left\{ \dfrac{n^2}{n+1} : n \in \mathbb{Z}^+ \right\}$. Show that X does not have a supremum.

As we noted in section 6.5, a proof of non-existence is frequently accomplished using proof by contradiction, so we will attempt such a proof here: assume that X does have a supremum, $\sup X = M$ say, and deduce a contradiction. Although we will need to work with the general assumption, $\sup X = M$, in order to see the kind of argument we need, it may be helpful first to understand why some particular value is not a supremum.

So, suppose $\sup X = 500$. Then, according to definition 6.2, 500 is an upper bound and any value less than 500 is not an upper bound. To show that this is false, only one of the two conditions needs to fail. If 500 is an upper bound for X then, for all $n \in \mathbb{Z}^+$,

$$\frac{n^2}{n+1} \le 500 \quad \Rightarrow \quad n^2 \le 500n + 500$$
$$\Rightarrow \quad n^2 - 500n \le 500$$
$$\Rightarrow \quad n(n-500) \le 500.$$

Clearly this last inequality is *not* true for all $n \in \mathbb{Z}^+$; take $n = 501$, for example. This reasoning is sufficiently generic that we can attempt a proof by contradiction. However, we just need to be a little careful not to assume that the supposed supremum is an integer (as we have done in this example).

Proof. Let $X = \left\{ \dfrac{n^2}{n+1} : n \in \mathbb{Z}^+ \right\}$ and suppose that $\sup X$ exists; call it M.

Then M is an upper bound for X so that, for all $n \in \mathbb{Z}^+$,

$$\frac{n^2}{n+1} \le M \quad \Rightarrow \quad n^2 \le Mn + M$$
$$\Rightarrow \quad n^2 - Mn \le M$$
$$\Rightarrow \quad n(n-M) \le M.$$

This is a contradiction since, for example, when n is any positive integer satisfying $n \ge M + 1$, then $n - M \ge 1$, so $n(n-M) \ge n > M$.

Therefore X does not have a supremum. $\qquad\qquad\square$

Note that we have actually proved that X does not have an upper bound and this is the reason why it does not have a supremum.

3. Each of the two previous examples have been about particular sets. For our last example, we prove a result about infima and suprema for arbitrary (non-empty, bounded) sets.

Theorem 6.17

Let A and B be non-empty bounded[4] subsets of $\mathbb{R}$ such that $A \subseteq B$. Then $\inf A \ge \inf B$ and $\sup A \le \sup B$.

---

[4] In fact, we only need to assume that B is bounded. If B is bounded and $A \subseteq B$, then A is bounded — see exercise 6.5.7 (ii).

First note that the logical structure of the theorem is

$$A \subseteq B \implies (\inf A \geq \inf B) \wedge (\sup A \leq \sup B),$$

universally quantified over all non-empty bounded subsets of $\mathbb{R}$. As we noted in section 6.4, $P \implies (Q \wedge R)$ is logically equivalent to $(P \implies Q) \wedge (P \implies R)$. This means that the theorem is equivalent to

$$(A \subseteq B \implies \inf A \geq \inf B) \wedge (A \subseteq B \implies \sup A \leq \sup B)$$

and we may prove each of the two conjuncts, $A \subseteq B \implies \inf A \geq \inf B$ and $A \subseteq B \implies \sup A \leq \sup B$ separately.

Although we will give the complete *structure* of the proof below, we will only prove in detail one of the two conjuncts, leaving the other as an exercise.

Proof. Let A and B be non-empty bounded sets of real numbers such that $A \subseteq B$. We need to show both $\inf A \geq \inf B$ and $\sup A \leq \sup B$. We leave the first of these as an exercise — see exercise 6.5.7 (i).

Let $x \in A$. Then $x \in B$, since $A \subseteq B$. Now $\sup B$ is an upper bound for B, so $x \leq \sup B$. We have proved that

$$x \in A \implies x \leq \sup B,$$

which means that $\sup B$ is *an* upper bound for A.

Since the supremum is the *least* upper bound and $\sup B$ is *an* upper bound (for A), it now follows that $\sup A \leq \sup B$. □

Exercises 6.5

1. Let $(G, *)$ and $(H, \circ)$ be groups and let $\theta : G \to H$ be a morphism.

   Prove that:

   (i) $\theta(e_G) = e_H$ where e_G is the identity element of G and e_H is the identity element of H,

   (ii) $\theta\left(x^{-1}\right) = (\theta(x))^{-1}$ for all $x \in G$,

   (iii) $\ker \theta$ is a subgroup of G, and

   (iv) the image of θ, $\operatorname{im} \theta = \{\theta x : x \in G\}$ is a subgroup of H.

2. Let G be a group (written multiplicatively). Prove that G is Abelian if and only if $(gh)^{-1} = g^{-1}h^{-1}$ for all elements $g, h \in G$.

3. Let $T : V \to W$ be a linear transformation between vector spaces: see exercise 5.3.11. Prove that T is injective if and only if $\ker T = \{\mathbf{0}\}$.

4. Let $T : V \to V$ be a linear transformation from a vector space V to itself. Prove that $T \circ T = $ zero if and only if $\operatorname{im} T \subseteq \ker T$.

   Here $T \circ T$ is the usual composition of mappings and zero is the mapping $V \to V$ that sends every vector to the zero vector, $\operatorname{zero}(\mathbf{x}) = \mathbf{0}$ for all $\mathbf{x} \in V$.

5. Using definition 6.2 as a model, define the **infimum** or **greatest lower bound** of a non-empty set X of real numbers.

6. For each of the following sets of real numbers, prove that the supremum is as stated.

   (i) $X = \left\{ \dfrac{x-1}{x} : x \in \mathbb{R}^+ \right\}$, $\sup X = 1$.

   (ii) $X = \left\{ x \in \mathbb{R} : \dfrac{1}{x} > 2 \right\}$, $\sup X = \frac{1}{2}$.

   (iii) $X = \left\{ x \in \mathbb{R} : \dfrac{1}{x^2} > 2 \right\}$, $\sup X = \dfrac{1}{\sqrt{2}}$.

   (iv) $X = \left\{ \dfrac{n}{n+1} : n \in \mathbb{Z}^+ \right\} = \{ \frac{1}{2}, \frac{2}{3}, \frac{3}{4}, \frac{4}{5}, \ldots \}$, $\sup X = 1$.

   (v) $X = \left\{ \dfrac{m}{n} : m, n \in \mathbb{Z}^+ \text{ and } m < 2n \right\}$, $\sup X = 2$.

7. (i) Complete the proof of theorem 6.17; that is, prove that if A and B are non-empty bounded subsets of $\mathbb{R}$ such that $A \subseteq B$, then $\inf A \geq \inf B$.

   (ii) Prove the result in the footnote to theorem 6.17: if A and B are non-empty subsets of $\mathbb{R}$ such that B is bounded and $A \subseteq B$, then A is bounded.

8. Let $f : \mathbb{R} \to \mathbb{R}$ be a function and let $a \in \mathbb{R}$. Note that, if (a_n) is a sequence, then 'applying' the function f gives another sequence $(f(a_n))$.

   Prove that $\lim\limits_{x \to a} f(x) = \ell$ if and only if, for every sequence (a_n) such that $a_n \neq \ell$ for all $n \in \mathbb{Z}^+$ and $\lim\limits_{x \to \infty} a_n = a$, the sequence $(f(a_n))$ converges to ℓ, $\lim\limits_{x \to \infty} f(a_n) = \ell$.

   Hint: for the converse, suppose that for every sequence (a_n) such that $a_n \neq \ell$ for all $n \in \mathbb{Z}^+$ and $\lim\limits_{x \to \infty} a_n = a$, the sequence $(f(a_n))$ converges to ℓ, $\lim\limits_{x \to \infty} f(a_n) = \ell$. Then, to establish $\lim\limits_{x \to a} f(x) = \ell$ use a proof by contradiction: suppose $\lim\limits_{x \to a} f(x) = \ell$ and obtain a contradiction.

Chapter 7

Existence and Uniqueness

7.1 Introduction

So far in this book we have concerned ourselves with the proof of theorems which are propositions P or universally quantified propositional functions such as $\forall x \bullet P(x)$. For much of the current chapter we shall turn our attention to proofs of existence theorems; that is, theorems which assert the existence within the universe of an object or objects with a certain property, P. We can symbolise such a theorem by $\exists x \bullet P(x)$.

Some examples of theorems of this form are the following.

  (i) Some prime numbers are of the form $32n + 1$, where n is an integer.

 (ii) Some quadratic equations do not have real roots.

(iii) Not all real numbers are rational.

 (iv) There exist sets which have the same cardinality as some of their proper subsets.

  (v) There exist non-Abelian simple groups.

Typically, existence theorems are stated using the phraseology '*Some ...*' or '*There exist ...*'. Notice, however, that example (iii) above is expressed rather differently, as the negation of a universally quantified propositional function $\neg\forall x \bullet P(x)$. The rule for negating quantified propositions from section 2.4, page 63, tells us that this is logically equivalent to $\exists x \bullet \neg P(x)$, which is an existence theorem. In our example, the equivalent existentially quantified statement may be expressed as: 'some real numbers are irrational' or 'there exist irrational real numbers'. The manner in which these theorems are expressed seems to suggest that they are asserting the existence of *several* objects of the required type. However, this is merely convention and each of the theorems could be expressed in the form 'There exists at least one object with the required property.' To prove a theorem of this type it is sufficient to demonstrate the existence of a single object of the appropriate type whether or not there actually exist many such objects. In section 7.5, we shall consider how we might prove that there is only one object with the required property if this is the case.

7.2   Constructive existence proofs

The most obvious way to prove a theorem of the form $\exists x \bullet P(x)$ is to find a specific object a in the universe for which the proposition $P(a)$ is true. This method of proof is called **proof by construction** because we construct (or find) a specific object a with the required property. How we actually go about finding or constructing the desired object a will, of course, depend on the particular theorem under consideration. The proof of the proposition $P(a)$ may employ any of the methods we have discussed in the previous two chapters.

Examples 7.1

1. We begin by considering the first of the examples given in the introduction.

   Theorem 7.1
   Some prime numbers are of the form $32n + 1$, where n is an integer.

   Suppose we define the universe of discourse to be the integers and we define propositional functions

   $$P(x): \quad x \text{ is prime}$$
   $$Q(x): \quad x = 32n + 1 \text{ for some integer } n.$$

   The theorem asserts that there exists an integer satisfying both properties; this can be symbolised as $\exists x \bullet P(x) \wedge Q(x)$. We need, therefore, to find a specific integer a for which both $P(a)$ and $Q(a)$ are true. The simplest approach is to list (some of) the integers a for which $Q(a)$ is true and then find one of these for which $P(a)$ is also true. Alternatively, we could list integers a for which $P(a)$ is true — the prime numbers — and find one of these for which $Q(a)$ is true. However, it is easier to list integers of the form $32n + 1$ than it is to list primes, so we adopt the former approach.

   The positive integers of the form $32n+1$ are: $1, 33, 65, 97, 129, 161, 193, \ldots$. Which, if any, of these are also prime? We begin:

   $$1 \text{ is not prime (by definition)},$$
   $$33 = 3 \times 11, \text{ so } 33 \text{ is not prime},$$
   $$65 = 5 \times 13, \text{ so } 65 \text{ is not prime}.$$

   However, on testing 97 for factors we find that 97 is, indeed, prime. Using the result of exercise 6.4.9, page 260, this is straightforward; we only need to show that none of the prime numbers 2, 3, 5, and 7 is a factor of 97. Therefore we have found an object with the desired properties and we can now proceed to the proof itself.

Proof. Consider 97. First note that $97 = 32 \times 3 + 1$. None of the primes $2, 3, 5$, and 7 is a factor of 97 so 97 is prime.

Hence 97 is a prime number of the form $32n + 1$ for some integer n. $\quad\square$

The proof itself is, of course, very straightforward. It is often the case with constructive proofs that the proofs themselves are relatively simple. The hard work frequently lies in finding the required object a, but this does not show in the final proof. In our example, we have merely exhibited an object $a = 97$ with the desired properties.

2. For our second example, we prove the third example introduced in the previous section.

Theorem 7.2
Not all real numbers are rational.

We noted above that the theorem is equivalent to the proposition: there exists a real number which is irrational. In section 6.5, we proved theorem 6.12: $\sqrt{2}$ is irrational. With this theorem now included as part of our background knowledge of the real numbers, we have the following one-line proof of the current theorem.

Proof. The real number $\sqrt{2}$ is irrational, by theorem 6.12. $\quad\square$

The hard work is first discovering that $\sqrt{2}$ is a suitable real number to consider, and then proving that $\sqrt{2}$ is indeed irrational. Of course, this is hidden from view in our one-line proof.

There is also a nice non-constructive proof of this theorem; that is, a proof which does not actually produce any specific irrational real numbers. However, the proof does rely on rather more background knowledge than the proof given here; see example 7.3.2.

3. When dealing with multiplication (of real numbers for example), we tend to take the commutative law for granted. This says that, for all x and y in $\mathbb{R}$, $xy = yx$. It is used in establishing familiar identities such as $(xy)^2 = x^2 y^2$ and $(x+y)^2 = x^2 + 2xy + y^2$. However, just because it is a familiar property for multiplication of real numbers does not mean that it is necessarily satisfied in other contexts. For example, with matrix multiplication we have the following theorem.

Theorem 7.3

Matrix multiplication is not commutative. To be more precise, for any integer $n \geq 2$, multiplication of $n \times n$ matrices is not commutative.

The statement of the theorem means that it is not *always* the case that $\mathbf{AB} = \mathbf{BA}$ for $n \times n$ matrices $\mathbf{A}$ and $\mathbf{B}$. Of course, there may be examples of some matrices where $\mathbf{AB} = \mathbf{BA}$ but, to prove the theorem, we must find two matrices $\mathbf{A}$ and $\mathbf{B}$ with the property that $\mathbf{AB} \neq \mathbf{BA}$. Two such matrices are easily found using a little trial and error.

For simplicity, we consider 2×2 matrices. Suppose we try

$$\mathbf{A} = \begin{pmatrix} 1 & 2 \\ 3 & 4 \end{pmatrix} \text{ and } \mathbf{B} = \begin{pmatrix} -1 & 1 \\ 3 & 2 \end{pmatrix}.$$

Then
$$\mathbf{AB} = \begin{pmatrix} 1 & 2 \\ 3 & 4 \end{pmatrix} \begin{pmatrix} -1 & 1 \\ 3 & 2 \end{pmatrix} = \begin{pmatrix} 5 & 5 \\ 9 & 11 \end{pmatrix}$$

and
$$\mathbf{BA} = \begin{pmatrix} -1 & 1 \\ 3 & 2 \end{pmatrix} \begin{pmatrix} 1 & 2 \\ 3 & 4 \end{pmatrix} = \begin{pmatrix} 2 & 2 \\ 9 & 14 \end{pmatrix},$$

so $\mathbf{AB} \neq \mathbf{BA}$. Since we have found a suitable example, we can proceed directly to a proof.

Proof. Let $\mathbf{A} = \begin{pmatrix} 1 & 2 \\ 3 & 4 \end{pmatrix}$ and $\mathbf{B} = \begin{pmatrix} -1 & 1 \\ 3 & 2 \end{pmatrix}.$

Then
$$\mathbf{AB} = \begin{pmatrix} 1 & 2 \\ 3 & 4 \end{pmatrix} \begin{pmatrix} -1 & 1 \\ 3 & 2 \end{pmatrix} = \begin{pmatrix} 5 & 5 \\ 9 & 11 \end{pmatrix}$$

and
$$\mathbf{BA} = \begin{pmatrix} -1 & 1 \\ 3 & 2 \end{pmatrix} \begin{pmatrix} 1 & 2 \\ 3 & 4 \end{pmatrix} = \begin{pmatrix} 2 & 2 \\ 9 & 14 \end{pmatrix},$$

so $\mathbf{AB} \neq \mathbf{BA}$.   □

In this example, trial and error easily produces elements of the universe with the required property. Indeed, a random choice of matrices is quite likely to give the desired result. However, this will not always be the case and sometimes we really do need to be careful when constructing a suitable example. Staying with 2×2 matrices for simplicity, we could do this here by first considering products of arbitrary matrices and then making simple choices for the entries as follows.

First note that

$$\begin{pmatrix} a & b \\ c & d \end{pmatrix} \begin{pmatrix} u & v \\ w & x \end{pmatrix} = \begin{pmatrix} au + bw & av + bx \\ cu + dw & cv + dx \end{pmatrix}$$

and

$$\begin{pmatrix} u & v \\ w & x \end{pmatrix} \begin{pmatrix} a & b \\ c & d \end{pmatrix} = \begin{pmatrix} ua + vc & ub + vd \\ wa + xc & wb + xd \end{pmatrix}.$$

We wish to choose the entries of the two matrices so that the two products are different. Recall that two matrices are not equal if they differ in at least one entry. Now, since $au = ua$, we can ensure that the top-left entry of the products are different if $bw \neq vc$. A simple choice would be $b = w = 0$, $v = c = 1$. With these choices and any choice of a, d, u and x, we can construct matrices with the required property. Taking $a = d = u = x = 1$, for example, gives matrices

$$\mathbf{A} = \begin{pmatrix} 1 & 0 \\ 1 & 1 \end{pmatrix} \text{ and } \mathbf{B} = \begin{pmatrix} 1 & 1 \\ 0 & 1 \end{pmatrix},$$

such that $\mathbf{AB} \neq \mathbf{BA}$.

4. For our final example, we show that the set of rational numbers $\mathbb{Q}$ is 'dense' in the set of real numbers $\mathbb{R}$. This means that, between any two real numbers x and y, there is a rational number q. We may express this formally as follows.

Theorem 7.4 (Denseness of the rational numbers)
For all $x, y \in \mathbb{R}$ such that $x < y$, there exists a rational number $q \in \mathbb{Q}$ such that $x < q < y$.

To begin to understand how we might find $q \in \mathbb{Q}$ (given real numbers $x < y$), suppose we try to find a rational number between $\sqrt{2}$ and $\sqrt{2.1}$. The obvious first step is to evaluate these real numbers as decimals:

$$\sqrt{2} = 1.41421\ldots \quad \text{and} \quad \sqrt{2.1} = 1.44913\ldots\,.$$

Now let $q = 1.42$. Then q is rational, since $q = 142/100$, and it clearly satisfies $\sqrt{2} < q < \sqrt{2.1}$.

Now let's set a slightly harder task: find a rational number between

$$x = 0.12399987\ldots \quad \text{and} \quad y = 0.124\ldots\,.$$

Again, we may proceed as before: find a *terminating* decimal that lies between the two real numbers. Let $q = 0.1239999$. Then q is rational since $q = 1239999/10^7$ and $0.12399987\ldots < q < 0.124\ldots$ as required.

The essence of a constructive proof is contained in these examples and we are now able to proceed to a proof.

Proof. Let $x, y \in \mathbb{R}$ be such that $x < y$. Consider the decimal expansions of x and y and let $x = x_0 \cdot x_1 x_2 \ldots x_n x_{n+1} \ldots x_{n+k} x_{n+k+1} \ldots$ where

- x_n is *first* position where expansions of x and y differ,
- $x_{n+1} \ldots x_{n+k}$ is a (possibly empty) sequence of 9s, and
- $x_{n+k+1} \neq 9$.

Now let $q = x_0 \cdot x_1 x_2 \ldots x_n x_{n+1} \ldots x_{n+k} 9$; thus q has the same decimal expansion as x up to the $(n+k)$th decimal place, but then terminates with 9 in the $(n+k+1)$st decimal place. Firstly, since q is a terminating decimal, it is rational, $q \in \mathbb{Q}$. Since $x_{n+k+1} < 9$ it follows that $x < q$. Since x_n is the first position where the decimal expansions of x and y differ, the decimal expansion of y is $y = x_0 \cdot x_1 x_2 \ldots x_{n-1} y_n \ldots$ where $y_n > x_n$; therefore $q < y$.

We have constructed the decimal expansion of a rational number q satisfying $x < q < y$. $\qquad\square$

Commentary

Although the proof is constructive in nature, it does not provide an actual rational number. Of course, the proof could not provide a specific rational number q because the selection of q must depend on the real numbers x and y. The proof actually gives a general *method of construction* of the rational number q depending on the reals x and y. We can think of the construction given in the proof as an algorithm for producing $q \in \mathbb{Q}$ given an 'input' of real numbers x and y satisfying $x < y$.

Exercises 7.1

1. Prove each of the following existence theorems.

   (i) Not all prime numbers are odd.

   (ii) There exists an integer k such that k, $k+2$ and $k+4$ are all prime.

   (iii) There exist consecutive positive integers n and $n+1$ which are both the sum of squares of two positive integers, i.e., $n = a^2 + b^2$ and $n + 1 = c^2 + d^2$ for some $a, b, c, d \in \mathbb{Z}^+$.

   (iv) There exists a triple of consecutive positive integers n, $n+1$, $n+2$ each of which is the sum of squares of two positive integers $a^2 + b^2$ $(a, b \in \mathbb{Z}^+)$.

   (v) There exists a complex number z such that $z^4 = -1$.

   (vi) There exists an irrational number x such that x^2 is also irrational.

   (vii) There exist positive integers n which can be expressed as the sum of two squares (of positive integers) in two distinct ways: that is, $n = a^2 + b^2 = c^2 + d^2$ where $a, b, c, d \in \mathbb{Z}^+$ and $\{a, b\} \neq \{c, d\}$.

   (viii) There exist integers $n > 1$ which are both the square of a positive integer and the cube of a positive integer: $n = a^2$ and $n = b^3$ for some $a, b \in \mathbb{Z}^+$.

(ix) There exist positive integers n which can be expressed as the sum of two cubes (of positive integers) in two distinct ways: that is, $n = a^3 + b^3 = c^3 + d^3$ where $a, b, c, d \in \mathbb{Z}^+$ and $\{a, b\} \neq \{c, d\}$.

(x) Let a and b be rational numbers such that $a < b$. There exists a rational number q such that $a < q < b$.

(xi) Let n be an odd positive integer. Then there exists a natural number m such that $n^2 = 8m + 1$.

(xii) Let $x, y \in \mathbb{R}$ where $x \neq 0$. Prove that there exists $n \in \mathbb{Z}^+$ such that $n|x| > y$.

2. Prove that there exist functions $f : A \to B$ and $g : B \to C$ satisfying the following conditions.

(i) The function g is not injective, but the composite function $g \circ f$ is injective.

(ii) The function f is not surjective, but the composite function $g \circ f$ is surjective.

3. Prove each of the following existence theorems about matrices.

(i) There exists a matrix $\mathbf{A}$ not equal to the $n \times n$ zero matrix, $\mathbf{0}_{n \times n}$, for any n such that $\mathbf{A}^2 = \mathbf{0}_{n \times n}$.

(ii) There exists an $n \times n$ matrix $\mathbf{A}$ not equal to $\mathbf{I}_n$ such that $\mathbf{A}^2 = \mathbf{I}_n$.

(iii) There exists a matrix $\mathbf{A}$ not equal either to $\mathbf{0}_{n \times n}$ or to $\mathbf{I}_n$ for any n such that $\mathbf{A}^2 = \mathbf{A}$.

(iv) There exist matrices $\mathbf{A}, \mathbf{B}$ such that $\mathbf{AB} = \mathbf{I}_n$ but $\mathbf{BA} \neq \mathbf{I}_m$ for any m.

(v) Let $\mathbf{A} = \begin{pmatrix} 2 & 1 \\ 1 & 2 \end{pmatrix}$.

There exists a 2×2 matrix $\mathbf{B}$ such that $\mathbf{BAB} = \begin{pmatrix} 6 & 0 \\ 0 & 2 \end{pmatrix}$.

4. This question refers to the set $\mathbb{Z}_8 = \{0, 1, 2, 3, 4, 5, 6, 7\}$ with the operation $\times_8$, called **multiplication modulo 8**, defined by:

$$n \times_8 m = \text{ remainder when } nm \text{ is divided by 8.}$$

For example, $2 \times_8 6 = 4$, $3 \times_8 7 = 5$, etc.

A **multiplicative inverse** of an element $n \in \mathbb{Z}_8$ is an element m such that $n \times_8 m = 1$.

Prove each of the following.

(i) In $\mathbb{Z}_8$ there exists a multiplicative inverse of 5.

(ii) Not all elements of $\mathbb{Z}_8$ have multiplicative inverse.

(iii) The equation $3 \times_8 x = 2$ has a solution in $\mathbb{Z}_8$.

(iv) The equation $x \times_8 x = 1$ has a solution in $\mathbb{Z}_8$.

(v) There exist elements x and y in $\mathbb{Z}_8$, both different from 1, such that $x \times_8 y = 7$.

5. Prove each of the following existence theorems from group theory.

(i) Not all groups are cyclic.

A group G is **cyclic** if there exists an element $g \in G$ such that every element $x \in G$ can be expressed as a power of g, $x = g^n$ for some $n \in \mathbb{Z}$. Any such element g is called a **generator** for G. For finite groups we may restrict n to be a positive integer. More generally, we interpret g^0 as being equal to the identity element e, $g^0 = e$, and we interpret g^{-n} to be $\left(g^{-1}\right)^n$ for $n \in \mathbb{Z}^+$.

(ii) There exist cyclic groups with more than one generator.

(iii) There exist cyclic groups with only one generator.

(iv) There exist non-Abelian groups.

(v) There exist groups which have no proper subgroups.

Proper subgroups of a group were defined in exercise 5.3.3.

(vi) There exist non-Abelian groups with the property that all of their proper subgroups are Abelian.

7.3　Non-constructive existence proofs

There are methods of proof of an existence theorem $\exists x \bullet P(x)$ which do not identify any specific element a in the universe of discourse which has the property defined by the predicate P. Any such proof is termed a **non-constructive existence proof** or an **indirect existence proof**. The methods used to provide non-constructive existence proofs are generally either proof by contradiction or the use of counting arguments.

In a proof by contradiction, we assume the negation of the theorem, that is $\neg \exists x \bullet P(x)$, and show that this leads to a contradiction. From section 2.4, we know that

$$\neg \exists x \bullet P(x) \equiv \forall x \bullet \neg P(x).$$

Thus we assume that $\neg P(x)$ for all x in the universe and show that this leads to a contradiction.

Examples 7.2

1. Our first example is closely related to Euclid's theorem about infinitely many primes, theorem 6.13, which also provides the key result used in the proof.

   ### Theorem 7.5
   There exists a prime number greater than 10^{100}.

   Before embarking on a non-constructive proof, it is worth reflecting on the difficulties associated with a constructive proof. The number 10^{100} is enormous; it is greater than the number of atoms in the observable universe, which is usually estimated to be around 10^{80}.

   Despite the obvious difficulties, there are constructive proofs of the theorem. In other words, there are specific known primes larger than 10^{100}. For example, several very large so-called Mersenne numbers of the form $2^p - 1$ (where p is prime) are known to be prime. The smallest of these that is greater than 10^{100} and which is also prime is $2^{521} - 1$, which has 157 digits in its decimal notation. The largest of the currently known primes have millions of digits in their decimal expansions.[1]

   The proof that any one of these extremely large integers is prime requires a computer (a fast supercomputer in the case of the largest) to perform all the necessary calculations. This should be contrasted with our non-constructive proof which follows readily from Euclid's theorem 6.13. It is also worth noting that Euclid's theorem itself had a non-constructive proof.

   Proof. The proof is by contradiction.

   Suppose that *every* prime number p satisfies $2 \leq p \leq 10^{100}$. Then, since there are only finitely many integers between 2 and 10^{100}, there can be only finitely many primes. This contradicts theorem 6.13 that there are infinitely many primes.

   Therefore our initial supposition is incorrect and hence there are prime numbers greater than 10^{100}. □

2. For our second example, we consider the following theorem which is not expressed as a 'pure' existence theorem in the sense that it is not simply asserting the existence of a particular object with a particular property. The statement of the theorem involves both universal and existential quantification and may be symbolised as $\forall n \, \exists \theta \bullet P(n, \theta)$ for suitably defined P and universes for n and θ.

---

[1] In fact, there is an Internet community called GIMPS — the Great Internet Mersenne Prime Search — dedicated to finding these large primes: see www.mersenne.org.

Theorem 7.6

In any n-sided polygon there is an interior angle θ such that

$$\theta \leq \left(\frac{n-2}{n}\right)\pi.$$

Figure 7.1 shows an 8-sided polygon with an interior angle θ marked.

FIGURE 7.1: Interior angle of an n-sided polygon.

The statement of the theorem has two quantifiers: 'for all n-sided polygons, there exists an interior angle ...'. Of course our standard first step to prove a universally quantified proposition is to consider an *arbitrary* object of the appropriate type. In our case, we will start by assuming P is an arbitrary n-sided polygon. Then the result is a simple existence theorem about P, namely P has an interior angle of the appropriate type.

In the proof, we will need the following result:

the sum of the interior angles in any n-sided polygon is $(n-2)\pi$.

This result may be proved using the method of mathematical induction, which is the subject of the next chapter; see example 8.3.2. For the current proof, we regard this as part of our background knowledge.

Proof. Let P be an arbitrary n-sided polygon.

The proof is by contradiction, so suppose that every interior angle θ of P satisfies

$$\theta > \left(\frac{n-2}{n}\right)\pi.$$

Since P has n interior angles, the sum of interior angles satisfies

$$\sum \text{interior angles} > n \times \left(\frac{n-2}{n}\right)\pi = (n-2)\pi.$$

This contradicts the result given above as background knowledge. Therefore, at least one of the interior angles must satisfy

$$\theta \leq \left(\frac{n-2}{n}\right)\pi.$$

$\square$

3. The proof of our next example does not rely on the method of proof by contradiction. It is a quite ingenious proof of the existence of an irrational number a^b, where a and b are irrational. The proof considers two numbers of the form a^b and proves that one of them is the appropriate irrational number but *without showing which one it is.*

We have included the proof for two reasons. Firstly it shows that not every non-constructive existence proof follows the method of proof by contradiction or the use of counting arguments described below. As we have previously indicated, we will never be able to specify precise rules for how to prove results of a particular type. The second reason is to indicate that, however many techniques we introduce and guidelines we give, sometimes there is no substitute for a flash of insight or brilliance.

Theorem 7.7

There exist irrational numbers a and b such that a^b is rational.

Let's begin by considering what might be suitable candidates for a and b. Suppose we take $a = b = \sqrt{2}$ and consider $a^b = \sqrt{2}^{\sqrt{2}}$. We do not know (immediately) whether $\sqrt{2}^{\sqrt{2}}$ is rational or irrational, even though it probably seems unlikely that it is rational. However, if $\sqrt{2}^{\sqrt{2}}$ is rational, then the proof will be complete. On the other hand, if $\sqrt{2}^{\sqrt{2}}$ is irrational, then we may take this to be the value of a: let $a = \sqrt{2}^{\sqrt{2}}$ and $b = \sqrt{2}$. Then $a^b = (\sqrt{2}^{\sqrt{2}})^{\sqrt{2}} = (\sqrt{2})^{(\sqrt{2}\sqrt{2})} = (\sqrt{2})^2 = 2$, which is certainly rational.

So, although we do not know *which* of $a = \sqrt{2}$ or $a = \sqrt{2}^{\sqrt{2}}$ will provide a suitable example, when taken together with $b = \sqrt{2}$, we do know that one of them will. So we may now proceed to the proof.

Proof. Consider $\sqrt{2}^{\sqrt{2}}$. Either $\sqrt{2}^{\sqrt{2}}$ is rational or irrational.

If $\sqrt{2}^{\sqrt{2}}$ is rational, then we have proved the theorem: take $a = b = \sqrt{2}$ which we know to be irrational by theorem 6.12.

However, if $\sqrt{2}^{\sqrt{2}}$ is irrational, then let $a = \sqrt{2}^{\sqrt{2}}$ and $b = \sqrt{2}$. Then

$$a^b = \left(\sqrt{2}^{\sqrt{2}}\right)^{\sqrt{2}} = \left(\sqrt{2}\right)^{(\sqrt{2}\sqrt{2})} = \left(\sqrt{2}\right)^2 = 2,$$

which is rational but a and b are irrational. $\qquad\square$

Use of counting arguments

Although counting is apparently an elementary activity, many quite advanced theorems can be proved using counting arguments. In fact, counting can be a complex task and there is a branch of mathematics, called enumeration theory, devoted to techniques of counting. Our aim in this section is to show how some simple counting theorems can form the basis of non-constructive existence proofs.

Theorem 7.8 (The Subset Counting Theorem)
If A and B are finite sets such that $A \subseteq B$ and $|A| \neq |B|$, then there exists an element of B which does not belong to A.

Proof. Let A and B be finite sets.

Clearly, if $A = B$, then $|A| = |B|$. The contrapositive is: $|A| \neq |B| \Rightarrow A \neq B$.

Now suppose that $A \subseteq B$ and $|A| \neq |B|$. From the contrapositive above, we have $A \neq B$. Hence $A \subseteq B$ and $A \neq B$, which means that A is a proper subset of B. Therefore, there exists at least one element of B that is not an element of A. □

Theorem 7.8 provides a basis for non-constructive existence proofs. If we can establish $A \subseteq B$ and $|A| \neq |B|$ for two sets A and B, then the theorem ensures the existence of an element of $B - A$ without of course identifying any specific element. We illustrate the use of the Subset Counting Theorem in the following examples.

Examples 7.3

1. For our first application of the Subset Counting Theorem we prove that any group with an even number of elements must have an element that is its own inverse.

   Theorem 7.9
   Let $(G, *)$ be a finite group with an even number of elements. Then there exists an element $g \in G$ such that $g \neq e$ and g is self-inverse, $g^{-1} = g$.

   The idea for the proof is to compare the set G with its subset comprising the identity e together with those elements that are not self-inverse, $g^{-1} \neq g$: $X = \{x \in G : x = e \text{ or } x^{-1} \neq x\}$. Providing we can show that the subset X has a different number of elements, it will follow from the Subset Counting Theorem that there exists an element of G that is not in X. Any such element satisfies the required conditions: $g \neq e$ and $g^{-1} = g$.

Proof. Let $(G, *)$ be a finite group with an even number of elements.

Define $X \subseteq G$ to be the identity element together with those elements that are not self-inverse,

$$X = \{x \in G : x = e \text{ or } x^{-1} \neq x\}.$$

Apart from e, all the other elements in X may be grouped in pairs, x and x^{-1}. There are clearly an even number of elements that may be grouped together in pairs like this. Since X also contains the identity element, it must contain an odd number of elements in total.

However, G contains an even number of elements, so $|G| \neq |X|$. Therefore, there exists an element $g \in G$ that does not belong to X. Any such element $g \notin X$ is self-inverse, $g^{-1} = g$ and is not equal to the identity element e. $\square$

Commentary

It is worth remarking that the proof assumes that each element of the group has a unique inverse. This allows the elements to be paired off with their inverses, g and g^{-1}. We will consider uniqueness proofs in section 7.5, where we will also prove that the inverse of each element of a group is unique; see theorem 7.15.

2. For our second example, we give a non-constructive existence proof of the existence of irrational numbers, theorem 7.2. Part of our motivation for this is to highlight that the Subset Counting Theorem applies equally well to infinite sets, although a proper understanding of the cardinality of infinite sets is required for this. Recall the corollary to theorem 5.7 which states that if there is a bijective function between two finite sets then they have the same cardinality.

In fact this can be used as the *definition* of 'same cardinality' for arbitrary sets (whether finite or infinite): A and B are defined to have the **same cardinality**, $|A| = |B|$, if there exists a bijection $A \to B$. This simple definition leads to a sophisticated theory of cardinality of infinite sets, originally developed in the 1870s and 1880s by Georg Cantor, but which is beyond the scope of this book. However, we mention two aspects of this general theory. Firstly, unlike a finite set, it is possible for an infinite set to have the same cardinality as a proper subset; see exercise 7.2.5. Secondly, the sets of rational and real numbers have different cardinality, $|\mathbb{Q}| \neq |\mathbb{R}|$. The proof of this uses a clever argument, now known as 'Cantor's diagonal argument'; see Garnier and Taylor [6] for example.

We are now in a position to give an alternative proof of theorem 7.2 that asserts the existence of irrational real numbers.

Alternative proof of Theorem 7.2. First note that $\mathbb{Q} \subseteq \mathbb{R}$. As noted above, $\mathbb{Q}$ and $\mathbb{R}$ have different cardinalities, $|\mathbb{Q}| \neq |\mathbb{R}|$. Therefore, by the Subset Counting Theorem, there exists an element in $\mathbb{R}$ that is not in $\mathbb{Q}$. In other words, there exists an irrational real number. $\qquad\square$

There is a second 'counting theorem' that we will use to provide the basis for non-constructive existence proofs. This is commonly referred to as the 'Pigeonhole Principle' as this provides an easy context to visualise the theorem. From a more mathematical point of view, the theorem is more naturally expressed as a property of functions. We have chosen to state the theorem using the more colloquial language of pigeonholes, but use functions to give the proof.

Theorem 7.10 (Pigeonhole Principle)
If k objects are placed in n pigeonholes where $k > n$, then some pigeonhole contains more than one object.

Proof. Let A denote a set of k pigeonholes and B denote a set of n objects placed in the pigeonholes. Suppose $k > n$.

Define a function $f : A \to B$ by

$$f(\text{object}) = \text{pigeonhole in which it is placed.}$$

Recall theorem 5.7 (i) which says that if $f : A \to B$ is injective, then $|A| \leq |B|$. The contrapositive of this asserts that if $|A| > |B|$, then $f : A \to B$ is not injective. Since $k > n$ we may deduce that the function f defined above is not injective. This means that there exists objects $a_1, a_2 \in A$ such that $a_1 \neq a_2$, but $f(a_1) = f(a_2)$. In other words, there are distinct objects a_1 and a_2 that are placed in the same pigeonhole, proving the theorem. $\qquad\square$

Examples 7.4

1. For our first application of the Pigeonhole Principle, we prove the following simple theorem.

Theorem 7.11
Let A be a set of 6 distinct positive integers. Then there exists a pair of elements in A whose difference is a multiple of 5.

Proof. Let $A = \{a_1, a_2, \ldots, a_6\}$ be a set of six distinct positive integers. For each a_k, let r_k be the remainder when a_k is divided by 5. Formally, r_k is the unique integer satisfying $a_k = 5q_k + r_k$ where $q_k \in \mathbb{N}$ and $0 \leq r_k \leq 4$. This follows from the Division Algorithm; see exercise 7.4.5.

There are 6 remainders, $r_1, r_2, \ldots, r_6$, but only five possible *values* for the remainders, $0, 1, 2, 3, 4$. Therefore, by the Pigeonhole Principle, at least two of the remainders are equal, say $r_i = r_j$. Hence the difference $a_i - a_j = (5q_i + r_i) - (5q_j - r_j) = 5(q_i - q_j)$ is a multiple of 5, as required. □

Commentary

Note that we appealed to the Pigeonhole Principle without explicitly referring to 'objects' and 'pigeonholes'. It is quite common to do this, assuming that the reader can, if necessary, relate the argument to theorem 7.10. If we wanted to refer explicitly to the Pigeonhole Principle, we would have five pigeonholes labelled by the possible remainders $0, 1, 2, 3, 4$. Then each integer a_k would be placed in the pigeonhole labelled m if its remainder $r_k = m$.

Alternatively, we could have proved the theorem without any reference to theorem 7.10. Instead we could have modelled the argument in terms of functions given in the proof of the Pigeonhole Principle. Thus we would first define a function $f : A \to \{0, 1, 2, 3, 4\}$ by defining $f(a_k)$ to be the remainder when a_k is divided by 5. Then use the cardinalities of the sets to deduce that f is not injective and so on.

2. Our second example is most naturally a theorem in graph theory. We have formulated a statement of the theorem in non-graph-theoretic terms in order to avoid introducing more terminology. Any reader who has studied the basics of graph theory should have no trouble in reformulating the following theorem in graph theory terminology.

Theorem 7.12

Let S be a network of bus stations and bus routes where each bus route connects exactly two bus stations. Suppose that there are n bus stations and m bus routes where $m > \frac{1}{2}n(n-1)$. Then there exists a pair of bus stations connected by at least two distinct bus routes.

Proof. Since each bus route connects exactly two stations, we need to compare the number of bus routes with the number of *pairs* of bus stations.

Given that there are n bus stations, the number of pairs is $\frac{1}{2}n(n-1)$. However the number of bus routes m is greater than this. Therefore, by the Pigeonhole Principle, there exists a pair of bus stations connected by more than one bus route. □

Commentary

As in the previous example, we have referred to the Pigeonhole Principle without explicitly saying what are the objects (bus routes) and pigeonholes (pairs of bus stations).

We have assumed, as background knowledge, that the number of pairs of objects that can be formed from n distinct objects is $\frac{1}{2}n(n-1)$. Readers may be familiar with the result that says the number of ways of selecting a subset of r objects from a set of n objects is the so-called binomial coefficient

$$^{n}C_r = \binom{n}{r} = \frac{n!}{(n-r)!r!}.$$

The particular result we need then follows by setting $r = 2$. For readers not familiar with the general result, we may reason as follows. To select a pair of objects from a set of n distinct objects, there are n choices for the first selected object and then there are $n-1$ choices for the second selected object, giving $n(n-1)$ selections in total. However, each possible pair is counted twice: the pair $\{a, b\}$ is counted once when selecting a first and b second and once when selecting b first and a second. Hence the total number of pairs is $\frac{1}{2}n(n-1)$ as claimed.

Exercises 7.2

1. A polygon is **convex** if every interior angle θ is such that $\theta < \pi$.

   Prove that, in any n-sided non-convex polynomial there is an interior angle θ such that

   $$\theta < \left(\frac{n-1}{n-3}\right)\pi.$$

   This theorem appears to assert that every non-convex triangle has a *negative* internal angle. Explain this apparent contradiction.

2. (i) Let $\{a_1, a_2, \ldots, a_n\}$ be a set of non-zero integers such that $\sum_{k=1}^{n} a_k < n$.

   Prove that at least one of the integers in the set is negative.

   (ii) Let $\{b_1, b_2, \ldots, b_n\}$ be a set of integers such that $\sum_{k=1}^{n} b_k^2 < n$.

   Prove that at least one of the integers in the set is zero.

3. A tennis club has $2n+1$ members, where n is a positive integer. During one week, $n+1$ matches were played between members.

   Prove that some member played more than once during the week.

4. Let x be an irrational number, $x \in \mathbb{R} - \mathbb{Q}$. Show that, in the decimal expansion of x, at least one digit $n \in \{0, 1, 2, \ldots, 9\}$ occurs infinitely many times.

5. Using the definition of 'same cardinality' given in example 7.3.2, prove that there is an infinite set that has the same cardinality as a proper subset. In other words, prove that there is an infinite set A and a proper subset B, $B \subset A$, such that $|A| = |B|$.

6. Prove that, in any collection of 12 distinct integers selected from the set $\{1, 2, 3, \ldots, 30\}$, there exists a pair with common factor greater than 1.

7. (i) Prove that, for any set of five points located in a rectangle of dimensions 6 units by 8 units, there exists a pair which are no more than 5 units apart.

   (ii) Prove that, for any set of $n^2 + 1$ points located in a square of side n, there exists a pair which are no more than $\sqrt{2}$ units apart.

   (iii) Prove that, for any set of $n^2 + 1$ points located in an equilateral triangle of side n, there exists a pair in which the points are no more than 1 unit apart.

8. Let $(G, *)$ be a group with n elements, $|G| = n$, and let g be an element of G. Prove that $g^k = e$ for some positive integer $k \leq g$.

9. Prove the following generalisation of the Pigeonhole Principle.

   Theorem (Generalised Pigeonhole Principle)
   Let k, n and r be positive integers. If k distinct objects are placed in n pigeonholes where $k > rn$, then some pigeonhole contains more than r objects.

10. Use the Generalised Pigeonhole Principle (given in the previous exercise) to prove each of the following.

    (i) In any set of 750 people, there exist three people with the same birthday (day and month but not necessarily year of birth).

    (ii) If a pair of dice is rolled 45 times, there is a score that occurs at least 5 times.

    (iii) In a certain lottery, six distinct numbers are drawn randomly each week from the set $\{1, 2, 3, \ldots, 49\}$. Prove that, in a year of lottery draws, some number was drawn on at least seven occasions.

11. The following theorem is given as background knowledge.

Theorem (Interval theorem)
Let $f : [a, b] \to \mathbb{R}$ be a continuous function where $[a, b]$ is a closed interval. Then the image of f is also a closed interval, $\mathrm{im}\, f = [c, d]$ for some $c, d \in \mathbb{R}$.

Use the Interval Theorem to prove the following existence theorem.

Theorem (Intermediate Value Theorem)
Let $f : [a, b] \to \mathbb{R}$ be a continuous function where $[a, b]$ is a closed interval. Suppose that k lies between $f(a)$ and $f(b)$; in other words, k satisfies $f(a) \leq k \leq f(b)$ or $f(b) \leq k \leq f(a)$. Then there exists $c \in [a, b]$ such that $f(c) = k$.

12. The Intermediate Value Theorem itself (given in the previous exercise) can be used as the basis for many non-constructive existence proofs. Prove each of the following using the Intermediate Value Theorem.

   (i) The polynomial $p(x) = x^3 - 5x + 1$ has a root between $x = 0$ and $x = 1$.

   (ii) The equation $e^{x^2} = x + 10$ has a solution between $x = 1$ and $x = 2$.

   (iii) Let f and g be two functions, each continuous on $[a, b]$. Suppose that $f(a) < g(a)$ and $f(b) > g(b)$. Then $f(c) = g(c)$ for some $c \in (a, b)$.

   (iv) Let f be continuous on $[0, 1]$ be such that $f(0) = f(1)$. There exists $c \in \left[\frac{1}{2}, 1\right]$ such that $f(c) = f\left(c - \frac{1}{2}\right)$.
   Hint: consider the function g defined by $g(x) = f(x) - f\left(x - \frac{1}{2}\right)$.

7.4   Counter-examples

Our principal concern in this book has been with finding and understanding proofs of theorems. Of course, given a particular proposition, we will not know whether it really is a theorem until a proof has been found. Suppose we are presented with a proposition of the form $\forall x \bullet P(x)$, which may or may not be a theorem. If it turns out that the proposition is not a theorem, then all our techniques and strategies for finding a proof are bound to fail for the glaringly obvious reason that no proof exists! Unfortunately, there is no way of showing that a proposition is a theorem in advance of finding a proof — finding a proof is precisely how a proposition is shown to be a theorem.

Consider, for example, the proposition:

for all non-negative integers n, the integer $F_n = 2^{2^n} + 1$ is prime.

In 1640, Pierre de Fermat asserted his belief that this proposition was a theorem, although he was unable to supply a proof. These numbers are now called **Fermat numbers** in his honour. Was Fermat correct in his belief? The first stage in investigating the question is to look at some of the smaller examples.

$$F_0 = 2^{2^0} + 1 = 2^1 + 1 = 3,$$

$$F_1 = 2^{2^1} + 1 = 2^2 + 1 = 5,$$

$$F_2 = 2^{2^2} + 1 = 2^4 + 1 = 17,$$

$$F_3 = 2^{2^3} + 1 = 2^8 + 1 = 257,$$

$$F_4 = 2^{2^4} + 1 = 2^{16} + 1 = 65\,537,$$

$$F_5 = 2^{2^5} + 1 = 2^{32} + 1 = 4\,294\,967\,297,$$

$$F_6 = 2^{2^6} + 1 = 2^{64} + 1 = 18\,446\,744\,073\,709\,551\,617.$$

It is clear that F_0, F_1, and F_2 are prime and we can fairly quickly verify that F_3 is prime. With rather more work, F_4 can be shown to be prime but even with a standard scientific calculator, this would be a lengthy and tedious task. Beyond F_4 the Fermat numbers grow very rapidly indeed. In exercise 6.4.9, we gave a method for testing the primality of a positive integer n: test whether n has a prime factor p in the range $2 \leq p \leq \sqrt{n}$. We cannot imagine anyone wishing to use this method to test whether or not F_5 is prime aided only by a pocket calculator. Indeed, it was not until 1732, nearly one hundred years after Fermat proposed the conjecture, that Euler established that F_5 is composite by showing that

$$F_5 = 4\,294\,967\,297 = 641 \times 6\,700\,417.$$

Of course, this factorisation of F_5 shows that Fermat's conjecture is not a theorem.[2] The factorisation provides a 'counter-example' to the proposition above.

Let us consider again the general situation: suppose we are presented with a proposition which is a universally quantified propositional function $\forall x \bullet P(x)$. If we can find a single specific member a of the universe such that $P(a)$ is false, then $\forall x \bullet P(x)$ is not a theorem. Any element a in the universe such

---

[2] As an aside, it is interesting to note that what took the mathematical community nearly 100 years to achieve now takes a modest desktop computer no more than a few seconds. There are various computer algebra packages which will obtain these factors in a fraction of a second. Indeed, the factorisation of the next three Fermat numbers, F_6, F_7, and F_8, can be obtained in at most a few minutes. This is quite impressive since, for example, $F_8 = 1238926361552897 \times 93461639715357977769163558199606896584051237541638185580280321$.

that $P(a)$ is false is called a **counter-example** to the proposition $\forall x \bullet P(x)$. The method of finding the appropriate element a and showing $P(a)$ is false is often called **proof by counter-example**. Since the existence of a counter-example establishes that $\forall x \bullet P(x)$ is not a theorem, perhaps '*disproof* by counter-example' would be a better term.

Given a proposition $\forall x \bullet P(x)$, which may or may not be a theorem, we are faced with a dilemma. Do we search for a proof or do we try to find a counter-example? If $\forall x \bullet P(x)$ is a theorem and we opt to search for a counter-example, then our quest is bound to fail. On the other hand, if $\forall x \bullet P(x)$ is not a theorem, then any search for a proof will inevitably be unsuccessful. The choice of which path to take — proof or counter-example — is often based on experience, intuition, or pure instinct. In practice, the situation is not as bad as it appears. As we have seen, the first step in the search for a proof is frequently to look at some examples, and during this initial phase we may come across a counter-example anyway.

Actually, there is a third possibility which is rather disturbing. It may be impossible to find a proof of, or a counter-example to, the proposition $\forall x \bullet P(x)$. Essentially, we have defined a theorem to be a proposition which is provable from the axioms. There are some situations when neither $\forall x \bullet P(x)$ nor its negation $\neg \forall x \bullet P(x)$ is provable from the axioms. In other words, the given axiom system is not sufficiently powerful to determine the 'truth' of $\forall x \bullet P(x)$. In this case, we say that $\forall x \bullet P(x)$ is **undecidable** from the given axioms. Fortunately, such situations are rare and tend to crop up only in the more esoteric areas of mathematics.

Examples 7.5

1. In example 7.2.1 we introduced the Mersenne prime numbers; that is, prime numbers of the form $2^p - 1$ where p is prime. In example 6.1.3, we proved theorem 6.3, which asserts that 'p is prime' is a necessary condition for '$2^p - 1$ is prime'. We now consider whether it is a sufficient condition; in other words, is the statement '*for all p, if p is prime, then $2^p - 1$ is prime*' a theorem?

   We begin by testing some small examples:

$$p = 2: \quad 2^p - 1 = 2^2 - 1 = 3$$
$$p = 3: \quad 2^p - 1 = 2^3 - 1 = 7$$
$$p = 5: \quad 2^p - 1 = 2^5 - 1 = 31$$
$$p = 7: \quad 2^p - 1 = 2^7 - 1 = 127$$
$$p = 11: \quad 2^p - 1 = 2^{11} - 1 = 2047$$

   Of these, the first three are clearly prime. It is easy to check that $2^7 - 1 = 127$ is prime using the method given in exercise 6.4.9: test whether n

has a prime factor p in the range $2 \leq p \leq \sqrt{n}$. Using this method to test $2^{11} - 1 = 2047$ takes a little longer, but we soon find that $2^{11} - 1 = 2047 = 23 \times 89$, so that we have found a counter-example to the proposition '*for all primes p, the integer $2^p - 1$ is prime*'.

2. Find a counter-example to the proposition:

$$\text{for all } x, y \in \mathbb{R}, \text{ if } x \leq y \text{ then} |x| \leq |y|.$$

Solution

We need to find specific real numbers x and y such that the conditional '$x \leq y \Rightarrow |x| \leq |y|$' is false; in other words. we need to find x and y where $x \leq y$ ie true but $|x| \leq |y|$ is false. If a is non-negative then $|a| = a$; hence any counter-example must have at least one of x and y being negative. Taking both to be negative gives simple counter-examples. For example, if $x = -2$ and $y = -1$, then $x < y$. However, $|x| = 2 > 1 = |y|$, so $|x| \leq |y|$ is false. We have found our counter-example.

3. Either find a proof or find a counter-example to the proposition

$$\text{for all sets } A, B \text{ and } C, \ A \cup (B - C) = (A \cup B) - C.$$

Solution

In this case, we do not initially know whether we should be seeking a proof or a counter-example. We could start by considering some specific examples of sets A, B, and C and calculate both $A \cup (B - C)$ and $(A \cup B) - C$. However, for sets, we have a powerful visualisation in the form of Venn-Euler diagrams. Although one or more Venn-Euler diagrams will not constitute a proof of a theorem,[3] they can be extremely useful in pointing the way towards a proof or counter-example. Figure 7.2 represents the sets $A \cup (B - C)$ and $(A \cup B) - C$ in two separate Venn-Euler diagrams.

From the diagrams in figure 7.2, we can see that the 'difference' between the two sets is that $A \cup (B - C)$ contains $A \cap C$ as a subset whereas $(A \cup B) - C$ does not. This indicates that a counter-example will require the set $A \cap C$ to be non-empty.

Counter-example

Let $A = \{1, 2, 3, 4, 5\}$, $B = \{2, 4, 6\}$, and $C = \{2, 3, 5\}$.

---

[3] In fact, this is not strictly true. In her seminal 1994 book [10], Sun-Joo Shin develops two logical systems related to Venn diagrams where rigorous proofs *can* be carried out entirely diagrammatically. Since then, other diagrammatic logics have been developed, including those for Venn-Euler diagrams, where proofs are purely diagrammatic. However, our informal use of Venn-Euler diagrams only points the way towards proofs and they do not, in themselves, constitute a proof.

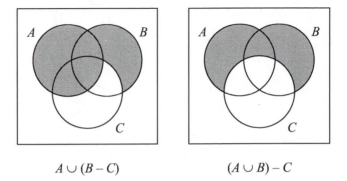

$$A \cup (B - C) \qquad\qquad (A \cup B) - C$$

FIGURE 7.2: Finding a set-theory counter-example.

Then $B - C = \{4, 6\}$, so $A \cup (B - C) = \{1, 2, 3, 4, 5, 6\}$.

However, $A \cup B = \{1, 2, 3, 4, 5, 6\}$, so $(A \cup B) - C = \{1, 4, 6\}$.

Therefore $A \cup (B - C) \neq (A \cup B) - C$ for these sets. $\square$

Commentary

In terms of proving a particular proposition to be false, any counter-example is as good as any other. However, simpler counter-examples are to be preferred to more complicated ones. A complicated example may be difficult to understand and can obscure the underlying reason why the particular proposition is false. A simpler example is more likely to come close to the heart of why the proposition is false and thus provide greater insight.

For instance, another counter-example to the proposition given above is provided by the sets $A = \mathbb{R}^+ = \{$positive real numbers$\}$, $B = \mathbb{Z} = \{$integers$\}$, and $C = \mathbb{R} - \mathbb{Q} = \{$irrational real numbers$\}$. (We leave it as an exercise to verify that these sets do, indeed, provide a counter-example; see exercise 7.3.3.) However, it is more difficult to evaluate the various sets involved here and we may wonder whether the reason that $A \cup (B - C) \neq (A \cup B) - C$ has something to do with the fact that the sets are infinite or involve irrational numbers.

4. Find counter-examples to the converse of each part of theorem 5.7. The theorem states that, for any function $f : A \rightarrow B$ between finite sets A and B:

    (i) if f is injective, then $|A| \leq |B|$;

    (ii) if f is surjective, then $|A| \geq |B|$.

Solution

In part (i), the converse is: if $|A| \leq |B|$, then f is injective. For a counter-example, we need to find a function $f : A \to B$ where $|A| \leq |B|$ but where f is not injective. Recall that a function is not injective if there exist different elements of the domain A with the same image in the codomain B; figure 3.20 (page 117) gives a diagrammatic representation of this situation.

In part (ii), the converse is: if $|A| \geq |B|$, then f is surjective. For a counter-example, we need to find a function $f : A \to B$ where $|A| \geq |B|$ but where f is not surjective. For a function to fail to be surjective, there must exist elements of the codomain B that do not belong to the image of f, im f; figure 3.13, page 109, illustrates this situation.

We can achieve these properties with functions like those shown in figure 7.3. The function in figure 7.3 (i) is not injective but has $|A| \leq |B|$, and the function in figure 7.3 (ii) is not surjective but has $|A| \geq |B|$ In fact, the figure essentially defines the counter-examples we are seeking.

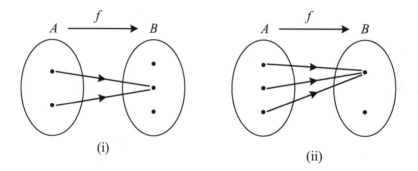

FIGURE 7.3: Counter-examples to theorem 5.7.

Counter-examples

(i) Let $A = \{a_1, a_2\}$ and $B = \{b_1, b_2, b_3\}$ and let $f : A \to B$ be defined by $f(a_1) = f(a_2) = b_2$.
   Then $|A| = 2 < 3 = |B|$. Note that f is not injective since $a_1 \neq a_2$ but $f(a_1) = f(a_2)$.

(ii) Let $A = \{a_1, a_2, a_3\}$ and $B = \{b_1, b_2\}$ and let $f : A \to B$ be defined by $f(a_1) = f(a_2) = f(a_3) = b_1$.
   Then $|A| = 3 < 2 = |B|$. Also f is not surjective since $b_2 \in B$ is not the image of any element of A, $b_1 \notin$ im f.

□

Commentary

These counter-examples certainly satisfy the 'simplicity test' that we discussed in the previous example. Some readers may be surprised that we have not defined the functions using formulae. However, recall from definition 3.1 that a function is any rule which associates, to each element $a \in A$, a unique element $b = f(a) \in B$. Explicitly defining $f(a_i)$ for each $a_i \in A$ defines such a rule.

Each of the counter-examples does, however, make implicit assumptions which, one might argue, should really be made explicit. These are that the elements of A are distinct and the elements of B are distinct. An alternative that avoids this issue is to define the sets A and B to contain familiar elements (rather that 'arbitrary' elements a_1, a_2 and so on). For example, in the counter-example in part (i), we could take $A = \{1, 2\}$ and $B = \{1, 2, 3\}$ and define $f(1) = f(2) = 2$; this is essentially equivalent to our example.

___

Exercises 7.3

1. Let $f(n) = n^2 + n + 41$. Then $f(0) = 41, f(1) = 43, f(2) = 47, f(3) = 53, f(4) = 61, \ldots$ are all prime.

   Find a counter-example to the proposition: for all non-negative integers n, $f(n)$ is prime.

   This formula, which does produce a long sequence of primes, was discovered by Euler. In fact, amongst all expressions of the form $n^2 + an + b$ where a and b are non-negative integers less than $10,000$, there is none which produces a longer sequence of primes.

2. Find a counter-example to each of the following propositions.

   (i) For all real numbers a, b, c, and d, if $a > b$ and $c > d$, then $(a - c) > (b - d)$.

   (ii) For all positive integers a, b, and c, if c is a factor of $a + b$, then c is a factor of a or c is a factor of b.

   (iii) $f(n) = n^2 - n + 17$ is prime for all positive integers n.

   (iv) $6^n + 4n^4$ is divisible by 5 for all positive integers n.

   (v) $3^n < 4n^4$ for all even positive integers n.

   (vi) $n^4 + 1$ is prime for all even positive integers n.

3. Verify that the sets $A = \mathbb{R}^+ = \{\text{positive real numbers}\}$, $B = \mathbb{Z} = \{\text{integers}\}$, and $C = \mathbb{R} - \mathbb{Q} = \{\text{irrational real numbers}\}$ provide a counter-example to the proposition: $A \cup (B - C) = (A \cup B) - C$ for all sets A, B, C, as claimed in example 7.5.3.

4. Prove or disprove each of the following propositions.

(i) If a and b are rational numbers, then ab is a rational number.

(ii) If a and b are irrational numbers, then ab is an irrational number.

(iii) If a and b are rational numbers and $b \neq 0$, then a/b is a rational number.

(iv) If a and b are irrational numbers, then a/b is an irrational number.

(v) (Compare with exercise 6.4.5 (ii).) Let α and β be irrational numbers. Then, for all non-zero integers m and n, $\alpha m + \beta n$ is irrational.

5. Find a counter-example to each of the following propositions where $\mathbf{A}$, $\mathbf{B}$, and $\mathbf{C}$ are 2×2 matrices with real number entries.

(i) If $\mathbf{AB} = \mathbf{AC}$ and $\mathbf{A}$ is not a zero matrix, then $\mathbf{B} = \mathbf{C}$.

(ii) The only matrices satisfying the equation $\mathbf{A}^2 = \mathbf{A}$ are $\mathbf{0} = \begin{pmatrix} 0 & 0 \\ 0 & 0 \end{pmatrix}$ and $\mathbf{I}_2 = \begin{pmatrix} 1 & 0 \\ 0 & 1 \end{pmatrix}$.

(iii) If $\mathbf{A}^4 = \begin{pmatrix} 1 & 0 \\ 0 & 1 \end{pmatrix}$, then $\mathbf{A}^2 = \begin{pmatrix} 1 & 0 \\ 0 & 1 \end{pmatrix}$.

(iv) If $\mathbf{A}$ and $\mathbf{B}$ are distinct matrices such that $\mathbf{AB} = \mathbf{BA}$, then either $\mathbf{A}$ or $\mathbf{B}$ is equal to $\mathbf{0} = \begin{pmatrix} 0 & 0 \\ 0 & 0 \end{pmatrix}$ or $\mathbf{I}_2 = \begin{pmatrix} 1 & 0 \\ 0 & 1 \end{pmatrix}$.

6. Prove or disprove each of the following propositions.

(i) For all positive integers a, b and n, if a and b are factors of n, and $ab < n$, then ab is a factor of n.

(ii) There exists a positive integer n such that $8^n + 1$ is prime.

7. For each of the following statements about functions $A \to B$ where A and B are subsets of $\mathbb{R}$, either find a proof or find a counter-example to the statement.

The definitions of increasing and decreasing functions are given in exercise 6.2.5.

(i) The function $f + g$ is increasing if and only if both f and g are increasing functions.

The function $f + g : A \to B$ is defined by $(f + g)(x) = f(x) + g(x)$.

(ii) The function fg is increasing if and only if both f and g are increasing functions.

The function $fg : A \to B$ is defined by $(fg)(x) = f(x)g(x)$.

(iii) Suppose that $g(x) \neq 0$ for all $x \in A$. Then the function f/g is increasing if and only if f is increasing and g is decreasing.

The function $f/g : A \to B$ is defined by $(f/g)(x) = f(x)/g(x)$.

8. (Compare with theorem 5.13.)

   Find a counter-example to the following proposition.

   Let S and T be subspaces of a vector space V. Their union $S \cup T$ is also a subspace of V.

9. The notion of a subgroup of a group was introduced in exercise 5.3.3. Find a counter-example to each of the following propositions.

   (i) Let H and K be subgroups of a group G. Their union $H \cup K$ is also a subgroup of G.

   (ii) If every proper subgroup of a group is Abelian, then the group itself is Abelian.

10. Prove or disprove each of the following propositions.

    (i) If A, B, and C are sets such that $A \subseteq B$ and $B \subseteq C$, then $A \subseteq C$.

    (ii) If A, B, and C are sets such that $A \nsubseteq B$ and $B \nsubseteq C$, then $A \nsubseteq C$. (The symbol $\nsubseteq$ means 'is not a subset of'.)

    (iii) If $\mathbf{A}$ and $\mathbf{B}$ are $n \times n$ matrices such that both $\mathbf{A}$ and $\mathbf{AB}$ are symmetric, then $\mathbf{B}$ is symmetric.

    A **symmetric matrix** is one that remains unchanged when its rows and columns are interchanged. The simplest way of formalising this is to first define the transpose of a matrix. Given an $m \times n$ matrix $\mathbf{X}$, its **transpose** $\mathbf{X}^{\mathrm{T}}$ is the $n \times m$ matrix whose rows are the columns of $\mathbf{X}$ written in the same order. Then a matrix $\mathbf{X}$ is symmetric if and only if it equals its transpose, $\mathbf{X} = \mathbf{X}^{\mathrm{T}}$.

    (iv) If G is a group and $g, h \in G$, then $(gh)^n = g^n h^n$ for all positive integers n.

    (v) If A, B, and C are sets such that $C \subseteq A \times B$, then $C = X \times Y$ for some $X \subseteq A$ and $Y \subseteq B$.

    Informally, this says that every subset of a Cartesian product is itself a Cartesian product.

11. (For readers who know about the elementary properties of continuous and differentiable functions.)

    Find a counter-example to each of the following propositions.

    (i) Every continuous function $f : A \to B$, where A and B are subsets of $\mathbb{R}$, is differentiable.

    (ii) Every continuous function $f : (a, b) \to \mathbb{R}$ is bounded, where (a, b) is an open interval.

    A real-valued function f is said to be **bounded** if its image, im f, is bounded. Referring to the definition of a bounded set given in section 6.6, this means that there exist real numbers m and M such that $m \leq f(x) \leq M$ for all x in the domain of the function.

Note: there is a theorem which states that every continuous function $f : [a, b] \to \mathbb{R}$ is bounded, where $[a, b]$ is a closed interval. Your counter-example shows that this theorem does not extend to open intervals.

(iii) If $f : \mathbb{R} \to \mathbb{R}$ is twice differentiable and f has a local maximum at $x = a$, then $f''(a) < 0$.

Similarly, if $g : \mathbb{R} \to \mathbb{R}$ is twice differentiable and g has a local minimum at $x = a$, then $g''(a) > 0$.

Note: your counter-examples indicate the limitations of what is frequently called the second derivative test for local maxima and minima.

7.5 Uniqueness proofs

Sometimes in mathematics, we wish to prove not only that an object with certain properties exists but also that there is only one such object; in other words, that the object is unique. The existence part of such a proof was discussed in sections 7.2 and 7.3. In this section, we focus on the uniqueness part. To see what is required in such a proof, suppose we define the natural number N to be the *number* of objects with the required property. An existence proof amounts to showing $N \geq 1$; in other words, that there is at least one such object. Given this, to establish uniqueness as well we must show that $N = 1$, so that there is exactly one object of the required type.

The method of proving uniqueness is to show that $N \leq 1$. This is generally done by assuming that there are two objects with the required property and then showing that they are equal. This sounds like a proof by contradiction but it is not (quite!). In everyday language, when we say 'I have two objects of a particular type', two apples say, we mean 'I have two *different* objects of the particular type'. In our proofs we will assume that there are two objects of the required type, a and b say, but we will not usually assume that they are different. The proof is completed by showing that $a = b$, so that there really is (at most) one object of the kind under consideration.

Examples 7.6

1. For our first example we will prove the uniqueness part of the following theorem.

Theorem 7.13

For every real number a, the equation $x^3 = a$ has a unique real solution.

The existence part of the proof is both subtle and difficult and amounts to proving that every real number a has a cube root $\sqrt[3]{a}$. Consider, for example, how we might prove that the real number $\sqrt[3]{2}$ exists? The answer to this question goes to the heart of what we mean by the real numbers. There are two approaches to describing the real numbers $\mathbb{R}$. We may start with a simpler system, usually the natural numbers $\mathbb{N}$, and from this 'build up' to the real numbers by constructing the integers $\mathbb{Z}$, the rational numbers $\mathbb{Q}$, and then the real numbers $\mathbb{R}$. The natural numbers themselves are generally described axiomatically by a system of five axioms called the Peano axioms; we shall consider one of the Peano axioms, called the Axiom of Induction, in section 8. The last step in this process, constructing the real numbers from the rational numbers, is subtle and there is more than one way of achieving this.

Alternatively, as we mentioned in chapter 4.5, the real numbers may be described axiomatically as a complete ordered field; thirteen axioms are required for this. The existence of real numbers such as $\sqrt[3]{2}$ follows from the Completeness Axiom for the real numbers, which says that every non-empty subset of $\mathbb{R}$ that is bounded above has a supremum. See definition 6.2 for the definition of the supremum of a non-empty subset of $\mathbb{R}$.

For the current theorem, we will not concern ourselves with how the existence of the real number $\sqrt[3]{a}$ is established — either through construction or via axioms. Instead, we will regard this as part of our background knowledge of the real numbers and concentrate only on the uniqueness part of the theorem.

Proof of uniqueness. Suppose that x and y are real numbers such that $x^3 = a$ and $y^3 = a$. Then $x^3 = y^3$, so we may reason as follows:

$$x^3 = y^3 \quad \Rightarrow \quad x^3 - y^3 = 0$$
$$\Rightarrow \quad (x - y)(x^2 + xy + y^2) = 0 \qquad \text{(factorising)}$$
$$\Rightarrow \quad x - y = 0 \quad \text{or} \quad x^2 + xy + y^2 = 0$$
$$\Rightarrow \quad x = y \quad \text{or} \quad x^2 + xy + y^2 = 0.$$

We are required to show that $x = y$. This would now follow if we could show that $x^2 + xy + y^2 \neq 0$. Unfortunately, we cannot show this because it is not true! However, from exercise 5.1.1 (iii) we know that

$$(x \neq 0 \text{ or } y \neq 0) \quad \Rightarrow \quad x^2 + xy + y^2 \neq 0.$$

Using De Morgan's rule (see Table 2.1 on page 39), the contrapositive of this is

$$x^2 + xy + y^2 = 0 \quad \Rightarrow \quad (x = 0 \text{ and } y = 0).$$

In particular, $x^2 + xy + y^2 = 0 \Rightarrow x = y$. From the argument above, it now follows that:

$$(x - y)(x^2 + xy + y^2) = 0 \quad \Rightarrow \quad x = y \quad \text{or} \quad x^2 + xy + y^2 = 0$$
$$\Rightarrow \quad x = y.$$

Therefore the equation $x^3 = a$ has a unique real solution. $\qquad\qquad\square$

Commentary

The proof follows the structure that we outlined above. That is, we assumed that x and y are both solutions to the equation and then we deduced that $x = y$. The proof itself is elementary in the sense that it uses little more than factorisation. However, there is a little complication in the middle where we need to show that 'if $x^2 + xy + y^2 = 0$, then $x = 0$ and $y = 0$', which requires us to be careful with the logical structure as we need to understand both the contrapositive and the negation of a disjunction using De Morgan's rule.

2. In this example, we first prove the following theorem about inverses of 2×2 matrices. For those readers who know group theory or who have read section 5.3, we then relate this to the uniqueness of inverses in any group.

Theorem 7.14

Let **A** be a 2×2 matrix. If $\det \mathbf{A} \neq 0$, then **A** has a unique inverse.

The **determinant** of a 2×2 matrix is given by

$$\det \begin{pmatrix} a & b \\ c & d \end{pmatrix} = ad - be.$$

This time we shall prove both the existence and uniqueness part of the theorem. The existence part is proved by construction, that is, given **A** we find a matrix **B** such that $\mathbf{AB} = \mathbf{BA} = \mathbf{I}_2$. We shall simply define the matrix **B** and show that it is the inverse of **A**. (For an explanation of where the matrix **B** comes from, a textbook covering basic matrix theory may be consulted.)

Proof. Let $\mathbf{A} = \begin{pmatrix} a & b \\ c & d \end{pmatrix}$ and suppose that $\det \mathbf{A} = ad - bc \neq 0$. Define **B** to be the 2×2 matrix

$$\mathbf{B} = \frac{1}{ad - bc} \begin{pmatrix} d & -b \\ -c & a \end{pmatrix} = \begin{pmatrix} d/(ad - bc) & -b/(ad - bc) \\ -c/(ad - bc) & a/(ad - bc) \end{pmatrix}.$$

Then

$$\mathbf{AB} = \begin{pmatrix} a & b \\ c & d \end{pmatrix} \times \frac{1}{ad-bc} \begin{pmatrix} d & -b \\ -c & a \end{pmatrix}$$

$$= \frac{1}{ad-bc} \begin{pmatrix} a & b \\ c & d \end{pmatrix} \begin{pmatrix} d & -b \\ -c & a \end{pmatrix}$$

$$= \frac{1}{ad-bc} \begin{pmatrix} ad-bc & -ab+ab \\ cd-cd & ad-bc \end{pmatrix}$$

$$= \frac{1}{ad-bc} \begin{pmatrix} ad-bc & 0 \\ 0 & ad-bc \end{pmatrix}$$

$$= \begin{pmatrix} 1 & 0 \\ 0 & 1 \end{pmatrix} = \mathbf{I}_2.$$

Therefore $\mathbf{AB} = \mathbf{I}_2$. Verifying that $\mathbf{BA} = \mathbf{I}_2$ is similar. Hence $\mathbf{B}$ is an inverse of $\mathbf{A}$, $\mathbf{B} = \mathbf{A}^{-1}$. This completes the existence part of the proof.

We now turn to the uniqueness part of the proof. We shall need to assume, as background knowledge, the associative property of multiplication of 2×2 matrices which states that $\mathbf{X(YZ)} = \mathbf{(XY)Z}$ for all 2×2 matrices $\mathbf{X, Y}$ and $\mathbf{Z}$.

Suppose that $\mathbf{B}$ and $\mathbf{C}$ are 2×2 matrices such that $\mathbf{AB} = \mathbf{BA} = \mathbf{I}_2$ and $\mathbf{AC} = \mathbf{CA} = \mathbf{I}_2$. Then

$$\begin{aligned}
\mathbf{B} &= \mathbf{BI}_2 && \text{(property of } \mathbf{I}_2\text{)} \\
&= \mathbf{B(AC)} && \text{(since } \mathbf{AC} = \mathbf{I}_2\text{)} \\
&= \mathbf{(BA)C} && \text{(associative property of matrix multiplication)} \\
&= \mathbf{I}_2\mathbf{C} && \text{(since } \mathbf{BA} = \mathbf{I}_2\text{)} \\
&= \mathbf{C} && \text{(property of } \mathbf{I}_2\text{).}
\end{aligned}$$

Therefore $\mathbf{B} = \mathbf{C}$, so the inverse of $\mathbf{A}$ is unique. $\square$

The proof of uniqueness clearly relies on the associative law for matrix multiplication. In fact, the uniqueness proof is valid for any associative binary operation with an identity element. In particular, the proof can be used to show that in any group the inverse of each element is unique, which we now do.

Theorem 7.15
The inverse of any element g of a group $(G, *)$ is unique.

Proof. Let G be a group which, for simplicity we will write multiplicatively; see section 5.3.

Let $g \in G$. Suppose that x and y are inverses of g. By definition, this means that $gx = xg = e$ and $gy = yg = e$ where e is the identity element of the group. Then

$$
\begin{aligned}
x &= xe && \text{(since e is the identity)} \\
&= x(gy) && \text{(since y is an inverse of g)} \\
&= (xg)y && \text{(associative property in G)} \\
&= ey && \text{(since x is an inverse of g)} \\
&= y && \text{(since e is the identity).}
\end{aligned}
$$

Therefore $x = y$, so the inverse of g is unique. $\qquad\square$

Note that the proof of theorem 7.15 is essentially identical to the uniqueness part of the proof of theorem 7.14. In fact, the set of all 2×2 invertible matrices — that is, matrices that have inverses — forms a group under the operation of matrix multiplication. Once we have established this, then theorem 7.14 is just a special case of theorem 7.15. This is an example of the economy of labour that we mentioned in section 4.5 as one advantage of using axiom systems. Having proved that inverses are unique in an (abstract) group, we may deduce this for any structure that is a group.

3. For our next example, we prove the uniqueness part of the Fundamental Theorem of Arithmetic. The proof of the existence part was outlined in section 4.2 and will be dealt with more rigorously in chapter 8.

Theorem 7.16 (Fundamental Theorem of Arithmetic)
Every integer greater than 1 can be expressed as a product of prime numbers in a manner which is unique apart from the ordering of the prime factors.

Proof. Let a be an integer greater than 1.

The existence of a prime factorisation for a was given in the Prime Factorisation Theorem 4.3 (page 132). We therefore only need to prove the uniqueness part.

Let

$$
a = p_1 p_2 \ldots p_m = q_1 q_2 \ldots q_n \tag{*}
$$

be two factorisations of a into prime factors. Without loss of generality, we may suppose that $m \geq n$.

Now q_n is a factor of a, so q_n divides the product $p_1 p_2 \ldots p_m$. If a prime number divides a product, it must divide one of the factors. (See exercise 6.2.3 for the case of two factors and exercise 8.1.7 for the extension to several factors.) Hence q_n divides one of the p's; suppose q_n divides

p_k. However p_k is prime, so it has no factors other than 1 and p_k itself. Since $q_n \neq 1$ we have $q_n = p_k$.

Dividing the equation (*) by $q_n = p_k$ and renumbering the p's (if necessary) gives

$$p_1 p_2 \dots p_{m-1} = q_1 q_2 \dots q_{n-1}.$$

Now consider q_{n-1} and repeat the same argument that we have just given: q_{n-1} must be equal to one of the p's, then divide by it. Continuing in this way using $q_{n-1}, q_{n-2}, \dots q_2$ in turn produces

$$p_1 p_2 \dots p_k = q_1, \tag{**}$$

where $k = m - n + 1 \geq 1$ (since $m \geq n$).

As before, q_1 divides the product $p_1 p_2 \dots p_k$ and so must divide one of the p's. If $k > 1$ so that there is more than one factor on the left-hand side of (**), then dividing by q_1 would leave a product of prime numbers equal to 1. This is impossible since every prime is strictly greater than 1. Therefore $k = 1$. This implies that $n = m$ so (**) is just $p_1 = q_1$.

In summary, we have shown that with a suitable re-numbering of the p's if necessary,

$$m = n \quad \text{and} \quad p_1 = q_1, p_2 = q_2, \dots, p_m = q_n.$$

Therefore the prime factorisation is unique apart from the ordering of the factors. $\qquad\square$

4. For our final example, we show that, for a non-empty subset of $\mathbb{R}$ that is bounded above, its supremum is unique. The definition of the supremum is given in definition 6.2.

Theorem 7.17 (Uniqueness of supremum)
Let X be a non-empty subset of $\mathbb{R}$ that is bounded above. Then the supremum of X, $\sup X$, is unique.

Proof. Let X be a non-empty subset of $\mathbb{R}$ that is bounded above. Let A and B be two suprema for X. This means that A and B both satisfy the conditions (i) and (ii) of definition 6.2.

Since A is a supremum for X and B is *an* upper bound for X, condition (ii) of the definition ensures that $A \leq B$.

Now reverse the roles of A and B. Since B is a supremum for X and A is *an* upper bound for X, condition (ii) of the definition ensures that $B \leq A$.

We have shown that $A \leq B$ and $B \leq A$, so $A = B$, as required. $\qquad\square$

Exercises 7.4

1. (i) Prove that the equation $ax = b$, where a and b are fixed real numbers and $a \neq 0$, has a unique solution.

   (ii) Prove that, if a is a positive real number, then the equation $x^2 = a$ has a unique positive solution.

   As in the example 7.6.1, you may assume the existence of $\sqrt{a}$ for any $a > 0$.

   (iii) Assuming the existence of a fifth root $\sqrt[5]{a}$ of any $a \in \mathbb{R}$, prove that, for every real number a, the equation $x^5 = a$ has a unique solution.

2. Prove that, if $a, b, c,$ and d are real numbers such that $ad - bc \neq 0$, then for all real numbers s, t there exists a unique solution (x, y) to the simultaneous equations

$$ax + by = s$$
$$cx + dy = t.$$

3. Prove that every integer $a > 2$ can be expressed uniquely as $a = 2^n b$ where n is an integer and b is an odd integer.

4. Prove that there is a unique prime number p for which $p^2 + 2$ is also prime.

5. Prove the following theorem. The theorem makes precise the idea that dividing a positive integer n by a positive integer m gives a 'quotient' q and 'remainder' r. For example, dividing 131 by 9 gives quotient 14 and remainder 5, which we can express as $131 = 14 \times 9 + 5$.

   Theorem (The Division Algorithm)
   Let m and n be positive integers. Then there exist unique $q, r \in \mathbb{N}$ such that $n = qm + r$ and $0 \leq r < m$.

   Hint: for the existence part, consider the set $\{n - qm : q \in \mathbb{N}\}$ and reason that this set must have a smallest non-negative element r.

6. Let $f : A \to B$ be a bijective function. Prove that the inverse function $f^{-1} : B \to A$ is unique.

7. Let $(G, *)$ be a group. Prove each of the following uniqueness results.

   (i) Uniqueness of the identity: the identity element e is unique.

   (ii) Uniqueness of the solution of equations (1): for all $g, h \in G$, the equation $gx = h$ has a unique solution for $x \in G$.

(iii) Uniqueness of the solution of equations (2): for all $g, h \in G$, the equation $xg = h$ has a unique solution for $x \in G$.

8. Let V be a vector space. Prove each of the following uniqueness results.

   (i) Uniqueness of the zero vector: $\mathbf{0}$ is unique.

   (ii) Uniqueness of negatives: for all $\mathbf{v} \in V$ the negative, $-\mathbf{v}$ is unique.

9. (i) Prove that $x = 0$ is the only real number that satisfies $|x| < \varepsilon$ for all $\varepsilon \in \mathbb{R}^+$.

   (ii) Let (a_n) be a convergent sequence. Prove that its limit is unique.

10. Let $f : A \to B$ be a function where A and B are subsets of R. Suppose that $\lim\limits_{x \to a} f(x)$ exists, where $a \in A$. Prove that the limit is unique.

Chapter 8

Mathematical Induction

8.1   Introduction

In this chapter we consider in some detail one particular method of proof, called proof by mathematical induction, which may be used to prove statements about the natural numbers or positive integers. In other words, the method will apply to the proofs of propositions of the form $\forall n \bullet P(n)$ where the universe for n is the set of natural numbers $\mathbb{N}$ or the set of positive integers $\mathbb{Z}^+$. The following are examples of the kinds of results that we may prove using the method of mathematical induction.

1. The sum of the first n positive integers is $\frac{1}{2}n(n+1)$; symbolically,

   for all $n \in \mathbb{Z}^+$, $1 + 2 + 3 + \cdots + n = \frac{1}{2}n(n+1)$.

2. For every $n \in \mathbb{N}$, the expression $5^n + 3$ is divisible by 4.

3. For all finite sets A, if A has cardinality n, then its power set $\mathbb{P}(A)$ has cardinality 2^n.

The first of these is clearly of the form $\forall n \in \mathbb{Z}^+ \bullet P(n)$ and the second is of the form $\forall n \in \mathbb{N} \bullet P(n)$. However, the third statement is not obviously of one of these two forms. Note that, to say that a set A is finite means it has cardinality $|A| = n$ where n is some natural number. This means we could rephrase the third statement as 'for all $n \in \mathbb{N}$, if A is a set with cardinality $|A| = n$, then its power set has cardinality $|\mathbb{P}(A)| = 2^n$' and this is of the form $\forall n \in \mathbb{N} \bullet P(n)$. This is not unusual. Sometimes a statement is not obviously one about $\mathbb{N}$ or $\mathbb{Z}^+$, but is more naturally about some other mathematical object such as a set, function, group, graph or whatever. But, by rethinking the statement, it may be rephrased to be one about say the positive integers; the advantage of this is that it brings into play the possibility of using the powerful method of mathematical induction.

Despite its name, any proof by mathematical induction uses deductive reasoning, which is the basis of all mathematical proofs; it does not rely on the kind of inductive reasoning described in section 1.3.

To motivate the idea behind the method, suppose we are given the following instruction.

- *If you write down an integer k, then you must also write down the integer $k + 1$.*

What do we need to do to satisfy the instruction? The simple answer is: nothing at all! The statement is a conditional of the form $W(k) \Rightarrow W(k + 1)$ where $W(k)$ stands for 'you write down k'. A conditional statement is true whenever its antecedent is false, so the statement above is true when you write nothing at all. Now let's add another instruction. Consider the following two instructions.

1. *Write down the number 1.*
2. *If you write down an integer k, then you must also write down the integer $k + 1$.*

What do we now need to do to satisfy these instructions? Firstly, to obey the first instruction, we need to write down '1'. Then the second instruction 'kicks in': because we have now written down '1' we then need to write down '2'. But once we have written down '2', to follow the second instruction, we also need to write down '3'. Then we have written '3' so, to follow the second instruction, we need to write '4', then '5', then '6', and so on. In other words, to follow both instructions, we would need to write down all the positive integers. Of course, since $\mathbb{Z}^+$ is infinite, we cannot actually do this. So, although we cannot actually carry out both instructions, they carry within them a 'process' for 'generating' all positive integers.

This idea is formalised within the natural numbers by the Axiom of Induction, which we give below. The natural numbers can be defined by five relatively simple axioms, now called the Peano Axioms, after the 19th-century Italian mathematician Giuseppe Peano, who presented them in a book in 1889. All of the properties of the arithmetic of the natural numbers may be developed from the five Peano axioms. The Axiom of Induction is one of Peano's axioms.

Axiom of Induction

Let S be a subset of $\mathbb{N}$ such that:

(i) $0 \in S$, and

(ii) for all k, $k \in S \Rightarrow (k + 1) \in S$.

Then S contains all the natural numbers, $S = \mathbb{N}$.

The Axiom of Induction gives a way of establishing that some property $P(n)$ holds for all natural numbers. Given a property $P(n)$, define $S = \{n \in \mathbb{N} : P(n)\}$ to be the set containing all those natural numbers for which $P(n)$ is true. Thus $n \in S$ if and only if $P(n)$ is true. Hence the two conditions in the Axiom of Induction become (i) $P(0)$ is true and (ii) if $P(k)$ is true, then $P(k+1)$ is true. This formulation is frequently called the Principle of Mathematical Induction.

Principle of Mathematical Induction

Let $P(n)$ be a propositional function with universe $\mathbb{N}$. Suppose that:

(i) $P(0)$ is true, and

(ii) for all $k \in \mathbb{N}$, if $P(k)$ is true, then $P(k+1)$ is true.

Then $P(n)$ is true for all $n \in \mathbb{N}$; that is, $\forall n \in \mathbb{N} \bullet P(n)$ is true.

8.2   Proof by induction

The Principal of Mathematical Induction provides a scheme for proving theorems that may be formulated as $\forall n \in \mathbb{N} \bullet P(n)$. There are two stages in the proof. Firstly, prove that $P(0)$ is true; in other words, prove that the result under consideration is true when $n = 0$. This is called the 'Base case' and will frequently amount to a simple checking of some condition of equation. For the second part, we need to prove that the conditional statement $P(k) \Rightarrow P(k+1)$ is true for all $k \in \mathbb{N}$. This is called the 'Inductive step' and will be proved using the method of direct proof; that is, we assume $P(k)$ and, from this, deduce $P(k+1)$. Assuming $P(k)$ is called the 'Inductive hypothesis' and it appears to be a slightly odd assumption to make. The reason is that we are attempting to prove $P(n)$ for all $n \in \mathbb{N}$, so *assuming* $P(k)$, for an arbitrary k, appears at first sight to be assuming what we are trying to prove. However, the assumption of $P(k)$ is *only* made in order to deduce $P(k+1)$. In other words, we only assume the truth of the result for some integer k *in order to deduce* the result for the 'next' integer $k+1$. We summarise this discussion in the following scheme for proving results about the natural numbers.

Method of Proof by Mathematical Induction

Let $P(n)$ be a propositional function with universe $\mathbb{N}$.

Base case   Prove that $P(0)$ is true.

Inductive step   Prove that, for all $k \in \mathbb{N}$, $P(k) \Rightarrow P(k+1)$.

To do this:
assume $P(k)$; this is the inductive hypothesis,
then deduce $P(k+1)$.

Conclusion:
$P(n)$ is true for all $n \in \mathbb{N}$; that is, $\forall n \in \mathbb{N} \bullet P(n)$ is true.

There is an obvious modification to the method if the universe is the positive integers $\mathbb{Z}^+$ rather than the natural numbers $\mathbb{N}$. In this situation the base case will be for the smallest positive integer rather than the smallest natural number; in other words, the base case will be to prove that $P(1)$ is true. We now illustrate the method, initially by proving the three results mentioned in the previous section.

Examples 8.1

1. One of the common contexts in which proof by induction is first encountered is to prove results that involve the summation of a finite number of terms. We illustrate the approach by proving that the sum of the first n positive integers is $\frac{1}{2}n(n+1)$.

Theorem 8.1
For all $n \in \mathbb{Z}^+$, $1 + 2 + 3 + \cdots + n = \frac{1}{2}n(n+1)$.

The base case is straightforward. For the inductive step we need to assume the result for $n = k$; that is, we assume

$$1 + 2 + 3 + \cdots + k = \tfrac{1}{2}k(k+1).$$

From this assumption, we then need to deduce the result for $n = k+1$

$$1 + 2 + 3 + \cdots + k + (k+1) = \tfrac{1}{2}(k+1)(k+2).$$

The left-hand side of this second equation is the same as the left-hand side of the previous equation, but with the addition of a single extra term $(k+1)$. Hence, to deduce the result for $n = k+1$ we will take the equation for $n = k$, add $(k+1)$ to both sides and then reorganise the new right-hand side. This right-hand side is $\frac{1}{2}k(k+1) + (k+1)$, which has a common factor of $k+1$. In fact, it is better to regard the common factor as being $\frac{1}{2}(k+1)$ by writing the expression as

$$\tfrac{1}{2}k(k+1) + (k+1) = \tfrac{1}{2}k(k+1) + 2 \times \tfrac{1}{2}(k+1) = \tfrac{1}{2}(k+1)(k+2).$$

This final expression is what we are seeking for the right-hand side of the equation for $k+1$. We are now in a position to organise these considerations into a proof.

Proof. The proof is by mathematical induction.

Base case. When $n = 1$, LHS $= 1$ and RHS $= \frac{1}{2} \times 1 \times 2 = 1$. Hence the result holds when $n = 1$.

Inductive step. Assume that $1 + 2 + 3 + \cdots + k = \frac{1}{2}k(k+1)$; this is the inductive hypothesis. Adding $(k+1)$ to both sides gives

$$
\begin{aligned}
1 + 2 + 3 + \cdots + k + (k+1) &= \tfrac{1}{2}k(k+1) + (k+1) \\
&= \tfrac{1}{2}k(k+1) + 2 \times \tfrac{1}{2}(k+1) \\
&= \tfrac{1}{2}(k+1)(k+2) \\
&= \tfrac{1}{2}(k+1)((k+1)k + 1).
\end{aligned}
$$

This is the result for $n = k+1$ and hence completes the inductive step.

Therefore, for all $n \in \mathbb{Z}^+$, $1 + 2 + 3 + \cdots + n = \frac{1}{2}n(n+1)$, by induction. □

Commentary

There are a number of points to make about the organisation and structure of the proof. Firstly, as with some other methods of proof such as proof by contradiction, it is usually a good idea to signal at the beginning that the proof uses the method of mathematical induction. The reader then knows to expect a particular structure to the proof.

To establish the base case, we just need to check that the equation is satisfied when $n = 1$. To verify an equation, it is good practice to evaluate *separately* the left- and right-hand sides. In this way, it will be clear what is being evaluated. Less clear would be to write '$1 = \frac{1}{2} \times 1 \times 2$' because it does not explicitly link this (trivial) equation with the result under consideration.

Finally, at the end of the proof, it is also good practice to conclude with the result that has been established and the reason that the conclusion is justified; the phrase 'by induction' signals this.

It is quite common to use the letter n, rather than k, within the body of the inductive step. Of course, the letter used is not important. We have chosen to use k rather than n to keep the reasoning within the inductive step slightly separate from the conclusion. Using n makes the inductive hypothesis look identical to the result that we are trying to establish. However, the choice of letter is a matter of personal taste on behalf of the proof writer.

2. In this example, we prove the following result which was the second example mentioned in the previous section. Here the universe is the natural numbers, so the base case will be $n = 0$.

Theorem 8.2
For every $n \in \mathbb{N}$, the expression $5^n + 3$ is divisible by 4.

Before commencing the proof, we just need to remind ourselves what it means to say that '$5^n + 3$ is divisible by 4'. This means that the expression is some integer multiple of 4, so there exists an integer, a say, such that $5^n + 3 = 4a$.

Proof. The proof is by mathematical induction.

Base case. When $n = 0$, $5^n + 3 = 5^0 + 3 = 1 + 3 = 4$, which is clearly divisible by 4. Hence the result holds when $n = 0$.

Inductive step. Assume that $5^k + 3$ is divisible by 4; then $5^k + 3 = 4a$ for some integer a. Hence $5^k = 4a - 3$. Now

$$
\begin{aligned}
5^{k+1} + 3 &= 5 \times 5^k + 3 \\
&= 5(4a - 3) + 3 \quad \text{from the inductive hypothesis} \\
&= 20a - 12 \\
&= 4(5a - 3) \quad \text{where } 5a - 3 \in \mathbb{Z}.
\end{aligned}
$$

Therefore $5^{k+1} + 3$ is divisible by 4. This completes the inductive step.

Therefore, for all $n \in \mathbb{N}$, the expression $5^n + 3$ is divisible by 4, by induction.   □

3. In section 5.2 we gave a direct proof of the following theorem. We also indicated a second proof using the Binomial Theorem; see exercise 5.2.2.

Theorem 5.5
For all finite sets A, if A has cardinality n, then its power set $\mathbb{P}(A)$ has cardinality 2^n.

As we indicated in the previous section, we may regard this theorem as being a candidate for proof by induction by writing it as 'for all $n \in \mathbb{N}$, if A is a set with cardinality $|A| = n$, then its power set has cardinality $|\mathbb{P}(A)| = 2^n$'. In our proof, we signal this at the beginning by stating that the proof uses induction 'on the cardinality of the set'.

Proof. The proof is by mathematical induction on the cardinality of A.

Base case. Let A be a set with cardinality 0; in other words A is the empty set, $A = \varnothing$. Then $\mathbb{P}(A) = \{\varnothing\}$, which is a set with one element (the empty set); see exercise 3.8.1. Hence, $|\mathbb{P}(a)| = 1 = 2^0$, so the result holds when $n = 0$.

Inductive step. Assume that, for all finite sets, if the set has cardinality k, then its power set has cardinality 2^k.

Let A be a set with cardinality $k + 1$. Select an element $a^* \in A$. We may divide the subsets of A into two types: those that contain a^* and those that don't.

Any subset $B \subseteq A$ that does not contain a^* is simply a subset of $A - \{a^*\}$, which is a set with k elements. Hence, by the inductive hypothesis, there are 2^k such subsets B.

Let $C \subseteq A$ be a subset that does contain a^*; then $C = B \cup \{a^*\}$, where again B is a subset of $A - \{a^*\}$. As before, $A - \{a^*\}$ is a set with k elements, so there are 2^k such subsets B, by the inductive hypothesis. Hence there are 2^k subsets $C = B \cup \{a^*\}$ that do contain a^*.

Therefore the total number of subsets of A is $2^k + 2^k = 2^{k+1}$. This completes the inductive step.

Therefore, for all $n \in \mathbb{N}$, if A is a set with cardinality $|A| = n$, then its power set has cardinality $|\mathbb{P}(A)| = 2^n$, by induction. $\qquad\square$

Commentary

The two previous proofs of theorem 5.5 involved counting arguments of one sort or another. The proof we gave on page 191 was based on counting the number of ways a subset of A could be 'built' by choosing whether to include or exclude each element of A from the subset. The proof outlined in exercise 5.2.2 proceeds by counting the number of r-element subsets for $r = 0, 1, 2, \ldots, n$. The proof we have given here is also based on a counting argument: counting the number of subsets that include a particular element a^* and also counting the number of subsets that exclude a^*.

There is one aspect of the proof that was implicit, which we now wish to make explicit. In concluding that there are 2^k subsets C that contain the element a^*, we are implicitly assuming that there is a bijective correspondence between the sets $C = B \cup \{a^*\}$ and the subsets $B \subseteq A$ that do not contain a^*. The bijection is, of course, defined by $B \cup \{a^*\} \mapsto B$, but we have not explicitly proved that this defines a bijection. We have also used implicitly the corollary to theorem 5.7 (page 198) to deduce that there are the *same number* of subsets $C = B \cup \{a^*\}$ containing a^* as there are subsets $B \subseteq A - \{a^*\}$ that do not contain a^*.

4. In example 6.1.3, we used the following theorem in the proof of theorem 6.3 which asserted that, for $n \in \mathbb{Z}^+$, if $2^n - 1$ is prime, then n is prime.

A **geometric progression** is a sequence of terms where each term is in a fixed constant ratio to the previous term; thus each term is obtained from the previous term by multiplying by a fixed real number, called the **common ratio**. If the first term is b and the common ratio is a, then the sequence of terms is $b, ba, ba^2, ba^3, \ldots$. In the theorem we evaluate the sum of the first n terms of a geometric progression with first term 1. Taking the first term equal to 1 is not a significant restriction as the sum of the general progression can be obtained from the theorem by multiplying both sides of the equation by b.

Theorem 8.3

For all real numbers $a \neq 1$ and $n \in \mathbb{Z}^+$,

$$1 + a + a^2 + \ldots + a^{n-1} = \frac{a^n - 1}{a - 1}.$$

Proof. The proof is by mathematical induction on n.

Base case. When $n = 1$, LHS $= 1$ and RHS $= \dfrac{a^1 - 1}{a - 1} = 1$. Hence the result holds when $n = 1$.

Inductive step. Assume that $1 + a + a^2 + \ldots + a^{k-1} = \dfrac{a^k - 1}{a - 1}$.

Adding the next term of the progression, a^k, to both sides gives

$$
\begin{aligned}
1 + a + a^2 + \ldots + a^{k-1} + a^k &= \frac{a^k - 1}{a - 1} + a^k \\
&= \frac{a^k - 1}{a - 1} + \frac{a^k(a - 1)}{a - 1} \\
&= \frac{a^k - 1 + a^{k+1} - a^k}{a - 1} \\
&= \frac{a^{k+1} - 1}{a - 1}.
\end{aligned}
$$

This is the result for $k + 1$ and hence completes the inductive step.

Therefore, for all $n \in \mathbb{N}$, $1 + a + a^2 + \ldots + a^{n-1} = \dfrac{a^n - 1}{a - 1}$, by induction. $\square$

5. Consider the following 'theorem' and 'proof'. Since the claim in the 'theorem' is patently false, there must be an error in the proof. The question is: What is wrong with the proof?

'Theorem'

All rectangles have the same area.

'*Proof.*' Let $P(n)$ be the statement: *in any set of n rectangles, all the rectangles have the same area.*

We shall prove that $P(n)$ is true for all positive integers n by induction. This will prove the theorem.[1]

Base case. The result $P(1)$ is obviously true since in any set containing only one rectangle, all the rectangles in the set have the same area.

Inductive step. Suppose that, $P(k)$ is true; in other words, in any set of k rectangles, all the rectangles have the same area.

Let $\{R_1, R_2, \ldots, R_{k+1}\}$ be an arbitrary set of $k + 1$ rectangles.

Since $\{R_1, R_2, \ldots, R_k\}$ is a set of k rectangles, all these have the same area. Similarly, since $\{R_2, R_3, \ldots, R_{k+1}\}$ is also a set of k rectangles, all these have the same area.

Hence, taking the union of these two sets, all the rectangles in $\{R_1, R_2, \ldots, R_{k+1}\}$ have the same area.

Hence $P(k + 1)$ is true and completes the inductive step.

Therefore, for all positive integers n, in any set of n rectangles, all the rectangles have the same area, by induction. □

Solution

The base case is uncontroversial. For any rectangle R, all rectangles in the set $\{R\}$ have the same area. The problem must lie in the inductive step. Let's examine one particular case of the inductive step, say, when $k = 4$. The inductive hypothesis in this case is that every set of four rectangles comprises rectangles with the same area. Now let $\{R_1, R_2, R_3, R_4, R_5\}$ be any set of five rectangles. By the inductive hypothesis, both the sets of four rectangles, $\{R_1, R_2, R_3, R_4\}$ and $\{R_2, R_3, R_4, R_5\}$, contain rectangles that all have the same area; hence all the rectangles in $\{R_1, R_2, R_3, R_4, R_5\}$ also all have the same area. This reasoning appears to be perfectly sound and *indeed it is!* The implication considered here, $P(4) \Rightarrow P(5)$, is a true proposition.

So what has gone wrong? Thinking about the truth of the result itself gives the clue. As we have indicated, the result is true for $n = 1$ (the base case) but clearly it is not true when $n = 2$. There are sets

---

[1] To see why this is equivalent to the stated theorem, suppose we have established that every finite set of rectangles contains rectangles which are all of the same area. The stated theorem then follows by contradiction, as follows. Assume that not all rectangles have the same area. Then there exist two rectangles, R and R', say, with different areas. Then any set of rectangles containing both R and R' contradicts the result that says all sets of rectangles contain only rectangles with the same area.

containing two rectangles $\{R_1, R_2\}$ that have different areas. In fact, it is the implication $P(1) \Rightarrow P(2)$ that is false. The inductive step $P(k) \Rightarrow P(k+1)$ relies on there being a non-empty intersection between the sets $\{R_1, R_2, \ldots, R_k\}$ and $\{R_2, R_3, \ldots, R_{k+1}\}$ so that the 'same area' property may be transferred to their union $\{R_1, R_2, R_3, \ldots, R_{k+1}\}$. When $k \geq 2$ this is, indeed, the case. However, when $k = 1$, removing the first and last elements gives two *disjoint* sets, $\{R_1\}$ and $\{R_2\}$, so the 'same area' property does not transfer to their union. In summary, we have: $P(1) \nRightarrow P(2) \Rightarrow P(3) \Rightarrow P(4) \Rightarrow P(5) \Rightarrow \cdots$.

This example illustrates the importance of showing that the inductive step $P(k) \Rightarrow P(k+1)$ is true for all k. In the present case, $P(1)$ is true and $P(k) \Rightarrow P(k+1)$ is true for all $k \geq 2$, but this is not sufficient.

Exercises 8.1

1. Prove each of the following by induction.

   (i) The sum of the first n odd positive integers is n^2:

   $$1 + 3 + 5 \cdots + (2n - 1) = n^2.$$

   (ii) The sum of the squares of the first n positive integers is $\frac{1}{6}n(n + 1)(2n + 1)$:

   $$1^2 + 2^2 + 3^2 \cdots + n^2 = \frac{1}{6}n(n + 1)(2n + 1).$$

   (iii) For all positive integers n,

   $$\frac{1}{1 \times 2} + \frac{1}{2 \times 3} + \frac{1}{3 \times 4} + \cdots + \frac{1}{n(n + 1)} = \frac{n}{n + 1}.$$

   (iv) The sum of the cubes of the first n positive integers is $\frac{1}{4}n^2(n+1)^2$:

   $$1^3 + 2^3 + 3^3 \cdots + n^2 = \frac{1}{4}n^2(n + 1)^2.$$

   (v) For all positive integers n,

   $$1 \times 3 + 2 \times 4 + 3 \times 5 + \ldots + n(n + 2) = \tfrac{1}{6}n(n + 1)(2n + 7).$$

   (vi) For all $n \in \mathbb{Z}^+$,

   $$1 \times 6 + 2 \times 7 + 3 \times 8 + \cdots + n(n + 5) = \tfrac{1}{3}n(n + 1)(n + 8).$$

(vii) Generalising the two previous examples, let m be a (fixed) positive integer. Then

$$1 \times m + 2 \times (m+1) + 3 \times (m+2) + \cdots + n(m+n-1)$$
$$= \tfrac{1}{6} n(n+1)(n+3m-2).$$

(viii) For all positive integers n,

$$\frac{1}{2!} + \frac{2}{3!} + \frac{3}{4!} + \cdots + \frac{n-1}{n!} = 1 - \frac{1}{n!}.$$

2. Prove each of the following by induction.

   (i) For all positive integers n, $2^n > n$.

   (ii) For all positive integers n, the expression $9^n + 7$ is divisible by 8.

   (iii) For all $n \in \mathbb{Z}^+$, the expression $7^n - 3^n$ is divisible by 4.

   (iv) For all $n \in \mathbb{Z}^+$, the expression $11^n - 4^n$ is divisible by 7.

   (v) Generalising the previous two examples, let a and b be two positive integers such that $a - b = m \in \mathbb{Z}^+$. Then, for all $n \in \mathbb{Z}^+$, the expression $a^n - b^n$ is divisible by m.

   (vi) For all positive integers n, the expression $2^{n+2} + 3^{2n+1}$ is divisible by 7.

   (vii) For all positive integers n, the expression $4^{2n+1} + 3^{n+2}$ is divisible by 13.

3. Prove each of the following results about products by mathematical induction. In each case, try to find an alternative proof that does not use mathematical induction.

   (i) For all $n \in \mathbb{Z}^+$, $1 \times 3 \times 5 \times \cdots \times (2n-1) = \dfrac{(2n)!}{2^n n!}$.

   (ii) For all $n \in \mathbb{Z}^+$, $\dfrac{2}{1} \times \dfrac{4}{3} \times \dfrac{6}{5} \times \cdots \times \dfrac{2n}{2n-1} > \sqrt{2n+1}$.

4. (i) Prove that, for all $n \in \mathbb{N}$ and all real numbers $x > -1$,

$$(1+x)^n \geq 1 + nx.$$

   (ii) Prove that, for all $n \in \mathbb{Z}^+$, if $x_1, x_2, \ldots, x_n$ are positive real numbers, then

$$(1+x_1)(1+x_2)\ldots(1+x_n) \geq 1 + (x_1 + x_2 + \cdots + x_n).$$

   (iii) Prove that, for all $n \in \mathbb{Z}^+$, if $x_1, x_2, \ldots, x_n$ are real numbers in the interval (0,1), (that is, $0 < x_r < 1$ for $r = 1, 2, \ldots, n$), then

$$(1-x_1)(1-x_2)\ldots(1-x_n) > 1 - (x_1 + x_2 + \cdots + x_n).$$

5. (i) Prove by induction that, for all $n \in \mathbb{Z}^+$, $n^3 - n$ is divisible by 6.

    (ii) Give a direct proof of the result in part (i). Which of the two proofs do you prefer?

    (iii) Using the result of part (i), give a direct proof that $n^4 - n^2$ is divisible by 12.

6. Let $\mathbf{R}(\theta)$ be the 2×2 matrix $\mathbf{R}(\theta) = \begin{pmatrix} \cos\theta & -\sin\theta \\ \sin\theta & \cos\theta \end{pmatrix}$.

    The matrix represents an anti-clockwise rotation of the plane $\mathbb{R}^2$ about the origin by an angle θ.

    Prove that, for all positive integers n, $\mathbf{R}(n\theta) = (\mathbf{R}(\theta))^n$.

    Hint: you may assume the following trigonometric identities.

    $$\sin(A + B) = \sin A \cos B + \cos A \sin B$$
    $$\text{and} \quad \cos(A + B) = \cos A \cos B - \sin A \sin B.$$

7. Prove that if $a_1, a_2, \ldots a_n$ are positive integers and p is a prime number that is a factor of the product $a_1 a_2 \ldots a_n$, then p is a factor of one of the integers a_r for some $r = 1, 2, \ldots, n$.

8. The edges of a triangle are divided into n equal segments by inserting $n - 1$ division points. Lines are drawn through each of these division points parallel to each of the three edges of the triangle, thus forming a set of small triangles as illustrated in the figure (for $n = 4$).

    Prove that there are n^2 small triangles.

    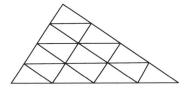

9. Prove that, for all positive integers n,

    $$\cos\theta + \cos 2\theta + \cdots + \cos n\theta = \frac{\sin\left((n + \tfrac{1}{2})\theta\right)}{2\sin\tfrac{1}{2}\theta} = \tfrac{1}{2}.$$

    Hint: you may assume the following trigonometric identity

    $$\sin A - \sin B = 2\cos\left(\frac{A + B}{2}\right)\sin\left(\frac{A - B}{2}\right).$$

10. Use mathematical induction to prove that, for all $n \in \mathbb{Z}^+$, $x + 1$ is a factor of $x^{2n-1} + 1$.

11. (i) Prove that, for all $n \in \mathbb{Z}^+$, a $2^n \times 2^n$ grid with a single corner square removed may be covered with L-shaped tiles like this one.

(ii) From part (i), prove that $2^{2n} - 1$ is divisible by 3 for all $n \in \mathbb{Z}^+$.

(iii) Prove the result in part (ii) by induction without reference to the result in part (i).

(iv) Using a similar approach to part (i), prove that for all $n \in \mathbb{Z}^+$, a $2^n \times 2^n \times 2^n$ cube with a single small cube (with side length 1) removed from one corner may be constructed using pieces like this one.

(v) Deduce a 'divisibility result' about numbers of the form $2^{3n} - 1$ from part (iv)?

Then give a standard proof by induction of this result.

12. Prove the following versions of De Morgan's Laws for n sets.

(i) For all sets $A_1, A_2, \ldots, A_n$,
$$\overline{A_1 \cup A_2 \cup \ldots \cup A_n} = \bar{A}_1 \cap \bar{A}_2 \cap \ldots \cap \bar{A}_n.$$

(ii) For all sets $A_1, A_2, \ldots, A_n$,
$$\overline{A_1 \cap A_2 \cap \ldots \cap A_n} = \bar{A}_1 \cup \bar{A}_2 \cup \ldots \cup \bar{A}_n.$$

13. Let $(G, *)$ be a group. Prove each of the following.

(i) Let $g, h \in G$. Then, for all $n \in \mathbb{Z}^+$, $\left(h^{-1}gh\right)^n = h^{-1}g^n h$.

(ii) Let $g_1, g_2, \ldots, g_n$ be n elements of G. Then $\left(g_1 g_2 \ldots g_n\right)^{-1} = g_n^{-1} \ldots g_2^{-1} g_1^{-1}$.

(iii) Suppose that $(G, *)$ is Abelian and let $g, h \in G$. Then, for all $n \in \mathbb{Z}^+$, $(gh)^n = g^n h^n$.

14. (i) Let $\mathbf{A} = \begin{pmatrix} 1 & 1 \\ 0 & 1 \end{pmatrix}$. Prove that, for all $n \in \mathbb{Z}^+$, $\mathbf{A}^n = \begin{pmatrix} 1 & n \\ 0 & 1 \end{pmatrix}$.

(ii) Let $\mathbf{A} = \begin{pmatrix} 1 & 0 \\ -1 & 2 \end{pmatrix}$. Prove that $\mathbf{A}^n = \begin{pmatrix} 1 & 0 \\ 1 - 2^n & 2^n \end{pmatrix}$ for all $n \in \mathbb{Z}^+$.

(iii) Let $\mathbf{A} = \begin{pmatrix} 5 & -2 & -2 \\ 2 & 1 & -2 \\ 2 & -2 & 1 \end{pmatrix}$. Prove that, for all $n \in \mathbb{Z}^+$,
$$\mathbf{A}^n = \begin{pmatrix} 2 \times 3^n - 1 & 1 - 3^n & 1 - 3^n \\ 3^n - 1 & 1 & 1 - 3^n \\ 3^n - 1 & 1 - 3^n & 1 \end{pmatrix}.$$

15. This example generalises theorem 5.15.

    Let $\mathbf{A}$ be an $m \times m$ matrix and let λ be an eigenvalue of $\mathbf{A}$. Prove that, for all $n \in \mathbb{Z}^+$, λ^n is an eigenvalue of $\mathbf{A}^n$.

    Show also that the eigenspace of λ for the matrix $\mathbf{A}$, $E_{\mathbf{A}}(\lambda)$, is a subspace of the eigenspace of λ^n for the matrix $\mathbf{A}^n$, $E_{\mathbf{A}^n}(\lambda^n)$.

16. Let f be a real-valued function such that $\lim\limits_{x \to a} f(x) = \ell$. For $n \in \mathbb{Z}^+$ define the function f^n by $f^n(x) = (f(x))^n$ for all x in the domain of f. Prove by induction that, for all $n \in \mathbb{Z}^+$, $\lim\limits_{x \to a} f^n(x) = \ell^n$.

    Note that you may assume, as background knowledge, the result of exercise 5.4.5.

8.3   Variations on proof by induction

In this section, we consider two variations to the method of proof by induction: the first variation modifies the base case and the second modifies the inductive step. The first modification is straightforward and just extends further the notion that the base case will vary depending on the result to be proved. We have already allowed two different base cases: $n = 0$ if the universe is $\mathbb{N}$ and $n = 1$ if the universe is $\mathbb{Z}^+$. We can readily extend this as follows.

Method of Proof by Mathematical Induction: Varying the Base Case

Let $P(n)$ be a propositional function with universe
$\mathbb{Z}_{\geq m} = \{n \in \mathbb{Z} : n \geq m\}$.

Base case   Prove that $P(m)$ is true.

Inductive step   Prove that, for all integers $k \geq m$, $P(k) \Rightarrow P(k+1)$.

To do this:
assume $P(k)$; this is the inductive hypothesis
then deduce $P(k+1)$.

Conclusion:
$P(n)$ is true for all integers $n \geq m$; that is, $\forall n \in \mathbb{Z}_{\geq m} \bullet P(n)$ is true.

Example 8.2

Since this modification is straightforward, we illustrate it with a single example.

Theorem 8.4

For all integers $n \geq 5$, $n^4 < 4^n$.

The following table compares the values of n^4 and 4^n for small positive integers n. Although the initial values give no clear pattern, the table suggests that, for $n \geq 5$, the values of 4^n increase more rapidly than do the values of n^4.

n	n^4	4^n
1	1	4
2	16	16
3	81	64
4	256	256
5	625	1024
6	1296	4096

Thinking about the proof, the table takes care of the base case which is, of course, $n = 5$. For the inductive step, we need to show that, for all $k \geq 5$, if $k^4 < 4^k$, then $(k+1)^4 < 4^{k+1}$. So, let's consider $(k+1)^4$. Expanding the expression, we have

$$(k+1)^4 = k^4 + 4k^3 + 6k^2 + 4k + 1.$$

Suppose we could show that the terms on the right-hand side were no greater than $k^4 + k^4 + k^4 + k^4 = 4 \times k^4$. Then the inductive step would allow us to reason: $4 \times k^4 < 4 \times 4^k = 4^{k+1}$. Now, considering the terms in the expression $k^4 + 4k^3 + 6k^2 + 4k + 1$, we have:

$$4k^3 < k^4 \qquad \text{since } 4 < k$$
$$6k^2 < k^4 \qquad \text{since } 6 < k^2 \text{ as } k \geq 5$$
$$4k + 1 < 4k + k = 5k < k^4 \qquad \text{since } 5 < k^3 \text{ as } k \geq 5.$$

From these 'pieces' we can deduce that $k^4 + 4k^3 + 6k^2 + 4k + 1 < 4k^4$, which then allows us to complete the inductive step. We can now marshal these reasoning fragments into a coherent proof as follows.

Proof. The proof is by mathematical induction.

Base case. When $n = 5$, $5^4 = 625$ and $4^5 = 1024$, so $n^4 < 4^n$ in this case. Hence the result holds when $n = 5$.

Inductive step. Assume that, for all integers $k \geq 5$, $k^4 < 4^k$.

Now $(k+1)^4 = k^4 + 4k^3 + 6k^2 + (4k+1)$ and we consider the terms on the right-hand side in turn. Since $k \geq 5$, we have
$$4k^3 < k \times k^3 = k^4$$
$$k^2 \geq 25 \text{ so } 6k^2 < 25k^2 \leq k^2 \times k^2 = k^4$$
$$k^3 \geq 125 \text{ so } 4k+1 < 4k+k = 5k < 125k \leq k^3 \times k = k^4.$$

Therefore
$$\begin{aligned}(k+1)^4 &= k^4 + 4k^3 + 6k^2 + (4k+1) \\ &< k^4 + k^4 + k^4 + k^4 \\ &= 4k^4 \\ &< 4 \times 4^k \qquad \text{by the inductive hypothesis} \\ &= 4^{k+1}.\end{aligned}$$

This shows that $(k+1)^4 < 4^{k+1}$, which is the result for $k+1$ and hence completes the inductive step.

Therefore, for all integers $n \geq 5$, we have $n^4 < 4^n$, by induction. $\qquad\square$

To motivate our second, more substantial, modification of the method of proof by induction, we look again at the Prime Factorisation Theorem (theorem 4.3): *every integer greater than 1 can be expressed as a product of prime numbers.* Let $P(n)$ be the propositional function:

$$P(n) : n \text{ can be expressed as a product of prime numbers.}$$

Then the Prime Factorisation Theorem may be symbolised as $\forall n \bullet P(n)$ where the universe is $\mathbb{Z}_{\geq 2} = \{n \in \mathbb{Z} : n \geq 2\}$. Hence this is a prime (excuse the pun) candidate for a proof by induction with base case $n = 2$. The base case itself is clear: 2 is already expressed as a product of primes, albeit in a rather trivial way.

Let's consider the inductive step. The inductive hypothesis is: assume that $k \geq 2$ may be expressed as

$$k = p_1 p_2 \ldots p_m \text{ where } p_1, p_2, \ldots, p_m \text{ are prime numbers.}$$

Now consider $k + 1$; we need to show that $k + 1$ can also be expressed as a product of primes. However, the factorisation of k does not give any information about the factorisation of $k + 1$. For example, 15 has factorisation 3×5 but 16 has factorisation $2 \times 2 \times 2 \times 2$; or 17 is itself prime but 18 has factorisation $2 \times 3 \times 3$. However, there are two possibilities for $k + 1$: either it is prime or it is composite. If it is prime, then there is nothing to prove; like 2, it is trivially expressed as a product of primes. If $k + 1$ is composite, then it has a factorisation

$$k + 1 = a_1 a_2 \text{ where } 2 \leq a_1, a_2 \leq k.$$

The inductive hypothesis only gives us information about k and not a_1 and a_2. In our previous, somewhat informal proof (page 134), we essentially applied the same reasoning to a_1 and a_2: each is either prime or it factorises. What is needed is a 'mathematical induction' version of this. Suppose that we had an inductive hypothesis that allowed us to assume $P(a_1)$ and $P(a_2)$ as well as $P(k)$. Then we may express each of a_1 and a_2 as a product of primes and hence we can also express $k+1 = a_1 a_2$ as a product of primes. An inductive hypothesis that allows us to assume $P(r)$ for all $r \leq k$ (rather than just $P(k)$ itself) is what is needed in this case.

Returning to our informal introduction to the Principle of Mathematical Induction expressed in terms of writing down integers, suppose we are given the following two instructions.

1. *Write down the number 1.*

2. *If you write down all of the integers $1, 2, \ldots, k$, then you must also write down the integer $k+1$.*

The effect is the same as following the original two instructions. Firstly, to obey the first instruction, we need to write down '1'. Then the second instruction 'kicks in': because we have now written down '1', we have (trivially) written down all the integers from 1 to 1; then need to write down '2'. But now we have written down *both* '1' and '2', so to follow the second instruction, we also need to write down '3'. Then we have written '1', '2', and '3' so, to follow the second instruction, we need to write '4' and so on. In other words, to follow both instructions, we again would need to write down all the positive integers. This is an informal justification for the following variation of the method of proof by induction.

Method of Proof by Mathematical Induction: Strong Form

Let $P(n)$ be a propositional function with universe $\mathbb{N}$.

Base case   Prove that $P(0)$ is true.

Inductive step   Prove that, for all integers $k \in \mathbb{N}$,

$$(P(0) \wedge P(1) \wedge \ldots \wedge P(k)) \Rightarrow P(k+1).$$

To do this:
assume $P(r)$ for all $r \leq k$ (and $r \in \mathbb{N}$);
this is the inductive hypothesis
then deduce $P(k+1)$.

Conclusion:
$P(n)$ is true for all integers $n \in \mathbb{N}$; that is, $\forall n \in \mathbb{N} \bullet P(n)$ is true.

We have stated this for the case where the universe is $\mathbb{N}$ to match our original statement of proof by induction but, of course, we may vary the base case appropriately to cover cases where the universe is $\mathbb{Z}^+$ or $\mathbb{Z}_{\geq m}$. The method is called the 'strong form' of mathematical induction because it allows us to make a stronger inductive hypothesis. In other words, we assume more than in the standard version of mathematical induction: instead of assuming the result for a single values k, we assume it for all values up to and including k. However, despite this stronger inductive hypothesis, the strong form is actually equivalent to the original method of proof by induction given in the previous section in the sense that anything that can be proved using the strong form can also be proved using the original version. Suppose that $P(n)$ is a propositional function with universe $\mathbb{N}$. Let $Q(n)$ be the propositional function

$$Q(n): \ P(0) \wedge P(1) \wedge \ldots \wedge P(n).$$

So $Q(n)$ asserts that $P(r)$ is true for all $0 \leq r \leq n$. Then two things are apparent:

- the propositions $\forall n \bullet P(n)$ and $\forall n \bullet Q(n)$ are logically equivalent: one is true if and only if the other is true; and
- the steps in the strong form of mathematical induction applied to $P(n)$ are identical to the steps in the original form of mathematical induction applied to $Q(n)$.

This demonstrates that any statement that can be proved using the strong form can actually be proved using the original form. Thus the strong form is a 'methodological convenience': it makes writing proofs easier, but is no stronger, logically, than the original form.

Examples 8.3

1. For our first example, we give an inductive proof of the Prime Factorisation Theorem.

   Theorem 4.3 (Prime Factorisation Theorem).
   Every integer greater than 1 can be expressed as a product of prime numbers.

   Proof. The proof is by mathematical induction, strong form.

   Base case. First note that $n = 2$ is trivially expressed as a product of primes since it is itself prime.

   Inductive step. Let $k \geq 2$, and suppose that every integer r such that $2 \leq r \leq k$ can be expressed as a product of prime numbers.

Consider $k+1$. Either $k+1$ is prime or it is composite. If it is prime, then there is nothing to do as it is expressed as a product of prime numbers. If $k+1$ is composite, then it has a factorisation

$$k+1 = a_1 a_2 \text{ where } 2 \le a_1, a_2 \le k.$$

By the inductive hypothesis, each of a_1 and a_2 may be expressed as a product of primes:

$$a_1 = p_1 p_2 \ldots p_s \quad \text{and} \quad a_2 = q_1 q_2 \ldots q_t$$

where $p_1, p_2, \ldots, p_s$ and $q_1, q_2, \ldots, q_t$ are all prime numbers. Therefore $k+1$ can be expressed as a product of prime numbers,

$$k+1 = p_1 p_2 \ldots p_s q_1 q_2 \ldots q_t.$$

This completes the inductive step.

Therefore, every integer greater than 1 can be expressed as a product of prime numbers, by induction. $\square$

2. Our second example is more geometric and concerns n-sided polygons. For $n = 3, 4, 5, 6$, and so on, these are, of course, just triangles, rectangles, pentagons, hexagons and so forth. Such a polygon is called **convex** if, for any two points of the polygon, the line segment joining those points is entirely contained within the polygon. This is illustrated in figure 8.1. Polygon A is convex: the line segment joining any two points of the polygon is entirely contained in the polygon. This is illustrated for two 'arbitrary' points in the figure. Polygon B is not convex: for the two points shown, the line segment joining them passes outside the polygon itself.

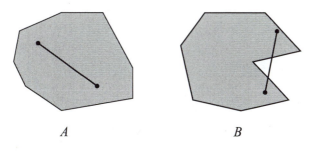

$$A \qquad\qquad\qquad B$$

FIGURE 8.1: Convex and non-convex polygons.

The following theorem was part of the 'background knowledge' that we needed for our non-constructive existence proof in example 7.2.2.

Theorem 8.5
The sum of the interior angles in any n-sided convex polygon is $(n-2)\pi$.

We are measuring angles in radians here. Were we to measure angles in degrees, the statement of the theorem would be 'the sum of the interior angles in any n-sided convex polygon is $180(n-2)$ degrees'.

The proof will be by induction on the number of sides of the polygon. The base case will therefore be $n = 3$ because a triangle is the polygon with the fewest possible number of sides. In fact, it is implicit in the statement of the theorem that n is an integer $n \geq 3$. We will make the strong inductive hypothesis that the result holds for all polygons with $r \leq k$ sides. The idea behind the inductive step is very simple: consider a $(k+1)$-sided polygon and 'cut it in two' to produce two polygons each with fewer sides; then apply the inductive hypothesis to the two smaller polygons.

Proof. The proof is by mathematical induction, strong form.

Base case. The base case is $n = 3$ since every (convex) polygon must have at least three sides. It is well known that the angle sum of any triangle is π radians.[2] Hence the base case is satisfied.

Inductive step. Suppose that $k \geq 3$ and that for every r-sided convex polygon where $3 \leq r \leq k$, the angle sum is $(r-2)\pi$.

Let P be an arbitrary $(k+1)$-sided polygon. Select any two non-adjacent vertices of P and draw the line segment joining them. Cutting along the line segment separates P into two convex polygons, Q and R, say. Of the $k + 1$ sides belonging to P, suppose that the cut separates the sides so that q of them belonging to Q and $k + 1 - q$ of them belonging to R. Since each of Q and R have an additional side (the cut itself), Q is a $(q + 1)$-sided polygon and R is a $(k + 2 - q)$-sided polygon. This is illustrated in figure 8.2 below.

Now, since there are at least two sides on each side of the cut, $2 \leq q \leq k-1$. Therefore $3 \leq q+1 \leq k$, so the inductive hypothesis applies to the polygon Q. Also, multiplying $2 \leq q \leq k-1$ by -1 gives $1-k \leq -q \leq -2$ and then adding $k + 2$ gives $3 \leq k + 2 - q \leq k$; hence the inductive hypothesis also applies to the polygon R.

From the inductive hypothesis we may deduce

$$\text{angle sum for } Q = (q + 1 - 2)\pi = (q - 1)\pi$$

$$\text{and} \quad \text{angle sum for } R = (k + 2 - q - 2)\pi = (k + q)\pi.$$

---

[2] As we noted on page 167, this is a theorem of Euclidean geometry that is not true in various non-Euclidean geometries. For the current proof, we will accept this result as part of our background knowledge. For a proof, see exercise 8.2.10.

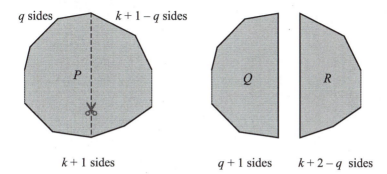

FIGURE 8.2: Illustrating the inductive step.

Consider the angle sum for P. Apart from each interior angle where the cut is made, each angle of P belongs either to Q or to R. For each of the two angles split by the cut, the sum of the angles of the two 'pieces' in Q and R equals the original angle in P. Hence

$$\text{angle sum for } P = (\text{angle sum for } Q) + (\text{angle sum for } R)$$
$$= (q-1)\pi + (k+q)\pi \quad \text{(from inductive hypothesis)}$$
$$= (k-1)\pi$$
$$= ((k+1) - 2)\,\pi.$$

This completes the inductive step.

Therefore, the sum of the interior angles in any n-sided convex polygon is $(n-2)\pi$, by induction. $\qquad\square$

Recursively defined sequences

In section 5.4, we considered the convergence behaviour of sequences (a_n). We now consider a way of describing sequences, that is often used where sequences are used to model real phenomena, and the role of mathematical induction in proving results about sequences defined in this way. Many systems in the real world that evolve over time can be described by measuring characteristics of the system only at specific time points, for example, every second, day, month, or year as appropriate. Such systems are called **discrete dynamical systems**, and they may be modelled using a sequence (a_n) where a_n represents the state of the system at the nth time point from the start. There are many systems that may be modelled in this way; these include the size of a population; the output of a factory or a whole economy; the temperature at a particular location or of the whole globe; the atmospheric pressure, humidity, wind speed at a particular place; and many others.

It is often the case that the state of the system at a particular time depends only on its state at the time point or points immediately beforehand. For example, a species whose growth is not constrained by environmental factors may be such that its population at time n depends only on its population at time $n - 1$. If a_n denotes the population at time n, then in this scenario, a_n is a function of a_{n-1}, $a_n = f(a_{n-1})$.

Another example is the famous **Fibonacci sequence**, which was introduced to Western mathematics in 1202 by Leonardo of Pica, known as Fibonacci, although this was known to Indian mathematics for centuries before this. In attempting to understand the growth of a rabbit population, Fibonacci proposed a model where the population at time n is the sum of the population at times $n - 1$ and $n - 2$. In this case, if a_n denotes the rabbit population, then Fibonacci's model is expressed as $a_n = a_{n-1} + a_{n-2}$.

These are examples of a **recursively defined sequence** which is one where each term in the sequence is defined in terms of the previous terms. We may express this more formally by saying that the nth term is a function of (some or all of) the terms $a_1, a_2, \ldots, a_{n-1}$,

$$a_n = f(a_1, a_2, \ldots, a_{n-1}).$$

This expression is called the **recursive relationship**. The first term or the first few terms will also need to be defined explicitly, and this will depend on which terms appear in the recursive relationship.

Examples 8.4

1. If the recursive relationship defined a_n only in terms of a_{n-1}, then it is only the initial term a_1 that also needs to be defined explicitly.

   A sequence (a_n) is defined by

   $$a_n = 3a_{n-1} + 2 \quad \text{for } n \geq 2,$$
   $$a_1 = 5.$$

   The first few terms of the sequence are:

   $$a_1 = 5 \qquad \text{(by definition)}$$
   $$a_2 = 3a_1 + 2 = 3 \times 5 + 2 = 17$$
   $$a_3 = 3a_2 + 2 = 3 \times 17 + 2 = 53$$
   $$a_4 = 3a_3 + 2 = 3 \times 53 + 2 = 158$$
   $$\vdots$$

2. For the Fibonacci sequence, where the recursive relationship defines a_n using both a_{n-1} and a_{n-2}, the first two terms are required explicitly.

These are both taken as having the value 1, so the Fibonacci sequence (a_n) is defined by

$$a_n = a_{n-1} + a_{n-2} \quad \text{for } n \geq 3,$$
$$a_1 = 1, \ a_2 = 1.$$

The first few terms of the sequence are:

$$
\begin{aligned}
a_1 &= 1 && \text{(by definition)} \\
a_2 &= 1 && \text{(by definition)} \\
a_3 &= a_2 + a_1 = 1 + 1 = 2 \\
a_4 &= a_3 + a_2 = 2 + 1 = 3 \\
a_5 &= a_4 + a_3 = 3 + 2 = 5 \\
a_6 &= a_5 + a_4 = 5 + 3 = 8 \\
a_7 &= a_6 + a_5 = 8 + 5 = 13
\end{aligned}
$$
$$\vdots$$

3. In the previous two examples, the nth term is defined in terms of a fixed *number* of previous terms — one or two in these examples. However, it is possible to use a variable number of terms. For example, a sequence (a_n) is defined by

$$a_n = a_1 + a_2 + \cdots a + a_{n-1} \quad \text{for } n \geq 2,$$
$$a_1 = 1.$$

In this case, the nth term is the sum of all of the previous terms, so the first few terms of the sequence are:

$$
\begin{aligned}
a_1 &= 1 && \text{(by definition)} \\
a_2 &= a_1 = 1 \\
a_3 &= a_1 + a_2 = 1 + 1 = 2 \\
a_4 &= a_1 + a_2 + a_3 = 1 + 1 + 2 = 4 \\
a_5 &= a_1 + a_2 + a_3 + a_4 = 1 + 1 + 2 + 4 = 8 \\
a_6 &= a_1 + a_2 + a_3 + a_4 + a_5 = 1 + 1 + 2 + 4 + 8 = 16
\end{aligned}
$$
$$\vdots$$

In the previous examples, we have only considered the process of *defining* sequences recursively. If we wish to *prove* properties of such sequences, then a form of mathematical induction is an obvious tool. In particular, it is often convenient to have an explicit formula for the nth term. For example, it would require a lot of work to use the recursive definition to calculate the 1000th term of each of the sequences defined in examples 8.4. In some cases it might be quite easy to conjecture a general formula; we suspect readers may 'spot' a formula for a_n in example 3 above. In other cases, this might be considerably more difficult; for the Fibonacci sequence, it would be almost impossible to 'guess' the following general formula

$$a_n = \frac{1}{\sqrt{5}} \left(\frac{1 + \sqrt{5}}{2} \right)^n - \frac{1}{\sqrt{5}} \left(\frac{1 - \sqrt{5}}{2} \right)^n.$$

However a conjecture is obtained, mathematical induction will be an appropriate method of proof. For those cases where several initial terms are defined explicitly, the result will need to be verified individually for these initial terms. Thus there may be several 'base cases' that need verifying and the inductive step will only apply where the recursive relationship is used. The general method, based on the strong form of induction, is described as follows.

Method of Proof by Mathematical Induction: Recursive Sequences

Let (a_n) be a sequence that is defined recursively as follows.

$$a_1, \ldots, a_m \text{ are defined explicitly}$$
$$a_n = f(a_1, \ldots, a_{n-1}) \text{ for } n \geq m + 1$$

Let $P(a_n)$ be a propositional function about the terms of the sequence.

Base cases   Prove that $P(a_1), \ldots, P(a_m)$ are true.

Inductive step   Prove that, for all integers $k \geq m + 1$,

$$(P(a_1) \wedge \ldots \wedge P(a_k)) \Rightarrow P(a_{k+1}).$$

To do this:
assume $P(a_r)$ for all $r \leq k$ (and $r \in \mathbb{Z}^+$);
this is the inductive hypothesis
then deduce $P(a_{k+1})$.

Conclusion:
$P(a_n)$ is true for all integers $n \in \mathbb{Z}^+$; that is, $\forall n \in \mathbb{Z}^+ \bullet P(a_n)$ is true.

Examples 8.5

1. Let (a_n) be the sequence in example 8.4.1 defined by

$$a_n = 3a_{n-1} + 2 \quad \text{for } n \geq 2,$$
$$a_1 = 5.$$

Then $a_n = 2 \times 3^n - 1$ for all $n \in \mathbb{Z}^+$.

Proof. The proof is by mathematical induction.

Base case. When $n = 1$, $a_1 = 5$ (by definition) and $2 \times 3^n - 1 = 2 \times 3^1 - 1 = 6 - 1 = 5$. Hence the result is true for $n = 1$.

Inductive step. Let $k \in \mathbb{Z}^+$. Suppose that, for every positive integer $r \leq k$, $a_k = 2 \times 3^k - 1$.

Consider a_{k+1}. Since $k + 1 \geq 2$, a_{k+1} is defined by the recursive relationship. So we have

$$\begin{aligned}
a_{k+1} &= 3a_k + 2 && \text{(definition of } a_{k+1}\text{)} \\
&= 3\left(2 \times 3^k - 1\right) + 2 && \text{(inductive hypothesis)} \\
&= 2 \times 3 \times 3^k - 3 + 2 \\
&= 2 \times 3^{k+1} - 1.
\end{aligned}$$

Hence a_{k+1} satisfies the given formula. This completes the inductive step.

Therefore, $a_n = 2 \times 3^n - 1$ for all $n \in \mathbb{Z}^+$, by induction. $\qquad\square$

Commentary

Alert readers will have noticed that, in this example, we have not actually used the strong form of the inductive hypothesis. The only place where the inductive hypothesis is used is to replace a_k with $2 \times 3^k - 1$. In this example, a proof using the 'standard' version of mathematical induction would be perfectly valid. Of course, the reason for this is that the recursive relationship is of the form $a_n = f(a_{n-1})$; in other words, a_n is defined only in terms of its immediate predecessor a_{n-1}.

2. Let (a_n) be the sequence defined by

$$\begin{aligned}
a_n &= 3a_{n-1} - 2a_{n-2} \quad \text{for } n \geq 3, \\
a_1 &= 0, a_1 = 2.
\end{aligned}$$

Then $a_n = 2^{n-1} - 1$ for all $n \geq 2$.

Proof. The proof is by mathematical induction.

Base cases. Since a_n is defined explicitly for $n = 1$ and $n = 2$, there are two base cases to consider.

When $n = 1$, $a_1 = 0$ (by definition) and $2^{n-1} - 1 = 2^0 - 1 = 1 - 1 = 0$.
When $n = 2$, $a_2 = 1$ (by definition) and $2^{n-1} - 1 = 2^1 - 1 = 2 - 1 = 1$.

Hence, the result is true for $n = 1$ and $n = 2$.

Inductive step. Let $k \geq 2$. Suppose that, for every integer $2 \leq r \leq k$, $a_k = 2^{k-1} - 1$.

Consider a_{k+1}. Since $k + 1 \geq 3$, a_{k+1} is defined by the recursive relationship. So we have

$$
\begin{aligned}
a_{k+1} &= 3a_k - 2a_{k-1} && \text{(definition of } a_{k+1}) \\
&= 3(2^{k-1} - 1) - 2(2^{k-2} - 1) && \text{(inductive hypothesis)} \\
&= 3 \times 2^{k-1} - 3 - 2 \times 2^{k-2} + 2 \\
&= 3 \times 2^{k-1} - 2^{k-1} - 1 \\
&= 2 \times 2^{k-1} - 1 \\
&= 2^k - 1.
\end{aligned}
$$

Hence a_{k+1} satisfies the given formula. This completes the inductive step.

Therefore, $a_n = 2^{n-1} - 1$ for all $n \geq 2$, by induction. $\qquad\square$

Commentary

In contrast to the previous example, the strong form of the inductive hypothesis is used in this proof. In the step labelled 'inductive hypothesis' we have replaced a_{k-1} with $2^{k-1} - 1$ and we have replaced a_{k-2} with $2^{k-2} - 1$. In other words, we used the inductive hypothesis for $r = k - 1$ and $r = k - 2$. This means that a simple inductive hypothesis would not have been sufficient in this case.

Exercises 8.2

1. Prove each of the following.

   (i) For all integers $n \geq 4$, $n! > 2^n$.

   (ii) For all integers $n \geq 7$, $n! > 3^n$.

   (iii) For all integers $n \geq 5$, $n^2 < 2^n$.

   (iv) For all integers $n \geq 4$, $n^3 < 3^n$.

   (v) For all integers $n \geq 10$, $n^3 < 2^n$.

2. Prove by induction that, for all integers $n \geq 2$,

$$
\frac{3}{4} \times \frac{8}{9} \times \frac{15}{16} \times \cdots \times \frac{n^2 - 1}{n^2} = \frac{n + 1}{2n}.
$$

3. Using the strong form of mathematical induction, prove that every integer $n \geq 6$ can be written as $n = 3a + 4b$ for some $a, b \in \mathbb{N}$.

   Proceed as follows.

   - Prove the three base cases $n = 6$, $n = 7$, $n = 8$ directly.
   - The inductive step applies for $k \geq 8$; the strong inductive hypothesis is that every integer r such that $6 \leq r \leq k$ can be expressed as $r = 3a + 4b$ for some $a, b \in \mathbb{N}$.
   - To complete the inductive step, write $k + 1 = (k - 2) + 3$ and use the inductive hypothesis.

4. Prove each of the following about recursively defined sequences.

   (i) Let (a_n) be the sequence defined by
   $$a_n = a_{n-1} + 3 \quad \text{for } n \geq 2,$$
   $$a_1 = 3.$$
   Then $a_n = 3n$ for all $n \in \mathbb{Z}^+$.

   (ii) Let (a_n) be the sequence defined by
   $$a_n = 3 - a_{n-1} \quad \text{for } n \geq 2,$$
   $$a_1 = 1.$$
   Then $a_n = \dfrac{3 + (-1)^n}{2}$ for all $n \in \mathbb{Z}^+$.

   (iii) Let (a_n) be the sequence defined by
   $$a_n = 5a_{n-1} - 6a_{n-2} \quad \text{for } n \geq 3,$$
   $$a_1 = 0, \ a_2 = 6.$$
   Then $a_n = 2 \times 3^n - 3 \times 2^n$ for all $n \in \mathbb{Z}^+$.

   (iv) Let (a_n) be the sequence defined by
   $$a_n = a_1 + a_2 + \cdots + a_{n-1} \quad \text{for } n \geq 2,$$
   $$a_1 = 1.$$
   Then $a_n = 2^{n-2}$ for all $n \geq 2$.

   (v) Let (a_n) be the sequence defined by
   $$a_n = 2a_{n-1} + a_{n-2} - 2a_{n-3} \quad \text{for } n \geq 4,$$
   $$a_1 = -3, \ a_2 = 1, \ a_3 = 3.$$
   Then $a_n = 2^n + (-1)^n - 4$ for all $n \in \mathbb{Z}^+$.

(vi) Let (a_n) be the sequence defined by

$$a_n = 2a_{n-1} + a_{n-2} + a_{n-3} \quad \text{for } n \geq 4,$$
$$a_1 = 1, \ a_2 = 2, \ a_3 = 3.$$

Then $a_n < 2^n$ for all $n \in \mathbb{Z}^+$.

5. Let (a_n) be the Fibonacci sequence defined in example 8.4.2. Prove each of the following.

  (i) $a_n < 2^n$ for all $n \in \mathbb{Z}^+$.

  (ii) $a_1 + a_2 + \cdots + a_n = a_{n+2} - 1$ for all $n \in \mathbb{Z}^+$.

  (iii) $a_{n+2}^2 - a_{n+1}^2 = a_n a_{n+3}$ for all $n \in \mathbb{Z}^+$.

  (iv) $a_1 + a_3 + \cdots + a_{2n-1} = a_{2n}$ for all $n \in \mathbb{Z}^+$.

  (v) $a_n = \dfrac{1}{\sqrt{5}} \left(\dfrac{1+\sqrt{5}}{2} \right)^n - \dfrac{1}{\sqrt{5}} \left(\dfrac{1-\sqrt{5}}{2} \right)^n$ for all $n \in \mathbb{Z}^+$.

  Hint: it will make the algebraic manipulation simpler if you first establish that

  $$\left(\frac{1 \pm \sqrt{5}}{2} \right)^2 = \frac{3 \pm \sqrt{5}}{2}.$$

6. It is well known that positive integers can be written using binary numerals. For example, $13 = 1101_2$ because $13 = 2^3 + 2^2 + 2^0$. That every $n \in \mathbb{Z}^+$ has a binary numeral representation is equivalent to the statement: every $n \in \mathbb{Z}^+$ can be expressed as a sum of distinct non-negative powers of 2.

   Use the strong form of induction to prove this statement.

7. The definition of 'convex' for a polygon is given in example 8.3.2. The definition applies to arbitrary subsets of the plane $\mathbb{R}^2$ or 3-dimensional space $\mathbb{R}^3$: a subset S of $\mathbb{R}^2$ or $\mathbb{R}^3$ is **convex** if, for all $x, y \in S$, the line segment joining x and y is entirely contained in S. For convenience, we regard the empty set $\varnothing$ and any singleton set $\{x\}$ as being convex.

   Prove that for all $n \geq 2$, if $A_1, A_2, \ldots, A_n$ are convex subsets of $\mathbb{R}^3$, then their intersection $A_1 \cap A_2 \cap \ldots \cap A_n$ is also a convex subset of $\mathbb{R}^3$.

8. Let $A = \{1, 5, 9, 13, \ldots\} = \{4n + 1 : n \in \mathbb{N}\}$ be the set of all positive integers whose remainder on division by 4 is 1.

  (i) Prove that A is closed under multiplication; that is, the product of any two elements of A is always an element of A.

Define an **A-prime** to be an element $p \in A$ such that $p > 1$ and the only factors of p which are elements of A are 1 and p.

For example, 9 is an A-prime, since the only divisors of 9 that are elements of A are 1 and 9.

(ii) Prove by induction on n that every element of A that is greater than 1 can be written as a product of A-primes.

(iii) Show that A-prime factorisation of the elements of A is not necessarily unique; in other words, it is possible for an element of A to be written as a product of A-primes in at least two different ways.

9. By induction on $|A|$, prove that, for all finite sets A and B, if $B \subset A$, then $|B| < |A|$.

10. Prove the base case in the proof of theorem 8.5: the angle sum in any triangle is π.

The following diagram provides the idea behind the proof.

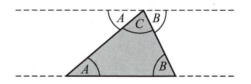

11. In the proof of theorem 8.5, where is the requirement that the polygon be convex used?

Is the theorem still true without the requirement that the polygon be convex? If so, how might the proof given in example 8.3.2 be modified to prove the more general theorem?

12. Let $P(n)$ be a propositional function with universe $\mathbb{Z}^+$. Suppose that the following are true propositions.

(I) $P(1)$;

(II) for all $n \in \mathbb{Z}^+$, $P(n) \Rightarrow P(2n)$;

(III) for all $n \in \mathbb{Z}^+$, $P(n+1) \Rightarrow P(n)$.

Prove that $P(n)$ is true, for all $n \in \mathbb{Z}^+$.

Hints and Solutions to Selected Exercises

Chapter 2

Exercises 2.1

1. (i)

P	Q	$\neg Q$	$P \vee \neg Q$	$\neg(P \vee \neg Q)$
T	T	F	T	F
T	F	T	T	F
F	T	F	F	T
F	F	T	T	F

(v)

P	Q	R	$P \Rightarrow R$	$R \Rightarrow Q$	$(P \Rightarrow R) \wedge (R \Rightarrow Q)$
T	T	T	T	T	T
T	T	F	F	T	F
T	F	T	T	F	F
T	F	F	F	T	F
F	T	T	T	T	T
F	T	F	T	T	T
F	F	T	T	F	F
F	F	F	T	T	T

2. (i) If the wind blows then, the sun doesn't shine or the rain falls (or both).

   (ii) The wind blows and the rain falls if and only if the sun doesn't shine.

   (v) If the rain doesn't fall or the temperature rises (but not both), then the sun shines and the wind doesn't blow.

3. (ii) $(S \wedge \neg W) \Rightarrow T$

   (iv) $(S \wedge T) \vee (W \wedge R)$

4. (iv)

P	R	$P \wedge Q$	$P \vee Q$	$\neg(P \vee Q)$	$(P \wedge Q) \wedge \neg(P \vee Q)$
T	T	T	T	F	F
T	F	F	T	F	F
F	T	F	T	F	F
F	F	F	F	T	F

Since all its truth values are F, $(P \wedge Q) \wedge \neg(P \vee Q)$ is a contradiction.

(v)

P	Q	R	$P \Rightarrow Q$	$Q \Rightarrow R$	$(P \Rightarrow Q) \vee (Q \Rightarrow R)$
T	T	T	T	T	T
T	T	F	T	F	T
T	F	T	F	T	T
T	F	F	F	T	T
F	T	T	T	T	T
F	T	F	T	F	T
F	F	T	T	T	T
F	F	F	T	T	T

Since all its truth values are T, $(P \Rightarrow Q) \vee (Q \Rightarrow R)$ is a tautology.

5. (i) (a), (b), (d), and (f) are the substitution instances.

For example, the substitution that gives the proposition in (f) is $P \longrightarrow (R \Rightarrow S)$ and $Q \longrightarrow S$.

(iii) (d) is a substitution instance of question 4 (v) and so is a tautology.

Exercises 2.2

1. (iii)

P	Q	$\neg Q$	$P \Rightarrow \neg Q$	$P \wedge Q$	$\neg(P \wedge Q)$
T	T	F	F	T	F
T	F	T	T	F	T
F	T	F	T	F	T
F	F	T	T	F	T

Since the truth values in the fourth and sixth columns are equal, $P \Rightarrow \neg Q \equiv \neg(P \wedge Q)$.

(iv)

P	Q	R	$P \wedge Q$	$(P \wedge Q) \vee R$	A $P \vee R$	B $Q \vee R$	$A \wedge B$
T	T	T	T	T	T	T	T
T	T	F	T	T	T	T	T
T	F	T	F	T	T	T	T
T	F	F	F	F	T	F	F
F	T	T	F	T	T	T	T
F	T	F	F	F	F	T	F
F	F	T	F	T	T	T	T
F	F	F	F	F	F	F	F

Since the truth values in the fifth and eighth columns are equal, $(P \wedge Q) \vee R \equiv (P \vee R) \wedge (Q \vee R)$.

(vi)

P	Q	R	$Q \Rightarrow R$	$(P \Rightarrow (Q \Rightarrow R)$	$P \wedge Q$	$(P \wedge Q) \Rightarrow R$
T	T	T	T	T	T	T
T	T	F	F	F	T	F
T	F	T	T	T	F	T
T	F	F	T	T	F	T
F	T	T	T	T	F	T
F	T	F	F	T	F	T
F	F	T	T	T	F	T
F	F	F	T	T	F	T

Since the truth values in the fifth and seventh columns are equal, $P \Rightarrow (Q \Rightarrow R) \equiv (P \wedge Q) \Rightarrow R$.

2. (ii)

P	Q	$P \Leftrightarrow Q$	$(P \Leftrightarrow Q) \wedge Q$
T	T	T	T
T	F	F	F
F	T	F	F
F	F	T	F

The only row where $(P \Leftrightarrow Q) \wedge Q$ is true is row 1 and P is also true in row 1.

Hence $(P \Leftrightarrow Q) \wedge Q \vDash P$.

(iv)

P	Q	R	$Q \vee R$	$P \Rightarrow (Q \vee R)$	$P \wedge R$	$P \vee R$
T	T	T	T	T	T	T
T	T	F	T	T	F	T
T	F	T	T	T	T	T
T	F	F	F	F	F	T
F	T	T	T	T	F	T
F	T	F	T	T	F	F
F	F	T	T	T	F	T
F	F	F	F	T	F	F

The propositions $P \Rightarrow (Q \vee R)$ and $P \wedge R$ are both true only in rows 1 and 3. In each of these rows, $P \vee R$ is also true.

Therefore $\{P \Rightarrow (Q \vee R), P \wedge R\} \vDash (P \vee R)$.

(vii)

P	Q	R	$\neg P$	$P \Rightarrow R$	$Q \Rightarrow R$	$\neg P \Rightarrow Q$	$\neg P \vee R$
T	T	T	F	T	T	T	T
T	T	F	F	F	F	T	F
T	F	T	F	T	T	T	T
T	F	F	F	F	T	T	F
F	T	T	T	T	T	T	T
F	T	F	T	T	F	T	T
F	F	T	T	T	T	F	T
F	F	F	T	T	T	F	T

The propositions $P \Rightarrow R$, $Q \Rightarrow R$ and $\neg P \Rightarrow Q$ are all true only in rows 1, 3, and 5. In each of these rows, $\neg P \vee R$ is also true.

Therefore $\{P \Rightarrow R, Q \Rightarrow R, \neg P \Rightarrow Q\} \vDash (\neg P \vee R)$.

3. (ii) $\neg(P \vee Q) \vee (\neg P \wedge Q)$

$$\equiv \quad (\neg P \wedge \neg Q) \vee (\neg P \wedge Q) \qquad \text{De Morgan's law}$$
$$\equiv \quad \neg P \wedge (\neg Q \vee Q) \qquad \text{Distributive law}$$
$$\equiv \quad \neg P \wedge \text{true} \qquad \text{Complement law}$$
$$\equiv \quad \neg P \qquad \text{Identity law}$$

(iv) $P \vee (Q \wedge (P \vee \neg Q))$

$$\equiv \quad P \vee ((Q \wedge P) \vee (Q \wedge \neg Q)) \qquad \text{Distributive law}$$
$$\equiv \quad P \vee ((Q \wedge P) \vee \text{false}) \qquad \text{Complement law}$$
$$\equiv \quad P \vee (Q \wedge P) \qquad \text{Identity law}$$
$$\equiv \quad P \vee (P \wedge Q) \qquad \text{Commutative law}$$
$$\equiv \quad P \qquad \text{Absorption law}$$

(vi) $(P \wedge Q) \vee (\neg P \vee R)$

$$\equiv \quad ((P \land Q) \lor \lnot P) \lor R) \qquad \text{Associative law}$$
$$\equiv \quad (\lnot P \lor (P \land Q)) \lor R \qquad \text{Commutative law}$$
$$\equiv \quad ((\lnot P \lor P) \land (\lnot P \lor Q)) \lor R \quad \text{Distributive law}$$
$$\equiv \quad (\text{true} \land (\lnot P \lor Q)) \lor R \qquad \text{Complement law}$$
$$\equiv \quad (\lnot P \lor Q) \lor R \qquad \text{Identity law}$$

(viii) $[(P \land Q) \lor (R \land S)]$
$$\equiv \quad ((P \land Q) \lor R) \land ((P \land Q) \lor S) \qquad \text{Distributive law}$$
$$\equiv \quad (R \lor (P \land Q)) \land (S \lor (P \land Q)) \qquad \text{Commutative law}$$
$$\equiv \quad (R \lor P) \land (R \lor Q) \land (S \lor P) \land (S \lor Q) \quad \text{Distributive law}$$
$$\equiv \quad (P \lor R) \land (Q \lor R) \land (P \lor S) \land (Q \lor S) \quad \text{Commutative law}$$
$$\equiv \quad (P \lor R) \land (P \lor S) \land (Q \lor R) \land (Q \lor S) \quad \text{Commutative law}$$

4. (ii) $P \Rightarrow (P \land Q)$
$$\equiv \quad \lnot P \lor (P \land Q) \qquad \text{Material Implication law}$$
$$\equiv \quad (\lnot P \lor P) \land (\lnot P \lor Q) \quad \text{Distributive law}$$
$$\equiv \quad \text{true} \land (\lnot P \lor Q) \qquad \text{Complement law}$$
$$\equiv \quad \lnot P \lor Q \qquad \text{Identity law}$$
$$\equiv \quad P \Rightarrow Q \qquad \text{Material Implication law}$$

(iv) $(P \lor Q) \Rightarrow R$
$$\equiv \quad \lnot (P \lor Q) \lor R \qquad \text{Material Implication law}$$
$$\equiv \quad (\lnot P \land \lnot Q) \lor R \qquad \text{De Morgan's law}$$
$$\equiv \quad R \lor (\lnot P \land \lnot Q) \qquad \text{Commutative law}$$
$$\equiv \quad (R \lor \lnot P) \land (R \lor \lnot Q) \qquad \text{Distributive law}$$
$$\equiv \quad (\lnot P \lor R) \land (\lnot Q \lor R) \qquad \text{Commutative law}$$
$$\equiv \quad (P \Rightarrow R) \land (Q \Rightarrow R) \qquad \text{Material Implication law}$$

Exercises 2.3

1. (ii) Universe: Footballers

Predicates: $O :$ ... *is overpaid*
$T :$ ... *is talented.*

Then symbolise as: $\exists x \bullet O(x) \land \lnot T(x).$

(iv) Universe: People (a specific group of people)

Predicates: $F :$ ... *shouted 'Fire'!*
$P :$ ... *panicked.*

Then symbolise as: $\exists x \bullet F(x) \land \forall x \bullet P(x).$

(vii) We interpret this as: *no-one likes **all** people who are rude.*

$$\begin{array}{rl} \text{Universe:} & \text{People} \\ \text{Predicates:} & L: \ \ldots \ likes \ \ldots \\ & R: \ \ldots \ is \ rude. \end{array}$$

Then symbolise as: $\quad \neg \exists x \, \forall y \bullet R(y) \Rightarrow L(x, y)$.

(xi)

$$\begin{array}{rl} \text{Universe:} & \text{People who went to the races} \\ \text{Predicates:} & C: \ \ldots \ was \ cold \\ & P: \ \ldots \ lost \ money. \end{array}$$

Then symbolise as: $\quad \neg \exists x \bullet C(x) \wedge \exists x \bullet L(x)$.

2. (ii) *If Pete is clever and honest, then he's likeable.*

(iv) *Some clever people are dishonest.*

(vi) *Not everyone is likeable and honest.*

(viii) *No one is likeable and dishonest.*

3. (iii) $\exists x \bullet T(x, \text{Statistics}) \wedge \neg P(x, \text{Statistics})$

(v) $\exists x \, \forall y \bullet T(x, y) \Rightarrow P(x, y)$

(vi) $P(\text{Gemma}, \text{Statistics}) \Rightarrow (\forall x \bullet T(x, \text{Statistics}) \Rightarrow P(x, \text{Statistics}))$

4. (iii) (a) *Poppy enjoys every course she takes.*

(b) $\exists y \bullet T(\text{Poppy}, y) \wedge \neg E(\text{Poppy}, y)$

(c) *Poppy take some courses that she doesn't enjoy.*

(v) (a) *Everyone passes some course.*

(b) $\exists x \, \forall y \bullet \neg P(x, y)$

(c) *Someone fails every course.*

(vii) (a) *Someone passes every course they take.*

(b) $\forall x \, \exists y \bullet T(x, y) \wedge \neg P(x, y)$

(c) *Everyone fails some course that they take.*

Exercises 2.4

1. (ii)
1.	$P \Rightarrow \neg Q$	premise
2.	$Q \vee R$	premise
3.	$\neg \neg Q \vee R$	2. Equivalence: involution law
4.	$\neg Q \Rightarrow R$	3. Equivalence: material implication law
5.	$P \Rightarrow R$	1,5. Hypothetical syllogism

(iv)
1. $(P \land Q) \Rightarrow (S \land T)$ premise
2. $Q \land P$ premise
3. $P \land Q$ 2. Equivalence: commutative law
4. $S \land T$ 1,3. Modus ponens
5. S 4. Simplification

(vi)
1. Q premise
2. $\neg S$ premise
3. $(P \land R) \Rightarrow S$ premise
4. $\neg(P \land R)$ 2,3. Modus tollens
5. $\neg P \lor \neg R$ 4. Equivalence: commutative law
6. $Q \lor P$ 1. Addition
7. $P \lor Q$ 6. Equivalence: De Morgan's law
8. $Q \lor \neg R$ 5,7. Resolution

(viii)
1. $P \lor Q$ premise
2. $R \Rightarrow \neg Q$ premise
3. $\neg P$ premise
4. $(\neg R \land Q) \Rightarrow S$ premise
5. Q 1,3. Disjunctive syllogism
6. $\neg(\neg Q)$ 5. Equivalence: involution law
7. $\neg R$ 2,6. Modus tollens
8. $\neg R \land Q$ 5,7. Conjunction
9. S 4,8. Modus ponens

(x)
1. $(P \Rightarrow Q) \land (R \Rightarrow Q)$ premise
2. $S \Rightarrow (P \lor R)$ premise
3. S premise
4. $P \lor R$ 2,3. Modus ponens
5. $P \Rightarrow Q$ 1. Simplification
6. $(R \Rightarrow Q) \land (P \Rightarrow Q)$ 1. Equivalence: commutative law
7. $R \Rightarrow Q$ 6. Simplification
8. $Q \lor Q$ 4,5,7. Constructive dilemma
9. Q 8. Equivalence: idempotent law

2. (i)
1. $P \Rightarrow Q$ premise
2. $Q \Rightarrow P$ premise
3. $(P \Rightarrow Q) \land (Q \Rightarrow P)$ 1,2. Conjunction
4. $P \Leftrightarrow Q$ 3. Equivalence: biconditional law

(iii)   1. $P \vee Q$ premise
      2. $P \Rightarrow R$ premise
      3. $Q \Rightarrow R$ premise
      4. $R \vee R$ 1,2,3. Constructive dilemma
      5. R 3. Equivalence: idempotent law

3. (iii) Let B: *the battery is flat*
              P: *the car is out of petrol*
              S: *the car won't start*
              L: *I'll be late for work.*

The argument has premises: $(B \vee P) \Rightarrow (S \wedge L)$ and $P \vee B$.
The conclusion is L.

    1. $(B \vee P) \Rightarrow (S \wedge L)$ premise
    2. $P \vee B$ premise
    3. $B \vee P$ 2. Equivalence: commutative law
    4. $S \wedge L$ 1,3. Modus ponens
    5. $L \wedge S$ 4. Equivalence: commutative law
    6. L 5. Simplification

(vi) Let V: *Peter is brave*
            B: *Peter is brainy*
            A: *Peter is bald.*

The argument has premises: $(V \vee B) \wedge (B \vee A)$ and $\neg B$.
The conclusion is $V \wedge A$.

    1. $(V \vee B) \wedge (B \vee A)$ premise
    2. $\neg B$ premise
    3. $V \vee B$ 1. Simplification
    4. $B \vee V$ 3. Equivalence: commutative law
    5. V 2,4. Disjunctive syllogism
    6. $(B \vee A) \wedge (V \vee B)$ 1. Equivalence: commutative law
    7. $B \vee A$ 6. Simplification
    8. A 2,7. Disjunctive syllogism
    9. $V \wedge A$ 5,8. Conjunction

(ix) Let G: *ghosts are a reality*
            S: *there are spirits roaming the Earth*
            F: *we fear the dark*
            I: *we have an imagination.*

The argument has premises: $(G \Rightarrow S) \wedge (\neg G \Rightarrow \neg F)$, $F \vee \neg I$ and $I \wedge G$.

The conclusion is S.

1.	$(G \Rightarrow S) \wedge (\neg G \Rightarrow \neg F)$	premise
2.	$F \vee \neg I$	premise
3.	$I \wedge G$	premise
4.	I	3. Simplification
5.	$\neg(\neg I)$	4. Equivalence: involution law
6.	$\neg I \vee F$	2. Equivalence: commutative law
7.	F	5,6. Disjunctive syllogism
8.	$(\neg G \Rightarrow \neg F) \wedge (G \Rightarrow S)$	1. Equivalence: commutative law
9.	$\neg G \Rightarrow \neg F$	8. Simplification
10.	$\neg(\neg F)$	7. Equivalence: involution law
11.	$\neg(\neg G)$	9,10. modus tollens
12.	G	7. Equivalence: involution law
13.	$G \Rightarrow S$	1. Simplification
14.	S	12,13. Modus ponens

4. (ii) Let P: *the project was a success*
  I: *Sally invested her inheritance*
  S: *Sally is sensible*
  B: *Sally is broke.*

The argument has premises: $\neg P \vee I$, $S \Rightarrow I$, P and $(\neg S \wedge \neg I) \Rightarrow B$. The conclusion is B.

Suppose that: P is true, I is true, S is true, and B is false.

Then the premises are all true and the conclusion is false. Therefore the argument is not valid.

  (iv) With the notation in question 3 (ix) above, the argument has premises: $(G \Rightarrow S) \vee (\neg G \Rightarrow \neg F)$, $F \vee \neg I$ and $I \vee G$.

The conclusion is S.

Suppose that G is true, S is false, F is true, and I is true.

Then the premises are all true and the conclusion is false. Therefore the argument is not valid.

5. (iii) Define the universe to be 'numbers' and symbolise the predicates as follows.

  Let E: *... is even*
  R: *... is rational*
  T: *... is divisible by two*
  F: *... is divisible by four.*

Then the argument has premises: $\forall x \bullet E(x) \Rightarrow (R(x) \wedge T(x))$ and $\exists x \bullet E(x) \wedge F(x)$. The conclusion is $\exists x \bullet T(x) \wedge F(x)$.

1.	$\forall x \bullet E(x) \Rightarrow (R(x) \wedge T(x))$	premise
2.	$\exists x \bullet E(x) \wedge F(x)$	premise
3.	$E(a) \Rightarrow (R(a) \wedge T(a))$	1. $\forall$-elimination
4.	$E(a) \wedge F(a)$	2. $\exists$-elimination
5.	$E(a)$	4. Simplification
6.	$R(a) \wedge T(a)$	3, 5. Modus ponens
7.	$T(a) \wedge R(a)$	6. Equivalence: commutative law
8.	$T(a)$	7. Simplification
9.	$F(a) \wedge E(a)$	4. Equivalence: commutative law
10.	$F(a)$	9. Simplification
11	$T(a) \wedge F(a)$	8, 10. Conjunction
12.	$\exists x \bullet T(x) \wedge F(x)$	11. $\exists$-introduction

(viii) Define the universe to be 'functions' and symbolise the predicates as follows.

Let P: ...*is a polynomial*
          D: ...*is differentiable*
          C: ...*is continuous.*

Then the argument has premises: $\neg \exists x \bullet P(x) \wedge \neg D(x)$ and $\forall x \bullet D(x) \Rightarrow C(x)$. The conclusion is $\forall x \bullet P(x) \Rightarrow C(x)$.

1.	$\neg \exists x \bullet P(x) \wedge \neg D(x)$	premise
2.	$\forall x \bullet D(x) \Rightarrow C(x)$	premise
3.	$\forall x \bullet \neg (P(x) \wedge \neg D(x))$	1. Equivalence: negating quantified propositions
4.	$D(a) \Rightarrow C(a)$	2. $\forall$-elimination
5.	$\neg (P(a) \wedge \neg D(a))$	3. $\forall$-elimination
6.	$\neg P(a) \vee \neg (\neg D(a))$	5. Equivalence: De Morgan's law
7.	$\neg P(a) \vee D(a)$	6. Equivalence: involution law
8.	$P(a) \Rightarrow D(a)$	7. Equivalence: material implication
9.	$P(a) \Rightarrow C(a)$	4, 8. Hypothetical syllogism
10.	$\forall x \bullet P(x) \Rightarrow C(x)$	9. $\forall$-introduction

Chapter 3

Exercises 3.1

1. (ii) $\{3, 6, 9, 12, \ldots\}$
(iv) $\left\{\frac{1}{3}, -2\right\}$

(vi) $\{-2\}$

(viii) $\{\frac{1}{2}, 1, \frac{3}{2}, 2, \frac{5}{2}, 3, \dots\}$

2. (ii) $\{m : m = 2n$ for some $n \in \mathbb{Z}$ and $1 \leq n \leq 50\}$ or, more simply, $\{2n : n \in \mathbb{Z}$ and $1 \leq n \leq 50\}$

   (vi) $\{m : m = 5n+2$ for some $n \in \mathbb{N}\}$ or, more simply, $\{5n+2 : n \in \mathbb{N}\}$

   (ix) $\{n \in \mathbb{Z}^+ : n = 2^m$ for some $m \in \mathbb{Z}\}$

3. (iii) True   (iv) False   (vi) False   (vii) True.

4. (ii) $|\{x : x$ is an integer and $2/3 < x < 17/3\}| = |\{1, 2, 3, 4, 5\}| = 5$

   (v) $|\{x \in \mathbb{R} : x^2 \leq 2\}| = \infty$

   (vii) $|\{2, 4, \{\{6, 8\}, 10\}\}| = 3$

   (ix) $|\{1, \{1\}, \{\{1\}\}, \{\{\{1\}\}\}\}| = 4$

5. (ii) $x \in A$ (iv) $x \in A$ and $x \subseteq A$ (vi) neither.

6. (ii) $\{\{a, b\}, \{a, c\}, \{a, d\}, \{b, c\}, \{b, d\}, \{c, d\}\}$

   (iv) $\{\{b\}, \{a, b\}, \{b, c\}, \{b, d\}, \{a, b, c\}, \{a, b, d\}, \{b, c, d\}, \{a, b, c, d\}\}$

7. (ii) Both   (iv) $B \subseteq A$ (vi) neither.

Exercises 3.2

1. (iii) 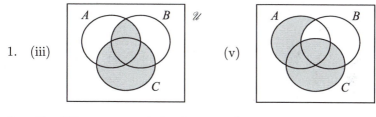   (v)

2. (i) $\{2\}$ (iii) $\{4, 6, 8, 10\}$
   (v) $\{3, 5, 7\}$ (vii) $\{3, 5, 7\}$

3. (iii) None.

   (vi) $X \subseteq Y$

   (ix) $Y \subseteq X$ and $X \cap Y = \emptyset$ (since $Y = \emptyset$).

4. (i) *Proof.* Let A, B, and C be sets such that $C \subseteq A$ and $C \subseteq B$.

   Let $x \in C$. Then $x \in A$ (since $C \subseteq A$) and $x \in B$ (since $C \subseteq B$). Therefore $x \in A \cap B$.

   We have shown that $x \in C \Rightarrow x \in A \cap B$, which means that $C \subseteq A \cap B$. $\square$

5. (i) (a) $A * B = \{2, 6, 10, 12, 16\}$
    (c) $A * (B \cap C) = \{2, 6, 8, 10, 16\}$
    (e) $(A * B) \cap (A * C) = \{2, 6, 10, 16\}$

   (ii) (c) $(A * B) \cap (A * C)$

| $A * B$ | $A * C$ | $(A*B) \cap (A*C)$ |

6. (iii) $\{\varnothing, \{\{1\}\}, \{\{2, 3\}\}, \{\{1\}, \{2, 3\}\}\}$
   (iv) $\{\varnothing, \{\{1, 2, 3\}\}\}$

7. $\mathbb{P}(A) = \{\varnothing, \{a\}, \{b\}, \{a, b\}\}$.

   $\mathbb{P}(B) = \{\varnothing, \{b\}, \{c\}, \{d\}, \{b, c\}, \{b, d\}, \{c, d\}, \{b, c, d\}\}$.

   $$\begin{aligned}
   \mathbb{P}(A \cup B) &= \mathbb{P}(\{a, b, c, d\}) \\
   &= \{\varnothing, \{a\}, \{b\}, \{c\}, \{d\}, \{a, b\}, \{a, c\}, \{a, d\}, \{b, c\}, \{b, d\}, \\
   &\quad \{c, d\}, \{a, b, c\}, \{a, b, d\}, \{a, c, d\}, \{b, c, d\}, \{a, b, c, d\}\}.
   \end{aligned}$$

   $\mathbb{P}(A) \cup \mathbb{P}(B) = \{\varnothing, \{a\}, \{b\}, \{c\}, \{d\}, \{a, b\}, \{b, c\}, \{b, d\}, \{c, d\}, \{b, c, d\}\}$.

   $\mathbb{P}(A) \cup \mathbb{P}(B) \subseteq \mathbb{P}(A \cup B)$.

   This relationship will always hold.

   Reason: an element $X \in \mathbb{P}(A) \cup \mathbb{P}(B)$ is either a subset of A or a subset of B. In either case, $X \subseteq A \cup B$, so $X \in \mathbb{P}(A \cup B)$. Hence every element of $\mathbb{P}(A) \cup \mathbb{P}(B)$ is an element of $\mathbb{P}(A \cup B)$, so $\mathbb{P}(A) \cup \mathbb{P}(B) \subseteq \mathbb{P}(A \cup B)$.

Exercises 3.3

1. (i) $A \times (X \cap Y) = \{(1, c), (2, c), (3, c)\}$
   (iii) $(A \times X) \cap (B \times Y) = \{(2, c), (3, c)\}$
   (v) $(A - B) \times X = \{(1, a), (1, b), (1, c)\}$
   (vii) $(A \cup B) \times (X \cap Y) = \{(1, c), (2, c), (3, c), (4, c)\}$

2. (iii) Not every subset of $X \times Y$ is of the form $A \times B$ for some $A \subseteq X$ and $B \subseteq Y$.

   Let $X = \{a, b, c\}$ and $Y = \{1, 2, 3, 4\}$. The set $S = \{(a, 4), (b, 1)\}$ is a subset of $X \times Y$ but it is not of the form $A \times B$ (for *any* sets A and B).

(iv) Suppose sets A, B, X, and Y are such that $A \times B \subseteq X \times Y$.

Provided both A and B are non-empty, it *does* follow that $A \subseteq X$ and $B \subseteq Y$.

However, if $A = \varnothing$, then for *any* set B, we have $\varnothing \times B = \varnothing \subseteq X \times Y$, so it does *not* follow that $B \subseteq Y$.

Similarly, if $B = \varnothing$, then it does not follow that $A \subseteq X$.

3. (i) $(A \cap B) \times (X \cap Y) = \{2, 4\} \times \{y\} = \{(2, y), (4, y)\}$ and
    $(A \times X) \cap (B \times Y)$

    $$= \{(1, x), (2, x), (3, x), (4, x), (1, y), (2, y), (3, y), (4, y)\}$$
    $$\cap \{(2, y), (4, y)(6, y), (2, z), (4, z), (6, z)\}$$
    $$= \{(2, y), (4, y)\},$$

    so $(A \cap B) \times (X \cap Y) = (A \times X) \cap (B \times Y)$.

    (ii) The following diagrams illustrate $(A \cap B) \times (X \cap Y) = (A \times X) \cap (B \times Y)$.

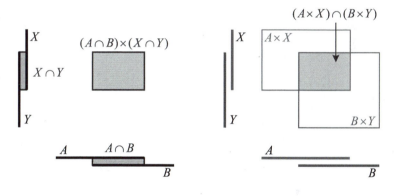

Exercises 3.4

1. (ii) $g\left(\frac{1}{2}\right) = \dfrac{2 \times \frac{1}{2}}{\left(\frac{1}{2}\right)^2 - 3} = \dfrac{1}{-\frac{11}{4}} = -\dfrac{4}{11}$.

    (iv) $(f \circ f)(-2) = f(f(-2)) = f(|-3|) = f(3) = 2$.

    (vi) $(h \circ h)(x) = h(h(x)) = h(2x + 1) = 2(2x + 1) + 1 = 4x + 3$.

    (viii) $(f \circ g)(x) = f(g(x)) = f\left(\dfrac{2x}{x^2 - 3}\right) = \left|\dfrac{2x}{x^2 - 3} - 1\right| = \left|\dfrac{2x - x^2 + 3}{x^2 - 3}\right|$.

2. (i) $(f \circ g)(x) = f(g(x)) = f\left(\dfrac{1}{x^2 + 1}\right)$

    $$= \dfrac{2}{x^2 + 1} + 1 = \dfrac{2 + (x^2 + 1)}{x^2 + 1} = \dfrac{x^2 + 3}{x^2 + 1}.$$

(iii) $(g \circ h)(x) = g(h(x)) = g\left(\sqrt{x^2+1}\right) = \dfrac{1}{\left(\sqrt{x^2+1}\right)^2 + 1} = \dfrac{1}{x^2+2}.$

(v) From part (i), $(f \circ g)(x) = \dfrac{x^2+3}{x^2+1}.$

Hence $\quad ((f \circ g) \circ h)(x) = (f \circ g)(h(x))$

$$= (f \circ g)\left(\sqrt{x^2+1}\right)$$

$$= \frac{\left(\sqrt{x^2+1}\right)^2 + 3}{\left(\sqrt{x^2+1}\right)^2 + 1}$$

$$= \frac{x^2+4}{x^2+2}.$$

4.   (i) $\operatorname{im} f = \{0, 1, 2, 3, 4\}.$

   (ii) $\operatorname{im} f = \{n^2 : n \in \mathbb{Z}\} = \{0, 1, 4, 9, 16, 25, 36, 49, \ldots\}.$

  (iii) $\operatorname{im} f = \{\text{capital cities of the world}\}.$

  (iv) $\operatorname{im} f = \{\varnothing, \{a\}\}.$

5. (iii) By definition, $\operatorname{im} f = \{f(x) : x \in \mathbb{R}\} = \{1/(x^2+2) : x \in \mathbb{R}\}.$

For all $x \in \mathbb{R}$, $x^2+2 \geq 2$, so $\dfrac{1}{x^2+2} \leq \frac{1}{2}$. Also $\dfrac{1}{x^2+2} > 0.$

Therefore $\operatorname{im} f \subseteq \left(0, \frac{1}{2}\right] = \left\{x \in \mathbb{R} : 0 < x \leq \frac{1}{2}\right\}.$

Conversely, if $0 < x \leq \frac{1}{2}$, then $\frac{1}{x} \geq 2$, so $\frac{1}{x} - 2 \geq 0$, in which case $\sqrt{\frac{1}{x} - 2} \in \mathbb{R}.$

Now $f\left(\sqrt{\frac{1}{x} - 2}\right) = \dfrac{1}{\left(\sqrt{\frac{1}{x} - 2}\right)^2 + 2} = \dfrac{1}{\left(\frac{1}{x} - 2\right) + 2} = \dfrac{1}{1/x} = x,$

so $x \in \operatorname{im} f.$

Hence $\left\{x \in \mathbb{R} : 0 < x \leq \frac{1}{2}\right\} \subseteq \operatorname{im} f.$

Therefore $\operatorname{im} f = \left(0, \frac{1}{2}\right] = \left\{x \in \mathbb{R} : 0 < x \leq \frac{1}{2}\right\}.$

(v) By definition, $\operatorname{im} f = \{f(x) : x \in \mathbb{R}\} = \left\{\sqrt{x^2+1} : x \in \mathbb{R}\right\}.$

For all $x \in \mathbb{R}$, $x^2+1 \geq 1$, so $\sqrt{x^2+1} \geq 1$; hence

$$\operatorname{im} f \subseteq \{x \in \mathbb{R} : x \geq 1\}.$$

Conversely, if $x \geq 1$, then $x^2 - 1 \geq 0$, so $\sqrt{x^2-1} \in \mathbb{R}$ and

$$f\left(\sqrt{x^2-1}\right) = \sqrt{\left(\sqrt{x^2-1}\right)^2 + 1} = \sqrt{(x^2-1) + 1} = \sqrt{x^2} = x,$$

since $x \geq 0$. Thus $x \in \operatorname{im} f$. Hence $\{x \in \mathbb{R} : x \geq 1\} \subseteq \operatorname{im} f).$

Therefore $\operatorname{im} f = \{x \in \mathbb{R} : x \geq 1\}.$

Exercises 3.5

1. (i) (a) For all $x, y \in \mathbb{R}$,

$$f(x) = f(y) \Rightarrow \tfrac{1}{2}x - 7 = \tfrac{1}{2}y - 7 \Rightarrow \tfrac{1}{2}x = \tfrac{1}{2}y \Rightarrow x = y,$$

so f is injective.

(b) Let $y \in \mathbb{R}$. Let $x = 2y + 14 \in \mathbb{R}$. Then

$$f(x) = f(2y + 14) = \tfrac{1}{2}(2y + 14) - 7 = y,$$

so f is surjective.

(iii) (a) $f(0) = |0-1| = |-1| = 1$ and $f(2) = |2-1| = 1$, so $f(0) = f(2)$; hence f is not injective.

(b) Let $y \in \mathbb{R}_{\geq 0}$. Let $x = y + 1$. Then $x \in \mathbb{R}_{\geq 0}$ and

$$f(x) = f(y + 1) = |(y + 1) - 1| = |y| = y.$$

Hence f is surjective.

(vi) (a) f is not injective. For example, $f(2) = 1 = f(3)$ but $2 \neq 3$.

(b) Let $m \in \mathbb{Z}$. Then $n = 2m \in \mathbb{Z}$ is even so

$$f(n) = f(2m) = \frac{2m}{2} = m.$$

Therefore f is surjective.

(ix) (a) $f(\{1, 2, 3\}) = \{1, 2, 3\} \cap \{1, 2\} = \{1, 2\}$ and $f(\{1, 2, 4\}) = \{1, 2, 4\} \cap \{1, 2\} = \{1, 2\}$.
Therefore $f(\{1, 2, 3\}) = f(\{1, 2, 4\})$ but $\{1, 2, 3\} \neq \{1, 2, 4\}$, so f is not injective.

(b) For all $X \in A$, we have $f(X) = X \cap \{1, 2\} \subseteq \{1, 2\}$.
Hence there is no $X \in A$ such that $f(X) = \{2, 3\}$, for example.
Therefore f is not surjective.

2. (i) Surjective but not injective.

(iii) Neither injective nor surjective.

(v) Injective and surjective.

3. (ii) (a) For all $x, y \in \mathbb{R} - \{-1\}$,

$$f(x) = f(y) \Rightarrow \frac{3x}{x + 1} = \frac{3y}{y + 1}$$
$$\Rightarrow 3xy + 3x = 3xy + 3y$$
$$\Rightarrow 3x = 3y$$
$$\Rightarrow x = y.$$

Therefore f is injective.

(b) Let $y \in \mathbb{R} - \{3\}$.

Define $x = -\dfrac{y}{y-3}$. Then $x \in \mathbb{R} - \{-1\}$ and

$$f(x) = f\left(-\frac{y}{y-3}\right) = \frac{3 \times -y/(y-3)}{-y/(y-3)+1}$$

$$= \frac{-3y}{-y+(y-3)} = \frac{-3y}{-3} = y.$$

Therefore f is surjective.

(c) The inverse function is

$$f^{-1} : \mathbb{R} - \{3\} \to \mathbb{R} - \{-1\}, \ f^{-1}(y) = -\frac{y}{y-3}.$$

(vi) (a) For all $(x, y), (a, b) \in \mathbb{R}^2$,

$$
\begin{aligned}
f(x,y) = f(a,b) \ &\Rightarrow \ (2x-1, 5y+3) = (2a-1, 5b+3) \\
&\Rightarrow \ 2x-1 = 2a-1 \text{ and } 5y+3 = 5b+3 \\
&\Rightarrow \ 2x = 2a \text{ and } 5y = 5b \\
&\Rightarrow \ x = a \text{ and } y = b \\
&\Rightarrow \ (x,y) = (a,b).
\end{aligned}
$$

Therefore f is injective.

(b) Let $(a, b) \in \mathbb{R}^2$.

Define $x = \frac{1}{2}(a+1)$ and $y = \frac{1}{5}(b-3)$. Then $(x, y) \in \mathbb{R}^2$ and

$$
\begin{aligned}
f(x,y) &= f\left(\tfrac{1}{2}(a+1), \tfrac{1}{5}(b-3)\right) \\
&= \left(2 \times \tfrac{1}{2}(a+1) - 1, 5 \times \tfrac{1}{5}(b-3) + 3\right) \\
&= ((a+1) - 1, (b-3) + 3) \\
&= (a, b).
\end{aligned}
$$

Therefore f is surjective.

(c) The inverse function is $f^{-1} : \mathbb{R}^2 \to \mathbb{R}^2, \ f^{-1}(a,b) = \left(\frac{1}{2}(a+1), \frac{1}{5}(b-3)\right)$.

4. (i) Let $B' = \operatorname{im} f$, the image of f.

(ii) For each $b \in \operatorname{im} f$, choose a single element $a \in A$ such that $f(a) = b$. Then define A' to be the set of all of these chosen elements a.

Chapter 4

Exercises 4.1

1. (ii) *Proof.* Suppose that n is prime. There are two cases: $n = 2$ or n is odd.

    If $n = 2$, then $n^2 + 5 = 9$, which is not prime.

    If n is odd, then n^2 is also odd — see example 1.3. Hence $n^2 + 5$ is even and so has a factor of 2. Thus $n^2 + 5$ is not prime in this case.

    In both cases $n^2 + 5$ is not prime. □

2. (i) *Proof.* Any three consecutive positive integers can be expressed as $n, n + 1, n + 2$ for some $n \in \mathbb{Z}^+$.

    The sum is $n + (n + 1) + (n + 2) = 3n + 3 = 3(n + 1)$, which is clearly divisible by 3. □

3. (i) *Proof.* Any three consecutive positive integers can be expressed as $n, n + 1, n + 2$ for some $n \in \mathbb{Z}^+$.

    Consider their product $n(n + 1)(n + 2)$. At least one of the factors is even (divisible by 2). Another factor is divisible by 3 as this is true of any three consecutive integers.

    Therefore the product has a factor of 6. □

4. (i) *Proof.* Any (two) consecutive positive integers can be expressed as $n, n + 1$ for some $n \in \mathbb{Z}^+$.

    Their sum of their squares is $n^2 + (n + 1)^2 = 2n^2 + 2n + 1 = 2(n^2 + n) + 1$ where $n^2 + n \in \mathbb{Z}^+$.

    Therefore $n^2 + (n + 1)^2$ is odd. □

Exercises 4.2

1. (ii) (a) The proof assumes that n is an *arbitrary* prime number. The proof immediately singles out a special case ($n = 2$) for separate consideration, but the initial assumption is just that n is (any) prime.

    (b) Some of the background knowledge assumed:
    - 9 is not prime;
    - the square of an odd integer is also odd;
    - any even integer (greater than 2) has a factor of 2 and hence is not prime.

(c) The initial statement is: n is prime.
The final concluding statement is: n^2+5 is odd (in both cases).

4. (i) (a) The proof implicitly assumes that a and b are *arbitrary* consecutive positive integers. It then makes the deduction that, if n is the smaller of the two, then the larger of the two is $n+1$. This initial assumption and deduction are wrapped together as the opening assumption that two *arbitrary* consecutive positive integers can be expressed as n and $n+1$.

(b) Some of the background knowledge assumed:
  • some algebraic manipulation including $(n+1)^2 = n^2+2n+1$
  • some elementary factorisation: $2n^2 + 2n = 2(n^2 + n)$
  • $n^2 + n$ is a positive integer when n is a positive integer
  • the oddness property of positive integers.

(c) The initial statement is: n and $n+1$ are consecutive positive integers.
The final concluding statement is: $n^2 + (n+1)^2$ is odd.

Exercises 4.3

1. (iii) *Proof.* Let x and y be real numbers. There are two cases to consider: $|x| \geq |y|$ and $|x| < |y|$.

From the triangle inequality for the modulus function, we have

$$|x| = |(x - y) + y| \leq |x - y| + |y|.$$

Hence $\quad\quad |x| - |y| \leq |x - y|.$ $\hspace{4cm}$ (1)

Swapping the roles of x and y in (1) gives

$$|y| - |x| \leq |y - x| = |x - y|.$$ $\hspace{3cm}$ (2)

We can now consider the two cases.

Case A: $|x| \geq |y|$.
In this case, $||x| - |y|| = |x| - |y| \leq |x - y|$ from (1).

Case B: $|x| < |y|$.
In this case, $||x| - |y|| = -(|x| - |y|) = |y| - |x| \leq |x - y|$ from (2).

Hence, in both cases, $||x| - |y|| \leq |x - y|.$ $\hspace{2cm}$ □

2. (iii) *Proof.* Let $\varepsilon > 0$.

Let $\delta = \min\left\{1, \dfrac{\varepsilon}{14}\right\}$.

Then $\quad 0 < |x + 2| < \delta$

$\Rightarrow \quad |x + 2| < 1$ and $|x + 2| < \dfrac{\varepsilon}{14}$

$$\Rightarrow \ |x^3 + x^2 + 4| = |(x+2)(x^2 - x + 2)|$$
$$= |x+2||(x+2)^2 - 5x - 2|$$
$$= |x+2||(x+2)^2 - 5(x+2) + 8|$$
$$\leq |x+2| \left(|x+2|^2 + 5|x+2| + 8\right)$$
$$< 14|x+2| \qquad \text{(since } |x+2| < 13)$$
$$< 14 \times \frac{\varepsilon}{14} = \varepsilon \qquad \text{(since } |x+2| < \frac{\varepsilon}{14}).$$

$\square$

4. (i) *Proof.* We need to show that d satisfies the conditions (M1), (M2), and (M3) of definition 4.1.

(M1) Let $x, y \in X$. By definition, $d(x, y) \geq 0$ and $d(x, y) = 0 \iff x = y$.
Hence d satisfies (M1).

(M2) Let $x, y \in X$.
If $x = y$, then $d(x, y) = 0 = d(y, x)$.
If $x \neq y$, then $d(x, y) = 1 = d(y, x)$.
Hence d satisfies (M2).

(M3) Let $x, y, z \in X$. We consider five cases.

Case A: $x = y = z$.
Then $d(x, z) = 0 = 0 + 0 = d(x, y) + d(y, z)$.

Case B: $x = y \neq z$.
Then $d(x, z) = 1 = 0 + 1 = d(x, y) + d(y, z)$.

Case C: $x \neq y = z$.
Then $d(x, z) = 1 = 1 + 0 = d(x, y) + d(y, z)$.

Case D: $x \neq y, y \neq z$ but $x = z$.
Then $d(x, z) = 0 < 1 + 1 = d(x, y) + d(y, z)$.

Case E: $x \neq y, y \neq z$ and $x \neq z$.
Then $d(x, z) = 1 < 1 + 1 = d(x, y) + d(y, z)$.

In each case $d(x, y) + d(y, z) \geq d(x, z)$, so d satisfies (M3).

Since d satisfies the three conditions (M1), (M2), and (M3), it is a metric on X. $\square$

(ii) $B(a, r) = \begin{cases} \{a\} & \text{if } 0 \leq r < 1 \\ X & \text{if } r \geq 1. \end{cases}$

(iii) Theorem 4.8 (ii) is true for the discrete metric.

Exercises 4.4

1. Here is an informal deduction in the style of formal deductions in chapter 2. We show that, if we assume $\Gamma \rightsquigarrow (P \Rightarrow Q)$, then Q may be deduced from Γ and P.

1.	$\Gamma \rightsquigarrow (P \Rightarrow Q)$	assumption
2.	Γ	assumption
3.	P	assumption
4.	$P \Rightarrow Q$	from 1 and 2.
5.	Q	from 3, 4: Modus ponens in propositional logic
6.	$\Gamma, P \rightsquigarrow Q$	2, 3, 5: Q has been deduced from Γ, P

2. Here is an informal deduction similar to that in question 1 above. We show that, if we assume both $\Gamma \rightsquigarrow \neg Q$ and $\Gamma, P \rightsquigarrow Q$, then $\neg P$ may be deduced from Γ.

1.	$\Gamma \rightsquigarrow \neg Q$	assumption
2.	$\Gamma, P \rightsquigarrow Q$	assumption
3.	Γ	assumption
4.	$\Gamma \rightsquigarrow (P \Rightarrow Q)$	from 2 by Conditional rule
5.	$P \Rightarrow Q$	from 3 and 4.
6.	$\neg Q \Rightarrow \neg P$	from 5 by Equivalence rule
7.	$\neg Q$	from 1 and 3.
8.	$\neg P$	from 6, 7: Modus ponens in propositional logic
9.	$\Gamma \rightsquigarrow \neg P$	3, 8: $\neg P$ has been deduced from Γ

3.

Formal structure	**'Real' proof**
Γ	We need to prove P and Q.
$\vdots$ (sequence of steps 1)	$\vdots$ (sequence of steps 1)
P	P
$\vdots$ (sequence of steps 2)	$\vdots$ (sequence of steps 2)
Q	Q
$P \wedge Q$ (Proof by conjunction rule)	Hence $P \wedge Q$.

Chapter 5

Exercises 5.1

1. (ii) *Proof.* Let x and y be non-negative real numbers such that $x^2 \geq y^2$.
    Then $x^2 - y^2 \geq 0$.

    If $x = y = 0$, then $x \geq y$.

    Otherwise at least one of x and y is non-zero so $x + y > 0$.

    Hence $\quad x^2 - y^2 \geq 0 \quad \Rightarrow \quad \dfrac{x^2 - y^2}{x + y} \geq 0 \qquad$ (since $x + y > 0$)

    $$\Rightarrow \quad \frac{(x+y)(x-y)}{x+y} \geq 0$$

    $$\Rightarrow \quad x - y \geq 0$$

    $$\Rightarrow \quad x \geq y.$$

    In both cases $x \geq y$. $\qquad \Box$

    (iv) *Proof.* Let $x, y \in \mathbb{R}$ be such that $x \neq 0$ and $y \neq 0$.

    Then $\left(x + \frac{1}{2}y\right)^2 \geq 0$ and $y^2 > 0$.

    Hence $x^2 + xy + y^2 = \left(x + \frac{1}{2}y\right)^2 + \frac{3}{4}y^2 > 0$. $\qquad \Box$

2. (i) *Proof.* Let $n \in \mathbb{Z}^+$.

    Then $n^3 - n = (n-1)n(n+1)$ is the product of three consecutive, non-negative integers. One of the integers $n-1, n, n+1$ is divisible by 3.

    Therefore the product $(n-1)n(n+1) = n^3 - n$ is also divisible by 3. $\qquad \Box$

3. *Proof.* Let $a \geq 1$.

    Then $\quad \left(\sqrt{a + \sqrt{a}} + \sqrt{a - \sqrt{a}}\right)^2$

    $$= (a + \sqrt{a}) + 2\sqrt{(a + \sqrt{a})(a - \sqrt{a})} + (a - \sqrt{a})$$

    $$= 2a + 2\sqrt{a^2 - a}$$

    $$\leq 2a + 2a \quad \text{since } \sqrt{a^2 - a} \leq \sqrt{a^2} = a$$

    $$= 4a.$$

    Therefore, taking square roots, $\sqrt{a + \sqrt{a}} + \sqrt{a - \sqrt{a}} \leq 2\sqrt{a}$. $\qquad \Box$

5. (i) *Proof.* The only integers in the range

    $$1, 2, \ldots, p, p + 1, p + 2, \ldots, 2p, 2p + 2, \ldots, 3p, \ldots p^2$$

which *are* divisible by p are $p, 2p, 3p, \ldots, p^2$. There are p such integers.

Hence $\phi(p^2) = p^2 - p = p(p - 1)$. $\qquad\qquad\qquad\qquad\qquad$ □

(iv) *Proof.* Let $n = p_1^{\alpha_1} p_2^{\alpha_2} \ldots p_k^{\alpha_k}$ be the prime factorisation of n.

Any pair of the terms $p_1^{\alpha_1}$, $p_2^{\alpha_2}$, $\ldots$, $p_k^{\alpha_k}$ are coprime. By the generalisation of the result of part (iii) to a product with k terms, we have

$$\phi(n) = \phi\left(p_1^{\alpha_1}\right) \phi\left(p_2^{\alpha_2}\right) \ldots \phi\left(p_k^{\alpha_k}\right).$$

Strictly, we need a proof by mathematical induction to justify this step; see chapter 8.

Now, using the result of part (ii), we have

$$\phi\left(p_1^{\alpha_1}\right) = p_1^{\alpha_1} - p_1^{\alpha_1-1} = p_1^{\alpha_1}\left(1 - \frac{1}{p_1}\right),$$

$$\vdots$$

$$\phi\left(p_k^{\alpha_k}\right) = p_k^{\alpha_k} - p_k^{\alpha_k-1} = p_k^{\alpha_k}\left(1 - \frac{1}{p_k}\right).$$

Piecing these results together gives

$$\phi(n) = \phi\left(p_1^{\alpha_1}\right) \phi\left(p_2^{\alpha_2}\right) \ldots \phi\left(p_k^{\alpha_k}\right)$$

$$= p_1^{\alpha_1}\left(1 - \frac{1}{p_1}\right) p_2^{\alpha_1}\left(1 - \frac{1}{p_2}\right) \ldots p_k^{\alpha_k}\left(1 - \frac{1}{p_k}\right)$$

$$= p_1^{\alpha_1} p_2^{\alpha_1} \ldots p_k^{\alpha_k}\left(\left(1 - \frac{1}{p_1}\right)\left(1 - \frac{1}{p_2}\right) \ldots \left(1 - \frac{1}{p_k}\right)\right)$$

$$= n\left(1 - \frac{1}{p_1}\right)\left(1 - \frac{1}{p_2}\right) \ldots \left(1 - \frac{1}{p_k}\right).$$

$\qquad\qquad\qquad\qquad\qquad\qquad\qquad\qquad\qquad\qquad\qquad\qquad$ □

6. (ii) *Proof.* Let p be prime.

Consider the sequence of factorials:

$$1!, 2!, 3!, \ldots, p!, (p + 1)!, (p + 2)!, \ldots, (2p)!, (2p + 1)!, \ldots.$$

As noted in the proof of theorem 5.3, the first term that p divides is $p!$.

In a similar way, p^2 does not divide any of the terms $(p+1)!, (p+2)!, \ldots, (2p-1)!$. However, p^2 does divide

$$(2p)! = 1 \times 2 \ldots \times \boldsymbol{p} \times (p+1) \times \ldots \times (\boldsymbol{2p}).$$

Therefore $(2p)!$ is the smallest factorial that p^2 divides, so $S(p^2) = 2p$. $\qquad\qquad\qquad\qquad\qquad\qquad\qquad\qquad\qquad\qquad\qquad$ □

(iii)

k	1	2	3	4	5	6	7	8	9	10
2^k	2	4	8	16	32	64	128	256	512	1024
$S(2^k)$	2	4	4	6	8	8	8	10	10	12

Exercises 5.2

4. *Proof.* Let A and B be sets.

There are two parts to the proof. We need to show both $\mathbb{P}(A \cap B) \subseteq \mathbb{P}(A) \cap \mathbb{P}(B)$ and $\mathbb{P}(A) \cap \mathbb{P}(B) \subseteq \mathbb{P}(A \cap B)$.

Let $C \in \mathbb{P}(A \cap B)$. Then $C \subseteq A \cap B$. Hence $C \subseteq A$ and $C \subseteq B$, which means $C \in \mathbb{P}(A)$ and $C \in \mathbb{P}(B)$. Therefore $C \in \mathbb{P}(A) \cap \mathbb{P}(B)$.

We have shown that $C \in \mathbb{P}(A \cap B) \Rightarrow C \in \mathbb{P}(A) \cap \mathbb{P}(B)$. Therefore $\mathbb{P}(A \cap B) \subseteq \mathbb{P}(A) \cap \mathbb{P}(B)$.

Now let $C \in \mathbb{P}(A) \cap \mathbb{P}(B)$. Then $C \in \mathbb{P}(A)$ and $C \in \mathbb{P}(B)$ which means $C \subseteq A$ and $C \subseteq B$. Hence $C \subseteq A \cap B$ — see exercise 3.2.4 (i). Therefore $C \in \mathbb{P}(A \cap B)$.

We have shown that $C \in \mathbb{P}(A) \cap \mathbb{P}(B) \Rightarrow C \in \mathbb{P}(A \cap B)$. Therefore $\mathbb{P}(A) \cap \mathbb{P}(B) \subseteq \mathbb{P}(A \cap B)$.

We have now shown both $\mathbb{P}(A \cap B) \subseteq \mathbb{P}(A) \cap \mathbb{P}(B)$ and $\mathbb{P}(A) \cap \mathbb{P}(B) \subseteq \mathbb{P}(A \cap B)$. Hence $\mathbb{P}(A \cap B) \subseteq \mathbb{P}(A) \cap \mathbb{P}(B)$. □

6. (ii) *Proof.* We need to show both $(A \cap B) \times (X \cap Y) \subseteq (A \times X) \cap (B \times Y)$ and $(A \times X) \cap (B \times Y) \subseteq (A \cap B) \times (X \cap Y)$.

   ($\subseteq$) Let $(x, y) \in (A \cap B) \times (X \cap Y)$. Then $x \in A \cap B$ and $y \in X \cap Y$. Therefore $x \in A$ and $y \in X$, so $(x, y) \in A \times X$. Also $x \in B$ and $y \in Y$ so $(x, y) \in B \times Y$.
   Hence $(x, y) \in (A \times X) \cap (B \times Y)$.
   We have shown that $(x, y) \in (A \cap B) \times (X \cap Y) \Rightarrow (x, y) \in (A \times X) \cap (B \times Y)$. Hence $(A \cap B) \times (X \cap Y) \subseteq (A \times X) \cap (B \times Y)$.

   ($\supseteq$) Let $(x, y) \in (A \times X) \cap (B \times Y)$. Then $(x, y) \in A \times X$, so $x \in A$ and $y \in X$. Also $(x, y) \in B \times Y$, so $x \in B$ and $y \in Y$.
   Hence $x \in A \cap B$ and $y \in X \cap Y$, so $(x, y) \in (A \cap B) \times (X \cap Y)$.
   We have shown that $(x, y) \in \Rightarrow (A \times X) \cap (B \times Y)(x, y) \in (A \cap B) \times (X \cap Y)$. Hence $(A \times X) \cap (B \times Y) \subseteq (A \cap B) \times (X \cap Y)$.

   Since $(A \cap B) \times (X \cap Y) \subseteq (A \times X) \cap (B \times Y)$ and $(A \times X) \cap (B \times Y) \subseteq (A \cap B) \times (X \cap Y)$, it follows that $(A \cap B) \times (X \cap Y) = (A \times X) \cap (B \times Y)$. □

7. (iii) *Proof.* For all sets A, B, C, we have
   $(A \cap B) * (A \cap C)$

$$= ((A \cap B) - (A \cap C)) \cup ((A \cap C) - (A \cap B))$$
(definition of $*$)

$$= ((A \cap B) \cap \overline{A \cap C}) \cup ((A \cap C) \cap \overline{A \cap B})$$
(definition of set difference)

$$= ((A \cap B) \cap (\bar{A} \cup \bar{C})) \cup ((A \cap C) \cap (\bar{A} \cup \bar{B}))$$
(De Morgan's law)

$$= (A \cap B \cap \bar{A}) \cup (A \cap B \cap \bar{C}) \cup (A \cap C \cap \bar{A}) \cup (A \cap C \cap \bar{B})$$
(Distributive law)

$$= \varnothing \cup (A \cap B \cap \bar{C}) \cup \varnothing \cup (A \cap C \cap \bar{B})$$
(Complement and commutative laws)

$$= (A \cap B \cap \bar{C}) \cup (A \cap C \cap \bar{B}) \text{ (Identity law)}$$

$$= A \cap ((B \cap \bar{C}) \cup (C \cap \bar{B})) \quad \text{(Distributive law)}$$

$$= A \cap ((B - C) \cup (C - B)) \quad \text{(definition of set difference)}$$

$$= A \cap (B * C) \qquad \text{(definition of $*$)}$$

$\square$

8. (ii) *Proof.* For all $x, y \in \mathbb{R} - \{-1\}$,

$$f(x) = f(y) \Rightarrow \frac{3x}{x+1} = \frac{3y}{y+1}$$
$$\Rightarrow 3xy + 3x = 3xy + 3y$$
$$\Rightarrow 3x = 3y$$
$$\Rightarrow x = y.$$

Therefore f is injective.

Now let $y \in \mathbb{R} - \{3\}$.

Define $x = -\dfrac{y}{y-3}$. Then $x \in \mathbb{R} - \{-1\}$ and

$$f(x) = f\left(-\frac{y}{y-3}\right) = \frac{3 \times -y/(y-3)}{-y/(y-3)+1} = \frac{-3y}{-y+(y-3)} = \frac{-3y}{-3} = y.$$

Therefore f is surjective.

Hence f is a bijection. $\square$

(v) *Proof.* For all $(x, y), (a, b) \in \mathbb{R}^2$,

$$f(x, y) = f(a, b) \Rightarrow (2x - 1, 5y + 3) = (2a - 1, 5b + 3)$$
$$\Rightarrow 2x - 1 = 2a - 1 \text{ and } 5y + 3 = 5b + 3$$
$$\Rightarrow 2x = 2a \text{ and } 5y = 5b$$
$$\Rightarrow x = a \text{ and } y = b$$
$$\Rightarrow (x, y) = (a, b).$$

Therefore f is injective.

Let $(a, b) \in \mathbb{R}^2$.

Define $x = \frac{1}{2}(a+1)$ and $y = \frac{1}{5}(b-3)$. Then $(x, y) \in \mathbb{R}^2$ and

$$
\begin{aligned}
f(x, y) &= f\left(\tfrac{1}{2}(a+1), \tfrac{1}{5}(b-3)\right) \\
&= \left(2 \times \tfrac{1}{2}(a+1) - 1, 5 \times \tfrac{1}{5}(b-3) + 3\right) \\
&= \left((a+1) - 1, (b-3) + 3\right) \\
&= (a, b).
\end{aligned}
$$

Therefore f is surjective.

Hence f is a bijection. $\qquad\square$

9. (iii) *Proof.* Let $x \in C$. Then $f(x) \in f(C)$ by the definition of $f(C)$. Hence $x \in f^{-1}(f(C))$ by the definition of $f^{-1}(D)$ with $D = f(C)$. We have shown that $x \in C \Rightarrow x \in f^{-1}(f(C))$. Therefore $C \subseteq f^{-1}(f(C))$. $\qquad\square$

   (iv) *Proof.* Suppose that f is injective.

   By part (iii), we only need to prove that $f^{-1}(f(C)) \subseteq C$.

   Let $x \in f^{-1}(f(C))$. Then $f(x) \in f(C)$ by the definition of $f^{-1}(D)$ with $D = f(C)$.

   By the definition of $f(C)$, this means that $f(x) = f(x')$ for some $x' \in C$. But f is injective, so $x = x'$, which means that $x \in C$.

   We have shown that $x \in f^{-1}(f(C)) \Rightarrow x \in C$. Therefore $f^{-1}(f(C)) \subseteq C$. $\qquad\square$

10. (ii) The required functions are defined by $g(x) = \frac{1}{2}(f(x) + f(-x))$ and $h(x) = \frac{1}{2}(f(x) - f(-x))$.

Exercises 5.3

3. *Proof.* Let H be a subset of G that satisfies the three given conditions.

   Condition (ii) says that $*$ is a binary operation on H, so we need to verify the three group axioms.

   (G1) The operation $*$ is associative on H because it is associative on G.

   (G2) Since H is non-empty, we may choose an $x \in H$. By condition (iii) its inverse is also in H, $x^{-1} \in H$.

   Therefore, by condition (ii), $x * x^{-1} = e \in H$. Hence H contains the identity element for $*$.

   (G3) Condition (iii) guarantees that every element of H has an inverse in H.

Therefore $(H, *)$ is a group and hence is a subgroup of $(G, *)$. ☐

5. *Proof.* Suppose that S is a set with binary operation $*$ satisfying (A1) and (A2).

Let $x, y \in S$.

By (A1), $e \in S$, so we may apply (A2) with elements e, x, y, which gives

$$x * y = (e * x) * y = e * (y * x) = y * x.$$

Therefore $*$ is commutative.

Now let $x, y, z \in S$. Then

$$\begin{aligned} (x * y) * z &= x * (z * y) \quad \text{by (A2)} \\ &= x * (y * z) \quad \text{by commutativity.} \end{aligned}$$

Therefore $*$ is associative. ☐

7. *Proof.* Let S and T be two subspaces of a vector space U. To prove $S + T$ is a subspace of U, we will use the Subspace Test, theorem 5.12.

First note that $\mathbf{0} \in S$ and $\mathbf{0} \in T$, so $\mathbf{0} = \mathbf{0} + \mathbf{0} \in S + T$. Therefore $S + T$ is non-empty.

Let $\mathbf{v}, \mathbf{w} \in S + T$. Then there exist vectors $\mathbf{x}_1, \mathbf{x}_2 \in S$ and $\mathbf{y}_1, \mathbf{y}_2 \in T$ such that $\mathbf{v} = \mathbf{x}_1 + \mathbf{y}_1$ and $\mathbf{w} = \mathbf{x}_2 + \mathbf{y}_2$.

Hence
$$\begin{aligned} \mathbf{v} + \mathbf{w} &= (\mathbf{x}_1 + \mathbf{y}_1) + (\mathbf{x}_2 + \mathbf{y}_2) \\ &= (\mathbf{x}_1 + \mathbf{x}_2) + (\mathbf{y}_1 + \mathbf{y}_2) \quad \text{(by vector space axioms} \\ &\qquad\qquad \text{(A1) and (A2)).} \end{aligned}$$

Now $\mathbf{x}_1 + \mathbf{x}_2 \in S$ since S is closed under addition of vectors and $\mathbf{y}_1 + \mathbf{y}_2 \in T$ since T is also closed under addition of vectors. Therefore $(\mathbf{x}_1 + \mathbf{x}_2) + (\mathbf{y}_1 + \mathbf{y}_2) \in S + T$. We have shown that $S + T$ is closed under addition.

Let $\mathbf{v} = \mathbf{x}_1 + \mathbf{y}_1 \in S + T$ be as above and let $\lambda \in \mathbb{R}$. Then

$$\lambda \mathbf{v} = \lambda(\mathbf{x}_1 + \mathbf{y}_1) = \lambda \mathbf{x}_1 + \lambda \mathbf{y}_1 \quad \text{(by vector space axiom (M2)).}$$

Now $\lambda \mathbf{x}_1 \in S$ since S is closed under scalar multiplication and $\lambda \mathbf{y}_1 \in T$ since T is also closed under scalar multiplication. Therefore $\lambda \mathbf{x}_1 + \lambda \mathbf{y}_1 \in S + T$. We have shown that $S + T$ is closed under scalar multiplication.

Hence $S + T$ satisfies the conditions of theorem 5.12 and is therefore a subspace of U. ☐

9. (i) *Proof.* Let λ be an eigenvalue of an $n \times n$ matrix $\mathbf{A}$.

By definition, there exists a non-zero vector $\mathbf{x}$ such that $\mathbf{A}\mathbf{x} = \lambda\mathbf{x}$.

Then
$$\begin{aligned} (\mathbf{A} + \mathbf{I}_n)\mathbf{x} &= \mathbf{A}\mathbf{x} + \mathbf{I}_n\mathbf{x} \\ &= \lambda\mathbf{x} + \mathbf{x} \\ &= (\lambda + 1)\mathbf{x}. \end{aligned}$$

Therefore $\lambda + 1$ is an eigenvalue of $\mathbf{A} + \mathbf{I}_n$ (with eigenvector $\mathbf{x}$). ☐

(ii) *Proof.* In part (i), we proved that $\mathbf{x} \in E_{\mathbf{A}}(\lambda) \Rightarrow \mathbf{x} \in E_{\mathbf{A}+\mathbf{I}_n}(\lambda+1)$.
This shows $E_{\mathbf{A}}(\lambda) \subseteq E_{\mathbf{A}+\mathbf{I}_n}(\lambda + 1)$ and we also need to prove the subset relation the other way around.

Now $\mathbf{x} \in E_{\mathbf{A}+\mathbf{I}_n}(\lambda + 1)$

$$\Rightarrow \quad (\mathbf{A} + \mathbf{I}_n)\mathbf{x} = (\lambda + 1)\mathbf{x}$$
$$\Rightarrow \quad \mathbf{A}\mathbf{x} + \mathbf{I}_n\mathbf{x} = \lambda\mathbf{x} + \mathbf{x}$$
$$\Rightarrow \quad \mathbf{A}\mathbf{x} + \mathbf{x} = \lambda\mathbf{x} + \mathbf{x}$$
$$\Rightarrow \quad \mathbf{A}\mathbf{x} = \lambda\mathbf{x}$$
$$\Rightarrow \quad \mathbf{x} \in E_{\mathbf{A}}(\lambda).$$

Therefore $E_{\mathbf{A}+\mathbf{I}_n}(\lambda + 1) \subseteq E_{\mathbf{A}}(\lambda)$.

Since we have established the subset relation both ways round, it follows that $E_{\mathbf{A}}(\lambda) = E_{\mathbf{A}+\mathbf{I}_n}(\lambda + 1)$. $\qquad\square$

10. Use the Subspace Test, theorem 5.12.

11. For each part, use the Subspace Test, theorem 5.12.

Exercises 5.4

1. (ii) *Proof.* Given $\varepsilon > 0$, let $N = \left\lfloor \dfrac{19}{16\varepsilon} \right\rfloor$.

Then

$$n > N \quad \Rightarrow \quad n > \frac{19}{16\varepsilon}$$

$$\Rightarrow \quad \left| \frac{n^3 + 3n}{4n^3 + 7n^2} - \frac{1}{4} \right| = \left| \frac{4(n^3 + 3n) - (4n^3 + 7n^2)}{4(4n^3 + 7n^2)} \right|$$

$$= \left| \frac{-7n^2 + 12n}{16n^3 + 28n^2} \right|$$

$$\leq \frac{7n^2 + 12n}{16n^3 + 28n^2}$$

$$\leq \frac{7n^2 + 12n^2}{16n^3} = \frac{19}{16n} < \varepsilon.$$

Hence $\displaystyle\lim_{n\to\infty} \frac{n^3 + 3n}{4n^3 + 7n^2} = \frac{1}{4}$ as claimed. $\qquad\square$

2. *Proof.* Let (a_n) be a convergent sequence with $\displaystyle\lim_{n\to\infty} a_n = \ell$ and let $\lambda \in \mathbb{R}$.

If $\lambda = 0$, then (λa_n) is the constant sequence with values 0. Hence (λa_n) converges to limit $0 = \lambda\ell$.

So suppose $\lambda \neq 0$.

Let $\varepsilon > 0$.

Then $\dfrac{\varepsilon}{|\lambda|} > 0$ and, since $\lim\limits_{n \to \infty} a_n = \ell$, there exists $N \in \mathbb{Z}^+$ such that

$$n > N \;\Rightarrow\; |a_n - \ell| < \frac{\varepsilon}{|\lambda|}.$$

Therefore, for $n > N$,

$$|\lambda a_n - \lambda \ell| = |\lambda||a_n - \ell| < |\lambda| \times \frac{\varepsilon}{|\lambda|} = \varepsilon.$$

Hence $\lim\limits_{n \to \infty} \lambda a_n = \lambda \ell$. $\qquad\square$

3. (iv) *Proof.* Let $\varepsilon > 0$. Let $\delta = \min\left\{1, \dfrac{\varepsilon}{14}\right\}$. Then

$$0 < |x - 1| < \delta \;\Rightarrow\; |x - 1| < 1 \text{ and } |x - 1| < \frac{\varepsilon}{14}$$

$$\begin{aligned}
\Rightarrow \quad |3x^2 + 5x - 8| &= |3x + 8||x - 1| \\
&= |3(x - 1) + 11||x - 1| \\
&\leq (3|x - 1| + 11)|x - 1| \quad \text{(triangle}\\
&\qquad\qquad\qquad\qquad\qquad\quad \text{inequality)} \\
&< 14|x - 1| \qquad\quad \text{since } |x - 1| < 1 \\
&< \varepsilon \qquad\qquad\qquad \text{since } |x - 1| < \frac{\varepsilon}{14}.
\end{aligned}$$

Therefore $\lim\limits_{x \to 1} 3x^2 + 5x = 8$. $\qquad\square$

(vi) *Proof.* Let $\varepsilon > 0$. Let $\delta = \min\left\{1, (1 + \sqrt{2})\varepsilon\right\}$.

First note that

$$|x - 2| < 1 \Rightarrow -1 < x - 2 < 1 \Rightarrow 1 < x < 3 \Rightarrow \sqrt{x} > 1. \qquad (*)$$

Then

$$0 < |x - 2| < \delta \;\Rightarrow\; |x - 2| < 1 \text{ and } |x - 2| < (1 + \sqrt{2})\varepsilon$$

$$\begin{aligned}
\Rightarrow \quad |\sqrt{x} - \sqrt{2}| &= \left|\frac{(\sqrt{x} - \sqrt{2})(\sqrt{x} + \sqrt{2})}{\sqrt{x} + \sqrt{2}}\right| \\
&= \frac{|x - 2|}{\sqrt{x} + \sqrt{2}} \\
&< \frac{|x - 2|}{1 + \sqrt{2}} \qquad \text{since } \sqrt{x} > 1 \text{ by } (*) \\
&< \varepsilon \qquad\qquad \text{since } |x - 1| < (1 + \sqrt{2})\varepsilon.
\end{aligned}$$

Therefore $\lim\limits_{x \to 2} \sqrt{x} = \sqrt{2}$. $\qquad\square$

4. *Proof.* Let $\varepsilon > 0$. Then $\dfrac{\varepsilon}{2} > 0$.

Therefore there exist $\delta_1 > 0, \delta_2 > 0$ such that

$$0 < |x - a| < \delta_1 \quad \Rightarrow \quad |f(x) - \ell| < \frac{\varepsilon}{2}$$

$$\text{and} \quad 0 < |x - a| < \delta_2 \quad \Rightarrow \quad |g(x) - m| < \frac{\varepsilon}{2}.$$

Now let $\delta = \min\{\delta_1, \delta_2\}$. Then

$$0 < |x - a| < \delta \Rightarrow 0 < |x - a| < \delta_1 \text{ and } 0 < |x - a| < \delta_2$$

$$\Rightarrow |f(x) - \ell| < \frac{\varepsilon}{2} \text{ and } |g(x) - m| < \frac{\varepsilon}{2}$$

$$\Rightarrow |f(x) + g(x) - (\ell + m)| = |(f(x) - \ell) + (g(x) - m)|$$

$$\leq |f(x) - \ell| + |g(x) - m|$$

$$< \frac{\varepsilon}{2} + \frac{\varepsilon}{2} = \varepsilon.$$

Therefore $\lim\limits_{x \to a} (f(x) + g(x)) = \ell + m$. $\qquad\qquad\square$

Chapter 6

Exercises 6.1

1. (ii) *Proof.* Let k be an odd integer. Then $k = 2a + 1$ for some $a \in \mathbb{Z}$. We prove the contrapositive: if n is odd, then $kn + (k + 1)$ is odd. So suppose that n is odd. Then $n = 2m + 1$ for some $m \in \mathbb{Z}$. Now

$$kn + (k + 1) = (2a + 1)(2m + 1) + (2a + 1) + 1$$

$$= 4am + 2a + 2m + 1 + 2a + 2$$

$$= 2(2am + 2a + m + 1) + 1$$

$$= 2M + 1 \text{ where } M = 2am + 2a + m + 1 \in \mathbb{Z}.$$

Therefore $kn + (k + 1)$ is odd, which completes the proof of the contrapositive.

Hence, if $kn + (k + 1)$ is even (and k is odd), then n is even. $\quad\square$

(iii) *Proof.* We prove the contrapositive: if n is not divisible by 3, then n^2 is not divisible by 3.

So suppose that n is an integer that is not divisible by 3.

There are two possibilities: $n = 3m + 1$ for some $m \in \mathbb{Z}$ or $n = 3m + 2$ for some $m \in \mathbb{Z}$.

If $n = 3m + 1$, then

$$n^2 = (3m + 1)^2 = 9m^2 + 6m + 1 = 3(3m^2 + 2m) + 1$$

where $3m^2 + 2m \in \mathbb{Z}$. Hence n^2 is not divisible by 3.

If $n = 3m + 2$, then

$$n^2 = (3m + 2)^2 = 9m^2 + 12m + 4 = 3(3m^2 + 4m + 1) + 1$$

where $3m^2 + 4m + 1 \in \mathbb{Z}$. Hence n^2 is not divisible by 3.

In both cases n^2 is not divisible by 3. Therefore, if n^2 is divisible by 3, then n itself is divisible by 3. □

2. (ii) *Proof.* Let k be a positive integer.

We prove the contrapositive, so suppose '$m \leq k$ or $n \leq k$' is false. Then $m > k$ and $n > k$ (by De Morgan's Law, page 39). Therefore $mn > k^2$ so, in particular, $mn \neq k^2$.

This completes the proof of the contrapositive. □

5. *Proof.* We prove the contrapositive, so let $f : A \to B$ and $g : B \to C$ be functions where g is *not* surjective. Then there exists $c \in C$ such that $c \neq g(b)$ for all $b \in B$.

Hence $c \neq g(f(a))$ for all $a \in A$, so the composite function $g \circ f$ is not surjective.

This completes the proof of the contrapositive. Hence if $g \circ f$ is surjective then so, too, is g. □

7. (i) *Proof.* We prove the contrapositive, so suppose that the statement '$A \subseteq X$ or $B \subseteq Y$' is false. Then, by De Morgan's law, $A \nsubseteq X$ *and* $B \nsubseteq Y$.

This means that there exists $a \in A$ such that $a \notin X$ and there exists $b \in B$ such that $b \notin Y$. Therefore $(a, b) \in A \times B$, but $(a, b) \notin X \times Y$. Hence $A \times B \nsubseteq X \times Y$.

This completes the proof of the contrapositive. Hence if $A \times B \subseteq X \times Y$, then $A \subseteq X$ or $B \subseteq Y$. □

Exercises 6.2

2. (ii) *Proof.* Let a, m, and n be positive integers.

$(\Rightarrow) \quad m|n \quad \Rightarrow \quad n = km \qquad \text{for some } k \in \mathbb{Z}^+$

$\Rightarrow \quad an = k(am) \quad (\text{where } k \in \mathbb{Z}^+)$

$\Rightarrow \quad am|an.$

$(\Leftarrow) \quad am|an \quad \Rightarrow \quad an = k(am) \quad \text{for some } k \in \mathbb{Z}^+$

$\Rightarrow \quad n = km \qquad \text{since } a \neq 0$

$\Rightarrow \quad m|n.$

$\qquad\qquad\qquad\qquad\qquad\qquad\qquad\qquad\qquad\qquad\qquad\qquad\qquad\square$

4. (i) *Proof.* Suppose that $f : A \to B$ is injective and let C_1 and C_2 be subsets of A. We need to prove that $f(C_1 \cap C_2) = f(C_1) \cap f(C_2)$, so we will show that each set is a subset of the other.

Firstly,

$$y \in f(C_1 \cap C_2)$$

$\Rightarrow \quad y = f(x) \text{ for some } x \in C_1 \cap C_2$

$\Rightarrow \quad y \in f(C_1) \text{ (since } x \in C_1) \text{ and } y \in f(C_2) \text{ (since } x \in C_2)$

$\Rightarrow \quad y \in f(C_1) \cap f(C_2).$

Hence $f(C_1 \cap C_2) \subseteq f(C_1) \cap f(C_2)$.

Secondly,

$$y \in f(C_1) \cap f(C_2)$$

$\Rightarrow \quad y \in f(C_1) \text{ and } y \in f(C_2)$

$\Rightarrow \quad y = f(x_1) \text{ for some } x_1 \in C_1 \text{ and}$
$\qquad y = f(x_2) \text{ for some } x_2 \in C_2$

$\Rightarrow \quad f(x_1) = f(x_2)$

$\Rightarrow \quad x_1 = x_2 \text{ since } f \text{ is injective}$

$\Rightarrow \quad y = f(x_1) \text{ where } x_1 \in C_1 \cap C_2$

$\Rightarrow \quad y \in f(C_1 \cap C_2).$

Hence $f(C_1) \cap f(C_2) \subseteq f(C_1 \cap C_2)$.

We have proved that each set is a subset of the other, so $f(C_1 \cap C_2) = f(C_1) \cap f(C_2)$.

Conversely, suppose that $f(C_1 \cap C_2) = f(C_1) \cap f(C_2)$ for all subsets C_1 and C_2 of A. We need to show that f is injective.

Let $x_1, x_2 \in A$. Then

$$f(x_1) = f(x_2)$$
$$\Rightarrow \quad f(x_1) \in f(\{x_1\}) \text{ and } f(x_1) \in f(\{x_2\})$$
$$\Rightarrow \quad f(x_1) \in f(\{x_1\}) \cap f(\{x_2\})$$
$$\Rightarrow \quad f(x_1) \in f(\{x_1\} \cap f\{x_2\})$$
$$\text{since } f(\{x_1\}) \cap f(\{x_2\}) = f(\{x_1\} \cap f\{x_2\})$$
$$\Rightarrow \quad \{x_1\} \cap f\{x_2\} \neq \varnothing$$
$$\Rightarrow \quad x_1 = x_2.$$

Hence f is injective. $\qquad\qquad\square$

5. (ii) *Proof.*

($\Rightarrow$) Suppose that $f : A \to B$ is an increasing function and $\alpha \in \mathbb{R}^+$. Let $x, y \in A$ be such that $x < y$. Then

$$f(x) \leq f(y) \quad \text{(since } f \text{ is increasing)}$$
$$\Rightarrow \quad \alpha f(x) \leq \alpha f(y) \quad \text{(since } \alpha > 0\text{)}.$$

Therefore αf is increasing.

($\Leftarrow$) Suppose that $\alpha f : A \to B$ is an increasing function where $\alpha \in \mathbb{R}^+$.
Let $x, y \in A$ be such that $x < y$. Then

$$\alpha f(x) \leq \alpha f(y) \quad \text{(since } \alpha f \text{ is increasing)}$$
$$\Rightarrow \quad f(x) \leq f(y) \quad \text{(since } \alpha > 0\text{)}.$$

Therefore f is increasing.

$\qquad\qquad\square$

Note that there is a slicker proof of the converse, as follows.

($\Leftarrow$) Suppose that $\alpha f : A \to B$ is an increasing function where $\alpha \in \mathbb{R}^+$.
Then $1/\alpha \in \mathbb{R}^+$, so by the first part, $f = (1/\alpha)(\alpha f)$ is also increasing.

6. *Proof.* Firstly, suppose that $n = p^4$ where p is prime. Then n has exactly 5 factors, namely $1, p, p^2, p^3$, and p^4.

Conversely, suppose that n is a positive integer that has exactly 5 factors. By the Prime Factorisation theorem 4.3, n may be expressed as a product of prime numbers

$$n = p_1^{\alpha_1} p_2^{\alpha_2} \dots p_m^{\alpha_m}$$

where $p_1, p_2, \dots, p_m$ are prime and each α_r is a positive integer. Furthermore, this expression is unique except for the ordering of the prime numbers — see theorem 7.16.

Any factor of n must therefore be of the form

$$p_1^{\beta_1} p_2^{\beta_2} \dots p_m^{\beta_m}$$

where, for $r = 1, \ldots, m$, $0 \le \beta_r \le \alpha_r$. Since any set of choices of the β_r gives a factor of n, there are $(\alpha_1 + 1)(\alpha_2 + 1) \ldots (\alpha_m + 1)$ factors altogether.

Since n has exactly five factors, $(\alpha_1 + 1)(\alpha_2 + 1) \ldots (\alpha_m + 1) = 5$. But 5 is a prime number, so we must have $m = 1$ and $\alpha_1 = 4$. Therefore $n = p_1^4$ (where p_1 is prime), as required.

$\square$

Exercises 6.3

1. (ii) *Proof.* If α and β are the roots of $x^2 + ax + b = 0$, then

$$x^2 + ax + b = (x - \alpha)(x - \beta) = x^2 - (\alpha + \beta)x + \alpha\beta.$$

If α and β are odd integers, then their sum $\alpha + \beta = a$ is even and their product $\alpha\beta = b$ is odd. $\square$

(iii) *Proof.* Firstly,

$$\begin{aligned} -1 \le x \le 4 \quad &\Rightarrow \quad -4 \le x - 3 \le 1 \\ &\Rightarrow \quad 0 \le (x - 3)^2 \le 16 \\ &\Rightarrow \quad 0 \le x^2 - 6x + 9 \le 16 \\ &\Rightarrow \quad -9 \le x^2 - 6x \le 7. \end{aligned}$$

Following a similar approach,

$$\begin{aligned} -1 \le x \le 4 \quad &\Rightarrow \quad 2 \le x + 3 \le 7 \\ &\Rightarrow \quad 4 \le (x + 3)^2 \le 49 \\ &\Rightarrow \quad 4 \le x^2 + 6x + 9 \le 49 \\ &\Rightarrow \quad -5 \le x^2 + 6x \le 40. \end{aligned}$$

$\square$

2. (iv) *Proof.* Let $(a, x) \in (A \cap B) \times (X \cap Y)$.
Then $a \in A \cap B$ and $x \in X \cap Y$.
Hence $a \in A$ and $x \in X$, so $(a, x) \in A \times X$. Also $a \in B$ and $x \in Y$, so $(a, x) \in B \times Y$. Therefore $(a, x) \in (A \times X) \cap (B \times Y)$.
Hence $(A \cap B) \times (X \cap Y) \subseteq (A \times X) \cap (B \times Y)$.

Now let $(a, x) \in (A \times X) \cap (B \times Y)$.
Then $(a, x) \in A \times X$, so $a \in A$ and $x \in X$. Also $(a, x) \in B \times Y$, so $a \in B$ and $x \in Y$. Hence $a \in A \cap B$ and $x \in X \cap Y$. Therefore $(a, x) \in (A \cap B) \times (X \cap Y)$.
Hence $(A \times X) \cap (B \times Y) \subseteq (A \cap B) \times (X \cap Y)$.

Since we have proved that each set is a subset of the other, we have $(A \cap B) \times (X \cap Y) = (A \times X) \cap (B \times Y)$. $\square$

3. (ii) *Proof.* Let $x \in (A \cup C) * (B \cup C)$. Then $x \in (A \cup C) - (B \cup C)$ or $x \in (B \cup C) - (A \cup C)$.

    In the first case,

    $$x \in A \cup C \text{ and } x \notin B \cup C$$
    $$\Rightarrow \quad (x \in A \text{ or } x \in C) \text{ and } (x \notin B \text{ and } x \notin C)$$
    $$\text{(since } x \in C \text{ and } x \notin C \text{ is impossible)}$$
    $$\Rightarrow \quad x \in A \text{ and } x \notin B \text{ and } x \notin C$$
    $$\Rightarrow \quad x \in A - B \text{ and } x \notin C$$
    $$\Rightarrow \quad x \in A * B \text{ and } x \notin C \quad \text{(since } A - B \subseteq A * B)$$
    $$\Rightarrow \quad x \in (A * B) - C.$$

    In the second case, interchanging A and B in the previous argument shows that $x \in (B * A) - C = (A * B) - C$.

    Therefore $x \in (A * B) - C$ in both cases, so that $(A \cup C) * (B \cup C \subseteq (A * B) - C$.

    Conversely, let $x \in (A * B) - C$. Then $x \in A * B$ and $x \notin C$.

    Now $x \in A * B$ means that $x \in A - B$ or $x \in B - A$.

    In the first case, we have $x \in A$, $x \notin B$, and $x \notin C$. Hence $x \in A$ and $x \notin B \cup C$. Since $A \subseteq (A \cup C)$, this means that $x \in A \cup C$ and $x \notin B \cup C$ so $x \in (A \cup C) - (B \cup C)$.

    In the second case, reversing the roles of A and B in the previous argument shows that $x \in (B \cup C) - (A \cup C)$.

    Putting the two cases together we have

    $$x \in ((A \cup C) - (B \cup C)) \cup ((B \cup C) - (A \cup C))$$
    $$= (A \cup C) * (B \cup C).$$

    Hence $(A * B) - C \subseteq (A \cup C) * (B \cup C)$.

    As we have shown that each set is a subset of the other, we have $(A \cup C) * (B \cup C) = (A * B) - C$. □

4. (iii) To prove $X \subseteq Y$: let $x \in X$. Clearly $x > 0$. To show that $x \le \frac{1}{2}$, use the approach given in the proof in example 5.1.

    To prove $Y \subseteq X$: let $0 < c \le \frac{1}{2}$. Then use the quadratic formula to show that the equation $x/(x^2 + 1) = c$ has a solution for $x \ge 1$.

Exercises 6.4

1. (i) *Proof.* Let x be rational and y be irrational.

    Suppose that $x + y$ is rational. Then $y = (x + y) - x$ is the difference of two rational numbers, which is therefore rational. This contradicts the fact that y is irrational.

    Therefore $x + y$ is irrational. □

3. Firstly, suppose that $ax^2 + bx + c = 0$ has a rational root (where a, b, and c are odd integers).

   Deduce that both roots must be rational — question 1 may be of use here.

   Then show that $ax^2 + bx + c = (px - q)(rx - s)$ for some integers p, q, r, s. Deduce that p, q, r, and s are all odd.

   The contradiction comes from the fact that the coefficient of x must then be even.

5. (i) *Proof.* Let m and n be integers where $n \neq 0$.

   Suppose that $m + \sqrt{2n}$ is rational; then $m + \sqrt{2n} = \dfrac{p}{q}$ where $p, q \in \mathbb{Z}$.

   Hence $\sqrt{2n} = \dfrac{p}{q} - m = \dfrac{p - qm}{q}$, so $\sqrt{2} = \dfrac{p - qm}{qn} \in \mathbb{Q}$. This is a contradiction since $\sqrt{2}$ is irrational.

   Therefore $m + \sqrt{2n}$ is irrational. □

6. *Proof.* Suppose that there is a rational number r such that $2^r = 3$.

   Now r may be expressed as $r = p/q$ where $p, q \in \mathbb{Z}$ and $q \neq 0$. Hence $2^{p/q} = 3$. Raising both sides to the power q gives $2^p = 3^q$. Now 2^p is even and 3^q is odd, so $2^p \neq 3^q$.

   Hence there is no rational number r such that $2^r = 3$. □

7. Suppose that x is the smallest real number such that $x > \frac{1}{2}$.

   Show that $y = \dfrac{x}{2} + \dfrac{1}{4}$ is smaller than x but still greater than $\frac{1}{2}$.

11. *Proof.* Suppose $\dfrac{m - 1}{m} \in A$ (where $m \in \mathbb{Z}^+$) is the largest element of A.

    Note that $\dfrac{m}{m + 1} = \dfrac{(m + 1) - 1}{m + 1} \in A$ and

    $$\dfrac{m}{m + 1} - \dfrac{m - 1}{m} = \dfrac{m^2 - (m + 1)(m - 1)}{m(m + 1)}$$
    $$= \dfrac{m^2 - (m^2 - 1)}{m(m + 1)} = \dfrac{1}{m(m + 1)} > 0.$$

    Hence $\dfrac{m}{m + 1} > \dfrac{m - 1}{m}$.

    In other words, $\dfrac{m}{m + 1}$ is an element of A that is greater than $\dfrac{m - 1}{m}$, which is therefore *not* the largest element of A.

    Hence A has no largest element. □

Exercises 6.5

1. (i) The identity element of H is the element e_H that satisfies $h \circ e_H = h = e_h \circ e_H$ for all $h \in H$.

    Show that $\theta(e_G)$ satisfies this property for any element h that is in the image of θ; $h = \theta(g)$ for some $g \in G$.

    (iii) *Proof.* Let $(G, *)$ and $(H, \circ)$ be groups and let $\theta : G \to H$ be a morphism.

    To prove that $\ker \theta$ is a subgroup of G, we use the Subgroup Test, theorem 5.16 given in exercise 5.3.3.

    Firstly, since $\theta(e) = e$, by part (i), it follows that $e \in \ker \theta$. (Note that we have dropped the subscripts on the identity elements to keep the notation simple.) Hence $\ker \theta \neq \varnothing$.

    Let $x, y \in \ker \theta$. Then, by definition of the kernel, $\theta(x) = e$ and $\theta(y) = e$. Therefore $\theta(x * y) = \theta(x) \circ \theta(y) = e \circ e = e$, so $xy \in \ker \theta$.
    By part (ii), $\theta(x^{-1}) = (\theta(x))^{-1} = e^{-1} = e$, so $x^{-1} \in \ker \theta$.

    We have shown that $\ker \theta$ satisfies the three conditions of the Subgroup Test. It therefore follows that $\ker \theta$ is a subgroup of G. □

2. *Proof.*

    ($\Rightarrow$) Suppose that G is Abelian.

    Let $g, h \in G$. Then

    $$(gh)^{-1} = h^{-1}g^{-1} \quad \text{(theorem 5.9)}$$
    $$= g^{-1}h^{-1} \quad \text{(G is Abelian.)}$$

    ($\Leftarrow$) Now suppose that $(gh)^{-1} = g^{-1}h^{-1}$ for all elements $g, h \in G$.
    In particular, since this equation is satisfied *all* elements of G, it is satisfied by g^{-1} and h^{-1}: $\left(g^{-1}h^{-1}\right)^{-1} = \left(g^{-1}\right)^{-1}\left(h^{-1}\right)^{-1}$.
    Hence we have

    $$gh = \left(g^{-1}\right)^{-1}\left(h^{-1}\right)^{-1}$$
    $$= \left(g^{-1}h^{-1}\right)^{-1}$$
    $$= \left(h^{-1}\right)^{-1}\left(g^{-1}\right)^{-1} \quad \text{(theorem 5.9)}$$
    $$= hg.$$

    Therefore G is Abelian.

    □

5. Let X be a non-empty subset of $\mathbb{R}$. A real number m satisfying:
    (i) m is a lower bound for X: $m \leq x$ for all $x \in X$ and

(ii) if $b > m$, then b is not a lower bound for X; if $b > m$ then, there exists $x \in X$ such that $x < b$

is called a **infimum** or **greatest lower bound** for X; it is denoted $m = \inf X$.

6. (i) *Proof.* First note that, for all $x \in \mathbb{R}^+$, $x - 1 < x$, so $\dfrac{x-1}{x} < 1$. Hence 1 is an upper bound for X.

Now let $a < 1$. Then $\dfrac{1}{1-a} > 0$.

Since there is no smallest positive real number (compare with exercise 6.4.7), there exists $x \in \mathbb{R}^+$ such that $x < \dfrac{a}{1-a}$. Hence

$$
\begin{aligned}
x < \frac{a}{1-a} \quad &\Rightarrow \quad x(1-a) < 1 \quad \text{(since } a-1 > 0\text{)} \\
&\Rightarrow \quad x - ax < 1 \\
&\Rightarrow \quad x - 1 < ax \\
&\Rightarrow \quad \frac{x-1}{x} < a \quad \text{(since } x > 0\text{).}
\end{aligned}
$$

In other words, there exists an element $(x-1)/x \in X$ that is less than a. Hence a is not an upper bound for X.

Therefore $\sup X = 1$. □

(v) *Proof.* Let $x \in X$. Then $x = m/n$ for some $m, n \in \mathbb{Z}^+$ such that $m < 2n$. Hence $x < 2$. Therefore 2 is an upper bound for X

Now let $a < 2$. Then $2 - a > 1$. Choose $n \in \mathbb{Z}^+$ such that $n > \dfrac{1}{2-a}$. Then

$$
n > \frac{1}{2-a} \quad \Rightarrow \quad 2 - a > \frac{1}{n} \quad \Rightarrow \quad a < 2 - \frac{1}{n} = \frac{2n-1}{n}
$$

where $\dfrac{2n-1}{n} \in X$.

Therefore a is not an upper bound for X.

Hence $\sup X = 2$. □

Chapter 7

Exercises 7.1

1. (iii) $73 = 3^2 + 8^2$ and $74 = 5^2 + 7^2$.

(vi) Both $1 + \sqrt{2}$ and $(1 + \sqrt{2})^2 = 3 + 2\sqrt{2}$ are irrational.

(ix) $1729 = 1^3 + 12^3 = 9^3 + 10^3$.

(xi) *Proof.* Let n be an odd positive integer.

Then $n = 2k + 1$ for some $k \in \mathbb{N}$. Hence $n^2 = (2k+1)^2 = 4k^2 + 4k + 1 = 4k(k+1) + 1$.

Now $k(k+1)$ is even — see theorem 1.2. Hence $k(k+1) = 2m$ for some $m \in \mathbb{N}$.

Therefore $n^2 = 4k(k+1) + 1 = 8m + 1$ for some $m \in \mathbb{N}$. $\square$

2. Try to construct examples where A, B, and C are small finite sets (at most three elements) and the functions f and g are defined by specifying explicitly the image of each element.

3. (ii) For example, $\mathbf{A} = \begin{pmatrix} -1 & 0 \\ 0 & 1 \end{pmatrix}$.

(iv) For example, $\mathbf{A} = \begin{pmatrix} 1 & 0 & -1 \\ 2 & -2 & -1 \end{pmatrix}$ and $\mathbf{B} = \begin{pmatrix} 1 & 1 \\ 1 & 0 \\ 0 & 1 \end{pmatrix}$.

Then $\mathbf{AB} = \mathbf{I}_2$ but $\mathbf{BA} \neq \mathbf{I}_3$.

5. The group $S(\triangle)$, defined in example 5.4.4, is a suitable example to establish parts (i), (ii), and (vi).

Exercises 7.2

2. (i) *Proof.* Let $\{a_1, a_2, \ldots, a_n\}$ be a set of non-zero integers such that $\sum_{k=1}^{n} a_k < n$ and suppose that none of the a_i are negative.

Then each $a_1 \geq 1$ so $\sum_{k=1}^{n} a_k \geq 1 + 1 + \cdots + 1(n \text{ terms}) = n$.

This is a contradiction. Hence at least one of the a_i is negative. $\square$

4. *Proof.* Suppose that $x \in \mathbb{R} - \mathbb{Q}$ is such that, in its decimal expansion, each digit $0, 1, 2, \ldots, 9$ occurs a finite number of times.

Then the decimal expansion of x terminates. Suppose the decimal expansion terminates after N decimal places, so that $x = x_0 \cdot x_1 x_2 \ldots x_N$ where $x_0 \in \mathbb{Z}$ and $x_1, \ldots, x_N \in \{0, 1, \ldots, 9\}$. Then $x \times 10^N$ is an integer, so x is rational. $\square$

6. Consider the prime numbers less than 30. Place each integer in the set $\{2, 3, \ldots, 30\}$ into a pigeonhole according to its *smallest* prime factor. Then use the Pigeonhole Principle to show that there are two integers with the same smallest prime factor.

7. (ii) *Proof.* Suppose that $n^2 + 1$ points are placed in a square of side n. Divide the square into n^2 smaller squares, each of side length 1 as shown.

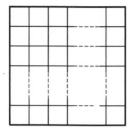

Since there are $n^2 + 1$ points placed in the (large) square and there are only n^2 small squares, there must be a small square that contains two points. These two points are no more than $\sqrt{2}$ units apart. ☐

10. (ii) If a pair of dice is rolled, there are 11 possible scores: $2, 3, \ldots, 12$. Since $45 > 4 \times 11$, if a pair of dice is rolled 45 times, there must be one score that occurs at least 5 times by the Generalised Pigeonhole Principle.

12. (iii) *Proof.* Let f and g be two functions, each continuous on $[a, b]$, with the property that $f(a) < g(a)$ and $f(b) > g(b)$.

Let h be the function $h = f - g$ defined on $[a, b]$; in other words $h(x) = f(x) - g(x)$ for all $x \in [a, b]$.

Then h is continuous on $[a, b]$, $h(a) = f(a) - g(a) < 0$ and $h(b) = f(b) - g(b) > 0$. Therefore, by the Intermediate Value Theorem, there exists $c \in (a, b)$ such that $h(c) = 0$. Hence $f(c) = g(c)$. ☐

Exercises 7.3

2. (i) For example, take $a = 6, b = 5, c = 4, d = 1$.

   (iii) $f(17) = 17^2$ is not prime.

   (v) The smallest counter-example is $n = 10$, where $3^{10} = 59049 > 40000 = 4 \times 10^4$.

4. (i) This is true. If $a = p/q$ and $b = r/s$ where $p, q, r, s \in \mathbb{Z}$ and $q \neq 0$, $s \neq 0$, then $ab = (pr)/(qs)$ where $pr, qs \in \mathbb{Z}$ and $qs \neq 0$.

   (ii) This is false. A counter-example is $a = b = \sqrt{2}$.

   (v) This is false. A counter-example is $\alpha = \beta = \sqrt{2}$, $m = 1$, and $n = -1$.

6. (ii) This is false: $8^n + 1$ factorises as $(2^n + 1)(4^n - 2^n + 1)$.

7.   (i) This is false. For example, let f and g be functions $\mathbb{R} \to \mathbb{R}$ defined by $f(x) = 2x$ and $g(x) = -x$. Then $f + g$ is increasing but g is not increasing.

Note that the implication 'f and g are increasing $\Rightarrow f + g$ is increasing' is true and has a simple direct proof.

9.   (i) Let $G = \{e, r, v, h\}$ be the Klein 4-group defined in example 5.4.3 and let $H = \{e, r\}$ and $K = \{e, v\}$. Then H and K are subgroups of G but $H \cup K$ is not a subgroup of G (because $r, v \in H \cup K$ but $r * v = h \notin H \cup K$).

(ii) The group $(S\triangle)$, defined in example 5.4.4, is a suitable example.

10.   (i) This is true and has a simple direct proof.

(ii) This is false. For example, let $A = \{1\}$, $B = \{2\}$, and $C = \{1, 3\}$. Then $A \not\subseteq B$ and $B \not\subseteq C$ but $A \subseteq C$.

(v) This is false. Let $A = \{1, 2\}$, $B = \{a, b\}$, and $C = \{(1, a), (2, b)\}$. Clearly $C \subseteq A \times B = \{\{(1, a), (1, b), (2, a)(2, b)\}\}$.

We need to show that C is not the Cartesian product of any two sets X and Y, which we do by contradiction.

So suppose that $C = X \times Y$ for some sets X and Y. Since $(1, a) \in C$, we have $1 \in X$; similarly, since $(2, b) \in C$, we have $b \in Y$. Then $(1, b) \in X \times Y = C$, which is a contradiction. Therefore C is not the Cartesian product of any sets X and Y.

Exercises 7.4

1.   (ii) *Proof.* Assuming the existence of $\sqrt{a}$ as a real number, then $x = \sqrt{a}$ is a solution of $x^2 = a$.

Suppose that x and y are both positive real numbers such that $x^2 = a$ and $y^2 = a$. Then

$$x^2 = y^2 \Rightarrow x^2 - y^2 = 0 \Rightarrow (x + y)(x - y) = 0.$$

Now $x + y > 0$, so $x - y = 0$ so $x = y$.

Hence $x^2 = a$ has a unique positive solution.   □

2.   *Proof.* First let $x = \dfrac{ds - bt}{ad - bc}$ and $y = \dfrac{at - cs}{ad - bc}$.

Then $ax + by = \dfrac{a(ds - bt) + b(at - cs)}{ad - bc} = \dfrac{ads - bcs}{ad - bc} = s$

and $cx + dy = \dfrac{c(ds - bt) + d(at - cs)}{ad - bc} = \dfrac{-bct + adt}{ad - bc} = t.$

Hence $(x, y) = \left(\dfrac{ds - bt}{ad - bc}, \dfrac{at - cs}{ad - bc} \right)$ is a solution to the simultaneous equations.

For the uniqueness part, suppose that (x, y) and (x', y') are solutions to the simultaneous equations. In other words,

$$\begin{array}{ll} ax + by = s & (1) \\ cx + dy = t & (2) \end{array} \quad \text{and} \quad \begin{array}{ll} ax' + by' = s & (3) \\ cx' + dy' = t & (4). \end{array}$$

Subtracting equation (3) from equation (1) and subtracting equation (4) from equation (2) gives:

$$\begin{array}{ll} a(x - x') + b(y - y') = 0 & (5) \\ c(x - x') + d(y - y') = 0 & (6) \end{array} \quad \Rightarrow \quad (ab - cd)(y - y') = 0.$$

Since $ad - bc \neq 0$, it follows that $y - y' = 0$, so $y = y'$.

Then equations (5) and (6) simplify to $a(x - x') = 0$ and $c(x - x') = 0$ respectively. Now a and c cannot both be zero (as $ad - bc \neq 0$), so at least one of these equations now implies that $x - x' = 0$ so $x = x'$.

Hence $(x, y) = (x', y')$, so the simultaneous equations have a unique solution. $\qquad \square$

4. *Proof.* Note that if $p = 3$, then $p^2 + 2 = 11$, which is also prime. Hence there exists a prime p for which $p^2 + 2$ is also prime.

Now suppose that p is prime and $p \neq 3$. Then p is not divisible by 3, so $p = 3n + 1$ or $p = 3n + 2$ for some $n \in \mathbb{N}$.

If $p = 3n+1$, then $p^2+2 = (3n+1)^2+2 = 9n^2+6n+3 = 3(3n^2+2n+1)$.

If $p = 3n+2$, then $p^2+2 = (3n+2)^2+2 = 9n^2+12n+6 = 3(3n^2+4n+2)$.

In either case, $p^2 + 2$ is divisible by 3 and therefore is not prime.

Hence $p = 3$ is the only prime for which $p^2 + 2$ is also prime. $\qquad \square$

7. (ii) Suppose that $gx = h$ and $gy = h$ in G. Then $gx = gy$ and 'multiply' this equation on the left by g^{-1}.

9. (i) *Proof.* Clearly $x = 0$ is such that $|x| < \varepsilon$ for all $\varepsilon > 0$, $\varepsilon \in \mathbb{R}$.

Suppose $x \neq 0$. Then $|x| > 0$. Let $\varepsilon = \dfrac{|x|}{2}$. Then $\varepsilon > 0$ and $|x| > \varepsilon$.

Hence $x = 0$ is the only real number that satisfies $|x| < \varepsilon$ for all $\varepsilon \in \mathbb{R}^+$. $\qquad \square$

(ii) *Proof.* Let (a_n) be a convergent sequence and suppose that $\lim_{n \to \infty} a_n = \ell$ and $\lim_{n \to \infty} a_n = \ell'$.

Let $\varepsilon > 0$, $\varepsilon \in \mathbb{R}$.

Then $\varepsilon/2 > 0$, so as $\lim_{n \to \infty} a_n = \ell$, there exists $N_1 \in \mathbb{Z}^+$ such that

$n > N_1 \Rightarrow |a_n - \ell| < \varepsilon/2$. Similarly, since $\lim_{n\to\infty} a_n = \ell'$, there exists $N_2 \in \mathbb{Z}^+$ such that $n > N_1 \Rightarrow |a_n - \ell'| < \varepsilon/2$.

For $n > \max\{N_1, N_2\}$ we have

$$|\ell' - \ell| = |(a_n - \ell) - (a_n - \ell')| \leq |a_n - \ell| + |a_n - \ell'| < \frac{\varepsilon}{2} + \frac{\varepsilon}{2} = \varepsilon.$$

Hence $|\ell - \ell'| < \varepsilon$ for all $\varepsilon \in \mathbb{R}^+$. Therefore, by part (i), $\ell - \ell' = 0$ so $\ell = \ell'$. □

Chapter 8

Exercises 8.1

1. (vi) *Proof.* The proof is by mathematical induction.

   Base case. When $n = 1$, LHS $= 1 \times 6 = 6$ and RHS $= \frac{1}{3} \times 1 \times 2 \times 9 = 6$. Hence the result holds when $n = 1$.

   Inductive step. Assume that $1 \times 6 + 2 \times 7 + 3 \times 8 + \cdots + k(k+5) = \frac{1}{3}k(k+1)(k+8)$; this is the inductive hypothesis.

   The next term on the left-hand side is $(k+1)(k+6)$. Adding this to both sides gives

   $$
   \begin{aligned}
   1 \times 6 + 2 \times 7 &+ 3 \times 8 + \cdots + k(k+5) + (k+1)(k+6) \\
   &= \tfrac{1}{3}k(k+1)(k+8) + (k+1)(k+6) \\
   &= \tfrac{1}{3}(k+1)\left(k(k+8) + 3(k+6)\right) \\
   &= \tfrac{1}{3}(k+1)(k^2 + 11k + 18) \\
   &= \tfrac{1}{3}(k+1)(k+2)(k+9).
   \end{aligned}
   $$

   This is the result for $n = k+1$ and hence completes the inductive step.

   Therefore, for all $n \in \mathbb{Z}^+$, $1 \times 6 + 2 \times 7 + 3 \times 8 + \cdots + n(n+5) = \frac{1}{3}n(n+1)(n+8)$, by induction. □

   (viii) *Proof.* The proof is by mathematical induction.

   Base case. When $n = 1$, LHS $= 1/2! = \frac{1}{2}$ and RHS $= 1 - 1/2! = \frac{1}{2}$. Hence the result holds when $n = 1$.

   Inductive step. Assume that $\dfrac{1}{2!} + \dfrac{2}{3!} + \dfrac{3}{4!} + \cdots + \dfrac{k-1}{k!} = 1 - \dfrac{1}{k!}$; this is the inductive hypothesis.

The next term on the left -hand side is $\dfrac{k}{(k+1)!}$. Adding this to both sides gives

$$
\frac{1}{2!} + \frac{2}{3!} + \cdots + \frac{k-1}{k!} + \frac{k}{(k+1)!} = 1 - \frac{1}{k!} + \frac{k}{(k+1)!}
$$

$$
= 1 - \frac{k+1}{(k+1)!} + \frac{k}{(k+1)!}
$$

$$
= 1 - \frac{1}{(k+1)!}
$$

This is the result for $n = k+1$ and hence completes the inductive step.

Therefore, for all $n \in \mathbb{Z}^+$, $\dfrac{1}{2!} + \dfrac{2}{3!} + \dfrac{3}{4!} + \cdots + \dfrac{n-1}{n!} = 1 - \dfrac{1}{n!}$, by induction. $\qquad\square$

2. (iv) *Proof.* The proof is by mathematical induction.

Base case. When $n = 1$, $11^n - 4^n = 11 - 4 = 7$, which is clearly divisible by 7. Hence the result holds when $n = 1$.

Inductive step. Assume that $11^k - 4^k$ is divisible by 7; in other words, $11^k - 4^k = 7a$ for some integer a. Hence $11^k = 7a + 4^k$. Now

$$
\begin{aligned}
11^{k+1} - 4^{k+1} &= 11 \times 11^k - 4 \times 4^k \\
&= 11(7a + 4^k) - 4 \times 4^k \ \text{(by inductive hypothesis)} \\
&= 7 \times 11a + 7 \times 4^k \\
&= 7(11a + 4^k) \ \text{where } 11a + 4^k \in \mathbb{Z}.
\end{aligned}
$$

Therefore $11^{k+1} - 4^{k+1}$ is divisible by 7. This completes the inductive step.

Therefore, for all $n \in \mathbb{N}$, the expression $11^n - 4^n$ is divisible by 7, by induction. $\qquad\square$

3. (i) *Proof.* The proof is by mathematical induction.

Base case. When $n = 1$, LHS $= 1$ and RHS $= \dfrac{2!}{2^1 \times 1!} = 1$. Hence the result holds when $n = 1$.

Inductive step. Assume that $1 \times 3 \times 5 \times \cdots \times (2k-1) = \dfrac{(2k)!}{2^k k!}$; this is the inductive hypothesis.

The next term on the left -hand side is $2(k+1) - 1 = 2k + 1$.

Multiplying both sides by $2k + 1$ gives

$$1 \times 3 \times 5 \times \cdots \times (2k - 1) \times (2k + 1)$$
$$= \frac{(2k)!}{2^k k!} \times (2k + 1)$$
$$= \frac{(2k)! \times (2k + 1) \times (2k + 2)}{2^k k! \times (2k + 2)}$$
$$= \frac{(2k + 2)!}{2^k k! \times 2(k + 1)}$$
$$= \frac{(2(k + 1))!}{2^{k+1}(k + 1)!}.$$

This is the result for $n = k + 1$ and hence completes the inductive step.

Therefore, for all $n \in \mathbb{Z}^+$, $1 \times 3 \times 5 \times \cdots \times (2n - 1) = \dfrac{(2n)!}{2^n n!}$, by induction. $\qquad\square$

The following is an alternative, and simpler, direct proof.

Proof. Let $n \in \mathbb{Z}^+$. Then

$$1 \times 3 \times 5 \times \cdots \times (2n - 1)$$
$$= \frac{1 \times 2 \times 3 \times 4 \times 5 \times \cdots \times (2n - 1) \times (2n)}{2 \times 4 \times 6 \times \cdots \times (2n)}$$
$$= \frac{(2n)!}{2^n(1 \times 2 \times 3 \times \cdots \times n)}$$
$$= \frac{(2n)!}{2^n n!}.$$

$\qquad\square$

5. (i) *Proof.* The proof is by mathematical induction.

Base case. When $n = 1$, $n^3 - n = 0$, which is divisible by 6. Hence the result holds when $n = 1$.

Inductive step. Assume that $k^3 - k$ is divisible by 6; then $k^3 - k = 6a$ for some integer a. Now

$$(k + 1)^3 - (k + 1) = (k^3 + 3k^2 + 3k + 1) - (k + 1)$$
$$= k^3 - k + 3(k^2 + k)$$
$$= 6a + 3k(k + 1) \text{ (by inductive assumption)}.$$

Now $k(k + 1)$ is the product of an even and an odd integer, so is even — see theorem 1.2. Therefore $3k(k + 1)$ is divisible by 6.

Hence is $(k+1)^3 - (k+1) = 6a + k(k+1)$ is divisible by 6. This completes the inductive step.

Therefore, for all $n \in \mathbb{Z}^+$, $n^3 - n$ is divisible by 6, by induction. □

(ii) *Proof.* Let $n \in \mathbb{Z}^+$. Then $n^3 - n = n(n^2 - 1) = (n-1)n(n+1)$.

Therefore $n^3 - n$ is the product of three consecutive integers, one of which is divisible by 3 and (at least) one of which is even.

Therefore $n^3 - n$ is divisible by 6. □

8. For the inductive step, note that adding the $(k+1)$st row at the bottom of the large triangle adds $2k+1$ additional small triangles. Then use the inductive hypothesis and $(k+1)^2 = k^2 + (2k+1)$.

11. (i) *Proof.* The proof is by mathematical induction.

Base case. When $n = 1$, the grid comprises a single L-shaped tile so the result holds (trivially) when $n = 1$.

Inductive step. Assume that a $2^k \times 2^k$ grid with a single corner square removed may be covered with L-shaped tiles.

Consider a $2^{k+1} \times 2^{k+1}$ grid with a single corner square removed. Split the grid into four, as shown in the diagram below, and place a single L-shaped tile in the middle as indicated.

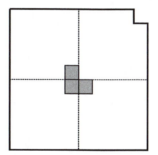

The remaining areas in each of the four quarters is a $2^k \times 2^k$ grid with a single corner square removed. By the inductive hypothesis, each of these quarters may be covered with L-shaped tiles. Therefore the $2^{k+1} \times 2^{k+1}$ grid with a single corner square removed may also be covered with L-shaped tiles. This completes the inductive step.

Therefore, for all $n \in \mathbb{Z}^+$, a $2^n \times 2^n$ grid with a single corner square removed may be covered with L-shaped tiles, by induction. □

(ii) The result follows by comparing areas. The area of the $2^n \times 2^n$ grid with a single corner square removed is $2^{2n} - 1$. Each tile has area 3. Hence, if the grid may be covered with tiles of area 3, it follows that 3 divides $2^{2n} - 1$.

(iii) *Proof.* The proof is by mathematical induction.

Base case. When $n = 1$, $2^{2n} - 1 = 2^2 - 1 = 3$, which is divisible by 3. Hence the result holds when $n = 1$.

Inductive step. Assume that $2^{2k} - 1$ is divisible by 3; then $2^{2k} - 1 = 3a$ for some integer a. Hence $2^{2k} = 3a + 1$. Now

$$
\begin{aligned}
2^{2(k+1)} - 1 &= 2^2 \times 2^{2k} - 1 \\
&= 4(3a + 1) - 1 \quad \text{(by inductive assumption)} \\
&= 3 \times 4a + 3 \\
&= 3(4a + 1) \qquad \text{where } 4a + 1 \in \mathbb{Z}.
\end{aligned}
$$

Hence $2^{2(k+1)} - 1$ is divisible by 3. This completes the inductive step.

Therefore, for all $n \in \mathbb{Z}^+$, $2^{2n} - 1$ is divisible by 3, by induction. $\square$

13. (i) *Proof.* The proof is by mathematical induction.

Base case. When $n = 1$, LHS $= h^{-1}gh =$ RHS. Hence the result holds when $n = 1$.

Inductive step. Assume that $\left(h^{-1}gh\right)^k = h^{-1}g^k h$. Then (assuming the associative law throughout so that terms may be regrouped as appropriate), we have

$$
\begin{aligned}
\left(h^{-1}gh\right)^{k+1} &= \left(h^{-1}gh\right)\left(h^{-1}gh\right)^k \\
&= \left(h^{-1}gh\right)\left(h^{-1}g^k h\right) \quad \text{(by inductive assumption)} \\
&= h^{-1}g\left(hh^{-1}\right)g^k h \\
&= h^{-1}gg^k h \qquad \text{(since } hh^{-1} = e) \\
&= h^{-1}g^{k+1}h.
\end{aligned}
$$

This completes the inductive step.

Therefore, for all $n \in \mathbb{Z}^+$ and for all $g, h \in G$, $\left(h^{-1}gh\right)^n = h^{-1}g^n h$, by induction. $\square$

14. (ii) *Proof.* The proof is by mathematical induction.

Base case. When $n = 1$, LHS $= \mathbf{A} = \begin{pmatrix} 1 & 0 \\ -1 & 2 \end{pmatrix}$ and

RHS $= \begin{pmatrix} 1 & 0 \\ 1 - 2^1 & 2^1 \end{pmatrix} = \begin{pmatrix} 1 & 0 \\ -1 & 2 \end{pmatrix}.$

Hence the result holds when $n = 1$.

Inductive step. Assume that $\mathbf{A}^k = \begin{pmatrix} 1 & 0 \\ 1 - 2^k & 2^k \end{pmatrix}.$

Then

$$\mathbf{A}^{k+1} = \mathbf{A}\mathbf{A}^k$$

$$= \begin{pmatrix} 1 & 0 \\ -1 & 2 \end{pmatrix} \begin{pmatrix} 1 & 0 \\ 1 - 2^k & 2^k \end{pmatrix} \text{ (by inductive assumption)}$$

$$= \begin{pmatrix} 1 & 0 \\ -1 + 2(1 - 2^k) & 2 \times 2^k \end{pmatrix}$$

$$= \begin{pmatrix} 1 & 0 \\ 1 - 2^{k+1} & 2^{k+1} \end{pmatrix}.$$

This completes the inductive step.

Therefore, for all $n \in \mathbb{Z}^+$, $\mathbf{A}^n = \begin{pmatrix} 1 & 0 \\ 1 - 2^n & 2^n \end{pmatrix}$, by induction. □

Exercises 8.2

1. (ii) *Proof.* The proof is by mathematical induction.

   Base case. For $n = 7$ we have $7! = 5040 > 2187 = 3^7$, so the result holds for $n = 7$.

   Inductive step. Assume that $k! > 3^k$ where $k \geq 7$. Then

   $$(k + 1)! = (k + 1) \times k! > (k + 1) \times 3^k > 3^{k+1}$$

   since $k + 1 > 3$. This is the result for $n = k + 1$ and hence completes the inductive step.

   Therefore, for all integers $n \geq 7$, $n! > 3^n$, by induction. □

2. *Proof.* The proof is by mathematical induction.

   Base case. When $n = 2$, LHS $= \frac{3}{4} =$ RHS. Hence the result holds when $n = 2$.

   Inductive step. Assume that

   $$\frac{3}{4} \times \frac{8}{9} \times \frac{15}{16} \times \cdots \times \frac{k^2 - 1}{k^2} = \frac{k + 1}{2k},$$

   where $k \geq 2$; this is the inductive hypothesis.

   The next term on the left-hand side is $\dfrac{(k + 1)^2 - 1}{(k + 1)^2}$. Multiplying both sides by this gives

$$\frac{3}{4} \times \frac{8}{9} \times \cdots \times \frac{k^2-1}{k^2} \times \frac{(k+1)^2-1}{(k+1)^2} = \frac{k+1}{2k} \times \frac{(k+1)^2-1}{(k+1)^2}$$

$$= \frac{k+1}{2k} \times \frac{k^2+2k}{(k+1)^2}$$

$$= \frac{k+1}{2k} \times \frac{k(k+2)}{(k+1)^2}$$

$$= \frac{k+2}{2(k+1)}.$$

This is the result for $n = k+1$ and hence completes the inductive step.

Therefore, for all $n \in \mathbb{Z}^+$, $\frac{3}{4} \times \frac{8}{9} \times \frac{15}{16} \times \cdots \times \frac{n^2-1}{n^2} = \frac{n+1}{2n}$, by induction. $\square$

4. (iii) *Proof.* The proof is by mathematical induction.

Base cases. Since a_n is defined explicitly for $n = 1$ and $n = 2$, there are two base cases to consider.

When $n = 1$, $a_1 = 0$ (by definition) and $2 \times 3^n - 3 \times 2^n = 2 \times 3 - 3 \times 2 = 0$.

When $n = 2$, $a_2 = 6$ (by definition) and $2 \times 3^n - 3 \times 2^n = 2 \times 9 - 3 \times 4 = 6$.

Hence the result is true for $n = 1$ and $n = 2$.

Inductive step. Let $k \geq 2$. Suppose that, for every integer $2 \leq r \leq k$, $a_r = 2 \times 3^r - 3 \times 2^r$.

Consider a_{k+1}. Since $k + 1 \geq 3$, a_{k+1} is defined by the recursive relationship. So we have

$$\begin{aligned}
a_{k+1} &= 5a_k - 6a_{k-1} & \text{(definition of } a_{k+1}) \\
&= 5(2 \times 3^k - 3 \times 2^k) - 6(2 \times 3^{k-1} - 3 \times 2^{k-1}) \\
& \qquad\qquad\qquad\qquad \text{(inductive assumption)} \\
&= 10 \times 3^k - 15 \times 2^k - 4 \times 2^k + 9 \times 2^k \\
&= 6 \times 3^k - 6 \times 2^k \\
&= 2 \times 3^{k+1} - 3 \times 2^{k+1}.
\end{aligned}$$

Hence a_{k+1} satisfies the given formula. This completes the inductive step.

Therefore, $a_n = 2 \times 3^n - 3 \times 2^n$ for all $n \geq 2$, by induction. $\square$

5. (ii) *Proof.* The proof is by mathematical induction.

Base cases. When $n = 1$, LHS $= a_1 = 1$ and RHS $= a_3 - 1 = 2 - 1 = 1$.

When $n = 2$, LHS $= a_1 + a_2 = 1 + 1 = 2$ and RHS $= a_4 - 1 = 3 - 1 = 2$.

Hence the result is true for $n = 1$ and $n = 2$.

Inductive step. Let $k \geq 2$ and suppose that $a_1 + a_2 + \cdots + a_k = a_{k+2} - 1$.

Adding a_{k+1} to both sides gives

$$a_1 + a_2 + \cdots + a_k + a_{k+1} = a_{k+2} - 1 + a_{k+1}$$
$$= a_{k+3} - 1.$$

since $a_{k+2} + a_{k+1} = a_{k+3}$ from the Fibonacci recurrence relation. This completes the inductive step.

Therefore, $a_1 + a_2 + \cdots + a_n = a_{n+2} - 1$, by induction. $\qquad\square$

(iii) This example may serve as a caution to assuming that proof by induction is always the best approach. Although a proof by induction is possible, it is complicated. The following is a much simpler direct proof.

Proof. Let $n \in \mathbb{Z}^+$. Then the Fibonacci recurrence relation gives both $a_{n+3} = a_{n+2} + a_{n+1}$ and $a_{n+2} = a_{n+1} + a_n$. Using these we have:

$$a_{n+2}^2 - a_{n+1}^2 = (a_{n+2} + a_{n+1})(a_{n+2} - a_{n+1})$$
$$= a_{n+3}(a_{n+1} + a_n - a_{n+1})$$
$$= a_{n+3}a_n.$$

$\qquad\square$

6. *Proof.* The proof is by mathematical induction.

Base case. When $n = 1$, $1 = 2^0$ can be expressed as a 'sum' of distinct non-negative powers of 2. Hence the result is true for $n = 1$.

Inductive step. Suppose that, for all integers $1 \leq r \leq k$, r can be expressed as a sum of distinct non-negative powers of 2.

Consider $k + 1$. We consider two cases: $k + 1$ is odd or $k + 1$ is even.

If $k + 1$ is odd, then k is even. By the inductive hypothesis, k can be expressed as a sum of distinct non-negative powers of 2. Since k is even, this expression does not include $2^0 = 1$. Hence adding 2^0 to the expression for k gives an expression for $k + 1$ as a sum of distinct non-negative powers of 2.

If $k + 1$ is even, then $k + 1 = 2r$ where $1 \leq r \leq k$. By the inductive hypothesis, r can be expressed as a sum of distinct non-negative powers of 2. Multiplying this expression for r by 2 gives an expression for $k+1$ as a sum of distinct non-negative powers of 2.

In either case, $k+1$ has an expression as a sum of distinct non-negative powers of 2. This completes the inductive step.

Therefore every $n \in \mathbb{Z}^+$ can be expressed as a sum of distinct non-negative powers of 2, by induction. $\qquad\square$

8. (i) *Proof.* Let $a, b \in A$. Then $a = 4m + 1$ and $b = 4n + 1$ for some integers m and n. Hence $ab = (4m + 1)(4n + 1) = 16mn + 4m + 4n + 1 = 4(4mn + m + n) + 1$ where $4mn + m + n \in \mathbb{Z}$. Therefore $ab \in A$. $\qquad\square$

   (ii) *Proof.* The proof is by mathematical induction.

   Base case. When $n = 1$, $5 = 4 \times 1 + 1$ is A-prime and so is a 'product' of A-primes. Hence the result is true for $n = 1$.

   Inductive step. Suppose that, for all integers $1 \le r \le k$, $4r + 1$ can be expressed as a product of A-primes.

   Consider $a = 4(k + 1) + 1$.

   If a is A-prime, then it is trivially expressed as a 'product' of A-primes.

   Otherwise a has a factorisation $a = a_1 a_2$ where both a_1 and a_2 are elements of A. By the inductive hypothesis, a_1 and a_2 can each be written as a product of A-primes

   $$a_1 = p_1 p_2 \ldots p_s \text{ and } a_2 = q_1 q_2 \ldots q_t$$

   where $p_1, p_2, \ldots, p_s$ and $q_1, q_2, \ldots, q_t$ are all A-primes. Therefore a can be expressed as a product of A-primes

   $$a = a_1 a_2 = p_1 p_2 \ldots p_s q_1 q_2 \ldots q_t.$$

   This completes the inductive step.

   Therefore every element of A that is greater than 1 can be written as a product of A-primes, by induction. $\qquad\square$

   (iii) $693 \in A$ has two distinct A-prime factorisations, $693 = 9 \times 77$ and $693 = 21 \times 33$.

9. *Proof.* The proof is by mathematical induction on $|A|$.

   Base case. When $n = 0$, $|A| = 0$ implies that $A = \varnothing$, which has no proper subsets. Hence the result is (trivially) true for $n = 0$.[3]

   ---

   [3] Readers who don't like this argument may wish to start the induction at $n = 1$. If $|A| = 1$ and $B \subset A$ it follows that $B = \varnothing$, so $|B| = 0 < 1 = |A|$ and the result holds for $n = 1$.

Inductive step. Suppose that, for all sets A with $|A| = k$, if $B \subset A$, then $|B| < |A|$.

Let A be a set with cardinality $|A| = k + 1$ and let $B \subset A$.

If $B = \varnothing$, then $|B| = 0 < k + 1 = |A|$.

Otherwise choose an element $b \in B$. Then $B' = B - \{b\}$ is a proper subset of $A' = A - \{b\}$, $B' \subset A'$. But $|A'| = |A| - 1 = k$ so, by the inductive hypothesis, $|B'| < |A'|$. Therefore $|B| = |B'| + 1 < |A'| + 1 = |A|$. This completes the inductive step.

Therefore, for all finite sets A and B, if $B \subset A$, then $|B| < |A|$, by induction. $\qquad\square$

Bibliography

[1] A. Cupillari. *The Nuts and Bolts of Proof, 4 th edition.* Elsevier Academic Press, 2013.

[2] P.J. Eccles. *An Introduction to Mathematical Reasoning: Numbers, Sets and Functions.* Cambridge University Press, 1997.

[3] J. Fauvell and J. Gray. *The History of Mathematics: A Reader.* Palgrave Macmillan, 1987.

[4] J. Franklin and A. Daoud. *Proof in Mathematics: An Introduction.* Key Books, 2010.

[5] R. Garnier and J. Taylor. *100% Mathematical Proof.* John Wiley & Sons, 1996.

[6] R. Garnier and J. Taylor. *Discrete Mathematics: Proofs, Structures and Applications, 3 rd edition.* Taylor and Francis, 2010.

[7] G.H. Hammack. *Book of Proof, 2 nd edition.* Creative Commons, 2013.

[8] G.H. Hardy. Mathematical proof. *Mind*, pages 1–25, 1929.

[9] G.H. Hardy. *A Mathematician's Apology (reprinted edition).* Cambridge University Press, 2012.

[10] S.-J. Shin. *The Logical Status of Diagrams.* Cambridge University Press, 1994.

[11] A.B. Slomson. Mathematical proof and its role in the classroom. *Mathematics Teaching*, 155:10–13, 1996.

[12] D. Solow. *How to Read and Do Proofs: An Introduction to Mathematical Thought Processes, 5 th edition.* John Wiley & Sons, 2010.

[13] D. Tall. Thinking through three worlds of mathematics. *Proceedings of the 28th Conference of the International Group for the Psychology of Mathematics Education, Bergen, Norway*, pages 281–288, 2004.

[14] D.J. Vellerman. *How to Prove It: A Structures Approach, 2 nd edition.* Cambridge University Press, 2006.

Index